Genetics, Phylogeny, and Evolution of Insects

Genetics, Phylogeny, and Evolution of Insects

Editor

Zhiteng Chen

MDPI

Basel • Beijing • Wuhan • Barcelona • Belgrade • Novi Sad • Cluj • Manchester

Editor
Zhiteng Chen
Jiangsu University of Science and Technology
Zhenjiang
China

Editorial Office
MDPI AG
Grosspeteranlage 5
4052 Basel, Switzerland

This is a reprint of articles from the Special Issue published online in the open access journal *Genes* (ISSN 2073-4425) (available at: https://www.mdpi.com/journal/genes/special_issues/8C5N4T0649).

For citation purposes, cite each article independently as indicated on the article page online and as indicated below:

Lastname, A.A.; Lastname, B.B. Article Title. *Journal Name* **Year**, *Volume Number*, Page Range.

ISBN 978-3-7258-1931-7 (Hbk)
ISBN 978-3-7258-1932-4 (PDF)
doi.org/10.3390/books978-3-7258-1932-4

Cover image courtesy of Zhiteng Chen

Contents

genes

MDPI

Editorial

Editorial for the "Genetics, Phylogeny, and Evolution of Insects" Special Issue

Zhiteng Chen

School of Grain Science and Technology, Jiangsu University of Science and Technology, Zhenjiang 212004, China; chenzhiteng@just.edu.cn

The rapid advancement of sequencing technologies has revolutionized our understanding of the phylogeny and evolution of insects, enabling researchers to generate extensive molecular data with unprecedented detail. The Special Issue "Genetics, Phylogeny, and Evolution of Insects" brings together 11 original research articles on various advancements of insect genetics and evolution. This Editorial aims to summarize the contributions of the papers published in this Special Issue, highlighting their significance and placing them in a broader context of entomological research.

One of the key themes in this Special Issue is the exploration of mitochondrial genomes, which serve as valuable markers for phylogenetic and genetic studies. Several studies in this collection exemplify this approach, focusing on different insect orders. Regarding Diptera, Bi et al. (2023) [1] sequenced and analyzed the complete mitochondrial genome of *Piophila casei* (Diptera: Piophilidae), a flesh-feeding fly, providing critical information for its genetic structure and phylogenetic position. This study not only aids in understanding the taxonomy and genetics of *P. casei* but also has implications for forensic science and food safety, given the species' relevance in decaying organic matter and foodstuffs. Similarly, Silva et al. (2024) [2] sequenced the first complete mitochondrial genome of *Orthopodomyia fascipes* (Diptera: Culicidae), revealing a 15,598 bp sequence that supports the monophyly of the family Culicidae and elucidates the evolutionary proximity between tribes Orthopodomyiini and Mansoniini. This study significantly contributes to the taxonomy and evolutionary understanding of the genus *Orthopodomyia* and offers insights into Culicidae, which includes many important vector species.

Regarding Hemiptera, the mitochondrial genomes of three species of *Egeirotrioza* (Hemiptera: Triozidae) were sequenced and analyzed by Aishan et al. (2024) [3], highlighting their evolutionary rates and phylogenetic relationships within Triozidae. This research provides valuable genomic information for the study of psyllid species and underscores the importance of mitochondrial data in resolving phylogenetic controversies. Zhu et al. (2023) [4] sequenced the mitochondrial genomes of three bamboo pests, namely *Notobitus meleagris*, *Macropes harringtonae*, and *Homoeocerus bipunctatus* (Hemiptera: Heteroptera), offering detailed insights into their genetic makeup and phylogenetic placement. This work improves the existing database of bamboo pests and supports the development of rapid identification techniques and pest control strategies.

For Coleoptera, Wang et al. (2023) [5] sequenced and compared the mitochondrial genomes of three tenebrionid species, revealing conserved genomic features and unique tRNA-like structures. Their study supports the monophyly of several subfamilies within Tenebrionidae and provides insights into the evolutionary and functional roles of tRNA-like sequences in these beetles. Li et al. (2023) [6] characterized the mitochondrial genome of *Luperomorpha xanthodera* (Coleoptera: Chrysomelidae: Galerucinae), revealing significant phylogenetic relationships within the family Chrysomelidae. This research contributes to resolving the classification status of *Luperomorpha* and understanding the evolutionary dynamics within the subfamily Galerucinae. Moreover, Yu et al. (2023) [7] sequenced the mitochondrial genomes of four species within the curculionid tribe Scolytoplatypodini,

Citation: Chen, Z. Editorial for the "Genetics, Phylogeny, and Evolution of Insects" Special Issue. *Genes* **2024**, *15*, 1000. https://doi.org/10.3390/genes15081000

Received: 17 July 2024
Accepted: 26 July 2024
Published: 30 July 2024

providing detailed phylogenetic analyses and supporting the monophyly of this group. This study enhances our understanding of the evolutionary relationships among scolytine beetles and offers a foundation for further research into their genetic diversity.

Regarding genetics, Liu et al. (2023) [8] explored the genetic diversity and migration patterns of the white-backed planthopper (WBPH) across Myanmar and Yunnan Province. Analyzing the mitochondrial DNA COI genes from 416 individuals, they identified 43 haplotypes, with a high gene flow being indicated by two common haplotypes shared across all populations. This study deduced that WBPH populations in Yunnan primarily migrated from northern and northeastern Myanmar. These findings contribute to the establishment of sustainable management strategies for this significant rice pest in Southeast Asia.

This Special Issue also includes significant contributions from transcriptomic and endosymbiont research, which provide deeper insights into insect biology and ecology. A comparative transcriptomic analysis of the rice leaf folder *Cnaphalocrocis medinalis* (Lepidoptera: Pyralidae) by Du et al. (2023) [9] revealed distinct chemosensory gene expression profiles between larvae and adults. This study highlights the specialized functions of these genes at different developmental stages and suggests potential molecular targets for pest management strategies. The research of Wang et al. (2023) [10] on the gut microbiota of the fall armyworm *Spodoptera frugiperda* (Lepidoptera: Noctuidae) demonstrated that its feeding on various host plants has significant effects on the structure and diversity of its gut bacterial community. These findings underscore the role of the gut microbiota in the adaptive evolution of this major agricultural pest and suggest new avenues for biological control. The study of Corpuz et al. (2023) [11] on *Wolbachia* endosymbionts in Hawaiian Drosophilidae documented the transmission patterns and strain diversity within this adaptive radiation. Their research highlights the complex interactions between *Wolbachia* and their hosts and suggests potential applications in conservation breeding programs for endangered species.

In conclusion, the articles included in this Special Issue entitled "Genetics, Phylogeny, and Evolution of Insects" cover a diverse array of topics and provide comprehensive insights to direct future research in insect genetics, phylogeny, and evolution. The extensive utilization of mitochondrial genomes, transcriptomic analyses, and studies on endosymbionts underscores the complexity and diversity of insect species. These findings highlight the importance of investigating insects not only at the molecular level, but also within the broader context of their ecological interactions and evolutionary histories. The findings presented in this Special Issue both advance our scientific knowledge and have practical implications for pest management, conservation, and biodiversity studies. We anticipate that this Special Issue will inspire further research, leading to a deeper understanding and innovative approaches in the study of insect biology and evolution.

Conflicts of Interest: The authors declare no conflicts of interest.

References

1. Bi, S.; Song, Y.; Liu, L.; Wan, J.; Zhou, Y.; Zhu, Q.; Liu, J. Complete Mitochondrial Genome of *Piophila casei* (Diptera: Piophilidae): Genome Description and Phylogenetic Implications. *Genes* **2023**, *14*, 883. [CrossRef] [PubMed]
2. Silva, F.S.d.; Nascimento, B.L.S.d.; Cruz, A.C.R.; Silva, S.P.d.; Aragão, C.F.; Dias, D.D.; Silva, L.H.d.S.e.; Reis, L.A.M.; Reis, H.C.F.; Chagas, L.L.d.; et al. Sequencing and Description of the Mitochondrial Genome of *Orthopodomyia fascipes* (Diptera: Culicidae). *Genes* **2024**, *15*, 874. [CrossRef] [PubMed]
3. Aishan, Z.; Mu, Z.-L.; Li, Z.-C.; Luo, X.-Y.; Huangfu, N. The First Three Mitochondrial Genomes for the Characterization of the Genus *Egeirotrioza* (Hemiptera: Triozidae) and Phylogenetic Implications. *Genes* **2024**, *15*, 842. [CrossRef] [PubMed]
4. Zhu, W.; Yang, L.; Long, J.; Chang, Z.; Gong, N.; Mu, Y.; Lv, S.; Chen, X. Characterizing the Complete Mitochondrial Genomes of Three Bugs (Hemiptera: Heteroptera) Harming Bamboo. *Genes* **2023**, *14*, 342. [CrossRef] [PubMed]
5. Wang, S.-H.; Hu, S.-Y.; Li, M.; Liu, M.; Sun, H.; Zhao, J.-R.; Chen, W.-T.; Yuan, M.-L. Comparative Mitogenomic Analyses of Darkling Beetles (Coleoptera: Tenebrionidae) Provide Evolutionary Insights into tRNA-like Sequences. *Genes* **2023**, *14*, 1738. [CrossRef] [PubMed]
6. Li, J.; Yan, B.; He, H.; Xu, X.; Ruan, Y.; Yang, M. Characterization of the Complete Mitochondrial Genome of a Flea Beetle *Luperomorpha xanthodera* (Coleoptera: Chrysomelidae: Galerucinae) and Phylogenetic Analysis. *Genes* **2023**, *14*, 414. [CrossRef] [PubMed]

7. Yu, G.; Lai, S.; Liao, S.; Cao, Y.; Li, W.; Long, C.; Tarno, H.; Wang, J. Complete Mitochondrial Genome of Scolytoplatypodini Species (Coleoptera: Curculionidae: Scolytinae) and Phylogenetic Implications. *Genes* **2023**, *14*, 162. [CrossRef] [PubMed]
8. Liu, Y.; Chan, K.N.; Li, X.; Zhao, X.; Chu, D.; Yin, Y.; Liu, Y.; Chen, A. The Genetic Diversity of White-Backed Planthoppers (*Sogatella furcifera*) between Myanmar and Yunnan Province of China. *Genes* **2023**, *14*, 2164. [CrossRef] [PubMed]
9. Du, H.-T.; Lu, J.-Q.; Ji, K.; Wang, C.-C.; Yao, Z.-C.; Liu, F.; Li, Y. Comparative Transcriptomic Assessment of Chemosensory Genes in Adult and Larval Olfactory Organs of *Cnaphalocrocis medinalis*. *Genes* **2023**, *14*, 2165. [CrossRef] [PubMed]
10. Wang, Y.-P.; Liu, X.; Yi, C.-Y.; Chen, X.-Y.; Liu, C.-H.; Zhang, C.-C.; Chen, Q.-D.; Chen, S.; Liu, H.-L.; Pu, D.-Q. The Adaptive Evolution in the Fall Armyworm *Spodoptera frugiperda* (Lepidoptera: Noctuidae) Revealed by the Diversity of Larval Gut Bacteria. *Genes* **2023**, *14*, 321. [CrossRef] [PubMed]
11. Corpuz, R.L.; Bellinger, M.R.; Veillet, A.; Magnacca, K.N.; Price, D.K. The Transmission Patterns of the Endosymbiont *Wolbachia* within the Hawaiian Drosophilidae Adaptive Radiation. *Genes* **2023**, *14*, 1545. [CrossRef]

genes

MDPI

Article

Complete Mitochondrial Genome of Scolytoplatypodini Species (Coleoptera: Curculionidae: Scolytinae) and Phylogenetic Implications

Guangyu Yu [1,†], Shengchang Lai [1,2,†], Song Liao [1], Yufeng Cao [1], Weijun Li [1], Chengpeng Long [3], Hagus Tarno [1,4] and Jianguo Wang [1,*]

1 Laboratory of Invasion Biology, School of Agricultural Sciences, Jiangxi Agricultural University, Nanchang 340045, China
2 Forest Protection, Forestry College, Nanjing Forest University, Nanjing 210036, China
3 College of Forestry and Biotechnology, Zhejiang Agricultural and Forestry University, Lin'an 311300, China
4 Department of Plant Pests and Diseases, Faculty of Agriculture, Universitas Brawijaya, Jl. Veteran, Malang 65145, Indonesia
* Correspondence: jgwang@jxau.edu.cn
† These authors contributed equally to this work.

Abstract: The complete mitochondrial genomes (mitogenomes) of beetles in the tribe Scolytoplatypodini (genus *Scolytoplatypus*) were sequenced and annotated. These included *Scolytoplatypus raja* (15,324 bp), *Scolytoplatypus sinensis* (15,394 bp), *Scolytoplatypus skyliuae* (15,167 bp), and *Scolytoplatypus wugongshanensis* (15,267 bp). The four mitogenomes contained 37 typical genes, including 13 protein-coding genes (PCGs), 22 transfer RNA genes (tRNAs), and 2 ribosomal RNA genes (rRNAs). The gene orientation and arrangement of the four mitogenomes were similar to other Coleoptera mitogenomes. PCGs mostly started with ATN and terminated with TAA. The Ka/Ks ratio of 13 PCGs in the four species revealed that *cox1* had the slowest evolutionary rate and *atp8* and *nad6* had a higher evolutionary rate. All tRNAs had typical cloverleaf secondary structures, but *trnS1* lacked dihydrouridine arm. Partial tRNAs lost the discriminator nucleotide. The *trnY* did not possess the discriminator nucleotide and also lost three bases, showing a special amino-acyl arm. Bayesian inference (BI) and maximum likelihood (ML) methods were conducted for phylogenetic analyses using 13 PCGs. Scolytoplatypodini was clustered with Hylurgini and Hylastini, and the monophyly of Scolytoplatypodini was supported. The four newly sequenced mitogenomes increase understanding of the evolutionary relationships of Scolytoplatypodini and other Scolytinae species.

Keywords: Scolytinae; Scolytoplatypodini; mitochondrial genome; phylogenetic analysis

Citation: Yu, G.; Lai, S.; Liao, S.; Cao, Y.; Li, W.; Long, C.; Tarno, H.; Wang, J. Complete Mitochondrial Genome of Scolytoplatypodini Species (Coleoptera: Curculionidae: Scolytinae) and Phylogenetic Implications. *Genes* **2023**, *14*, 162. https://doi.org/10.3390/genes14010162

Academic Editor: Zhiteng Chen

Received: 17 November 2022
Revised: 2 January 2023
Accepted: 4 January 2023
Published: 6 January 2023

1. Introduction

The Scolytoplatypodini is a tribe of wood-boring ambrosia beetles in the subfamily Scolytinae (Curculionidae). They farm fungi as the food source for larvae and adults in their gallery systems [1,2]. The females of most species have a unique mycangial structure located on the pronotum for carrying fungal spores [3]. However, this mycangial structure is absent in some species [4]. Except for *Scolytoplatypus bombycinus*, which is host-specific, other Scolytoplatypodini species have no selection propensity for hosts [4]. Scolytoplatypodini species are mostly secondary borers and usually attack small branches of dead trees. They do not usually attack healthy trees, but some species will attack healthy living trees [5]. There are two genera in Scolytoplatypodini: *Scolytoplatypus* Schaufuss 1891 and *Remansus* Jordal 2013, including 60 species. *Scolytoplatypus* occur mainly in the Afrotropical and Oriental regions, with a few occurrences in temperate regions of Japan and India, and *Remansus* are only found in Madagascar [6]. Phylogenetic analysis studies on Scolytoplatypodini have mostly focused on selected gene fragments. Jordal [6] conducted a phylogenetic analysis of the Scolytoplatypodini in Africa and Asia using four gene fragments and found that the Asian species of *Scolytoplatypus* clearly differ from the African species of *Scolytoplatypus*. Additional molecular data are needed to explore the taxonomic status and systematic relationship of *Scolytoplatypus* and Scolytoplatypodini in Scolytinae. Among various animal groups, complete mitochondrial genomes (mitogenomes) phylogenetic analyses were credible [7]. Mitogenomes also have been applied to evaluate population genetics, phylogeography, and systematics at different taxonomic levels [8]. Therefore, the mitogenome can be a powerful tool for determining the relationship of Scolytoplatypodini within the Scolytinae.

DNA sequencing technology has allowed the sequencing of many insect mitogenomes [9,10]. The typical mitogenome of insect has closed-circular and double-stranded DNA, containing 13 protein-coding genes (PCGs), 22 transfer RNA genes (tRNAs), two ribosomal RNA genes (rRNAs), and an A+T-rich region (D-loop), totaling 37 genes [8,11,12]. Mitogenomes are commonly used in phylogenetic analysis of insect lineages because of their maternal inheritance, rare recombination, relatively high evolutionary rate, and conserved gene components [11,13]. The mitogenome was used to reconstruct the phylogeny of weevils (Curculionoidea) and show that Scolytinae forms a separate lineage [7,14]. Representative bark and ambrosial beetles of Scolytinae have also been studied for phylogeny and taxonomy based on mitogenomes [15].

In this study, the mitogenome of *S. raja* (Blandford, 1893); *S. sinensis* (Tsai and Huang, 1965); *S. skyliuae* (Liao, Lai, and Beaver, 2022); and *S. wugongshanensis* (Liao, Lai, and Beaver, 2022) were provided and analyzed. These complete mitogenomes are first reported in the Scolytoplatypodini. We also explored the genome structure, nucleotide compositions, codon usage, gene overlaps, intergenic spacers, tRNA secondary structure, and the D-loop. Reconstructing the phylogenetic tree of Scolytoplatypodini based on 13 PCGs by Bayesian Inference (BI) and Maximum Likelihood (ML) methods.

2. Materials and Methods

2.1. Sample Collection and DNA Extraction

The samples for genome sequencing of *S. raja*, *S. sinensis*, *S. skyliuae*, and *S. wugongshanensis* were collected in China (Table 1) and identified according to the corresponding key [1]. All of the fresh samples were placed in absolute alcohol, preserved at −20 °C, and stored at the School of Agricultural Sciences, Jiangxi Agricultural University, Nanchang, Jiangxi 340045, China. The total genomic DNA was extracted from the leg muscle tissue of individual specimens using the Universal Genomic DNA Kit (Jiangsu, China), following manufacturer instructions. DNA was stored for sequencing at −20 °C.

Table 1. Voucher information of the specimens used for mitochondrial genomes sequencing.

Specimens	Location	Geographic Info.	Date of Collection
S. raja	Gaoligong National Nature Reserve, Yunnan	25.29 N, 98.80 E	27 July 2019
S. sinensis	Leibo County, Sichuan	28.41 N, 103.77 E	5 August 2021
S. skyliuae	Wuyishan National Nature Reserve, Jiangxi	27.88 N, 117.78 E	17 July 2017
S. wugongshanensis	Wugong Mountain, Jiangxi	27.58 N, 114.23 E	27 September 2017

2.2. Mitogenome Sequencing and Assembly

Following manufacturer recommendations, NEB Next® Ultra™ DNA Library Prep Kit was used to generate the sequencing library for Illumina (Lincoln, NE, USA) and add index code to each sample. Using Illumina PE Cluster Kit (Illumina, Lincoln, NE, USA), we performed the clustering of the index-coded samples on a cBot Cluster Generation System, according to manufacturer instructions. When the cluster was generated, we used the Illumina platform (NovaSeq 6000) to sequence the DNA libraries and generated 150 bp paired-end reads. After removing low-quality sequences, the reads were assembled into a complete mitogenome by MitoFlex [16].

2.3. Sequence Annotation and Analyses

MITOS [17] web server was used to determine 37 genes and all tRNA secondary structures under default parameters. According to the MITOS predictions, secondary structures for tRNAs were manually drawn with Microsoft PowerPoint 2017. The remaining PCGs and rRNA were manually corrected in Geneious 8.1.3 [18]. To draw the mitogenome circular map, we used OrganellarGenomeDRAW (OGDRAW) version 1.3.1 [19]. The nucleotide composition and relative synonymous codon usage (RSCU) were calculated by MEGA X [20] and PhyloSuite v 1.2.2 [21], respectively. Strand asymmetry was calculated by the formulas: GC-skew = [G − C]/[G + C] and AT-skew = [A − T]/[A + T] [22]. DnaSP v6.12.03 was used to calculate the nucleotide diversity (Pi) and nonsynonymous (Ka)/synonymous (Ks) mutation rate ratios of 13 PCGs [23].

2.4. Phylogenetic Analysis

A total of 34 mitogenomes (Table 2) were used to construct the phylogenetic tree. PhyloSuite v 1.2.2 [21] was used to download the data onto mitogenomes from GenBank (except *S. raja*, *S. skyliuae*, *S. sinensis*, and *S. wugongshanensis*). All sequences were standardized and extracted 13 PCGs by PhyloSuite v 1.2.2 [21]. The 13 PCGs (excluding the stop codons) of the 34 beetle species were aligned individually using codon-based multiple alignments with MAFFT v7.313 software with default settings [24]. Gblocks v 0.91b software was used to remove the intergenic gaps and ambiguous sites [25], and all PCGs genes were concatenated in PhyloSuite v 1.2.2 [21]. The best partitioning scheme and evolutionary models for constructing BI and ML trees were selected by PartitionFinder2 [26], with a greedy algorithm, BIC criterion, and the gene and codon model. The results are presented in Supplementary Table S1.

Table 2. Mitochondrial genome information used in this study.

Subfamily	Species	Length (bp)	GenBank Accession No.	References
Scolytinae	*Anisandrus dispar*	16,665	NC_036293	Unpublished
Scolytinae	*Xylosandrus crassiusculus*	16,875	NC_036284	Unpublished
Scolytinae	*Xylosandrus germanus*	16,099	NC_036280	Unpublished
Scolytinae	*Xylosandrus morigerus*	16,246	NC_036283	Unpublished
Scolytinae	*Cyclorhipidion bodoanus*	15,899	NC_036295	Unpublished
Scolytinae	*Euwallacea fornicates*	15,745	NC_059702	[27]
Scolytinae	*Dryocoetes autographus*	17,055	NC_036287	Unpublished
Scolytinae	*Dryocoetes hectographus*	16,040	NC_062125	[15]
Scolytinae	*Dryocoetes villosus*	15,859	NC_036282	Unpublished
Scolytinae	*Ips calligraphus*	19,144	NC_060365	[28]
Scolytinae	*Ips subelongatus*	16,040	MZ766130	[15]
Scolytinae	*Ips typographus*	16,793	MZ766131	[15]
Scolytinae	*Ips sexdentatus*	18,579	NC_036281	Unpublished
Scolytinae	*Orthotomicus erosus*	16,753	NC_060712	[15]
Scolytinae	*Orthotomicus laricis*	18,887	NC_036291	Unpublished
Scolytinae	*Pityogenes bidentatus*	18,781	NC_036289	Unpublished
Scolytinae	*Dendroctonus valens*	16,547	NC_061915	Unpublished
Scolytinae	*Hylastes attenuates*	17,409	NC_036290	Unpublished
Scolytinae	*Hylastes brunneus*	15,774	NC_036262	Unpublished
Scolytinae	*Scolytoplatypus raja*	15,324	OP719285	This study
Scolytinae	*Scolytoplatypus skyliuae*	15,166	OP719283	This study
Scolytinae	*Scolytoplatypus sinensis*	15,394	OP719284	This study
Scolytinae	*Scolytoplatypus wugongshanensis*	15,267	OP712675	This study
Scolytinae	*Trypodendron domesticum*	16,986	NC_036286	Unpublished
Scolytinae	*Trypodendron signatum*	16,909	NC_036292	Unpublished
Scolytinae	*Trypophloeus asperatus*	17,039	NC_036285	Unpublished
Scolytinae	*Gnathotrichus materiarius*	16,871	NC_036294	Unpublished
Scolytinae	*Pityophthorus pubescens*	17,316	NC_036288	Unpublished

Table 2. *Cont.*

Subfamily	Species	Length (bp)	GenBank Accession No.	References
Scolytinae	*Polygraphus poligraphus*	17,434	NC_060713	[15]
Scolytinae	*Phloeosinus perlatus*	17,054	NC_057470	Unpublished
Scolytinae	*Scolytus schevyrewi*	15,891	NC_046589	[29]
Scolytinae	*Scolytus seulensis*	16,396	NC_046588	Unpublished
Dryophthorinae	*Sitophilus oryzae*	17,602	NC_030765	[30]
Dryophthorinae	*Sitophilus zeamais*	18,531	MT294139	[30]

MrBayes v 3.2.6 [31] and IQ-TREE v.1.6.8 [32] software were employed in PhyloSuite v 1.2.2 to construct the BI and ML phylogenetic trees [21] (refer to Du et al. [15]). *Sitophilus oryzae* and *Sitophilus zeamais* were used as the out-group. ML phylogeny was inferred using IQ-TREE v.1.6.8 [32] under the Edge-linked partition model for 10,000 ultrafast bootstraps [33]. BI phylogeny was inferred using MrBayes 3.2.6 [31] under the partition model (two parallel runs, 2,000,000 generations), in which the initial 25% of sampled data were discarded as burn-in. A PSRF close to 1.0 and a standard deviation of split frequencies below 0.01 were accepted.

3. Results and Discussion

3.1. Mitogenome Organization and Nucleotide Composition

The complete mitogenome lengths of *S. raja*, *S. sinensis*, *S. skyliuae*, and *S. wugongshanensis* were 15,324 bp, 15,394 bp, 15,166 bp, and 15,267 bp, respectively. These mitogenomes have a similar structure; all mitogenomes exhibited the typical insect mitogenome structure, closed-circular and double-stranded DNA, containing 13 PCGs, 22 tRNAs, 2 rRNAs, and a D-loop (Figure 1). There were 23 genes encoded by the majority strand (J-strand), including 9 PCGs and 14 tRNAs. The remaining 14 genes were encoded by the minority strand (N-strand), including 4 PCGs, 8 tRNAs, and 2 rRNAs (Table S3).

The basic composition of *S. raja* was A = 39.4%, T = 36.4%, C = 14.8%, and G = 9.4; *S. sinensis* was A = 39.5%, T = 36.3%, C = 15.0%, and G = 9.2%; *S. skyliuae* was A = 39.6%, T = 36.7%, C = 15.5%, and G = 8.2%; *S. wugongshanensis* was A = 39.0%, T = 37.2%, C = 15.6%, and G = 8.2% (Table S2). In all four species, the nucleotide composition of the whole mitogenome exhibited a distinct A/T bias: 75.8% (*S. raja*), 75.8% (*S. sinensis*), 76.3% (*S. skyliuae*), and 76.2% (*S. wugongshanensis*). A higher A/T bias was also found in *Scolytinae* mitogenomes [15]. The AT-skew ranged from 0.02 (*S. wugongshanensis*) to 0.04 (*S. raja*, *S. sinensis*, and *S. skyliuae*), the GC-skew ranged from −0.31 (*S. wugongshanensis*) to −0.22 (*S. raja*) (Table S2).

3.2. Protein-Coding Genes

The total lengths of the 13 PCGs of *S. raja*, *S. sinensis*, *S, skyliuae*, and *S. wugongshanensis* were 11,070 bp, 11,062 bp, 11,023 bp, and 11,025 bp, respectively (Table S2). Of these 13 PCGs, 9 PCGs are located at the J-strand, the other 4 PCGs were encoded by the N-strand (Figure 1, Table S3). The whole 13 PCGs AT-skew and GC-skew were all negative; the AT-skew were −0.13 (*S. raja* and *S. skyliuae*) and −0.14 (*S. sinensis* and *S. wugongshanensis*), the GC-skew were −0.04 (*S. raja* and *S. sinensis*) and −0.07 (*S. skyliuae* and *S. wugongshanensis*). All PCGs used ATN as the initiation codons, except for *nad1* (*S. raja* and *S. sinensis*), which begins with TTG, and *atp6* (*S. raja*), which begins with GTG. Only *atp6* and *cox2* (*S. skyliuae*) had incomplete stop codon with T residue; other PCGs terminated with TAA/TAG. The incomplete termination codons are presumed to be filled by polyadenylation during the mRNA maturation process [34].

The RSCU of the four Scolytoplatypodini species was calculated (Figure 2). The codons that were most commonly used were UUA-Leu, UUU-Phe, AUU-Ile, and AUA-Met. This result indicated that UUA is the most preferred codon. Additionally, there was a strong A/T bias in the PCGs.

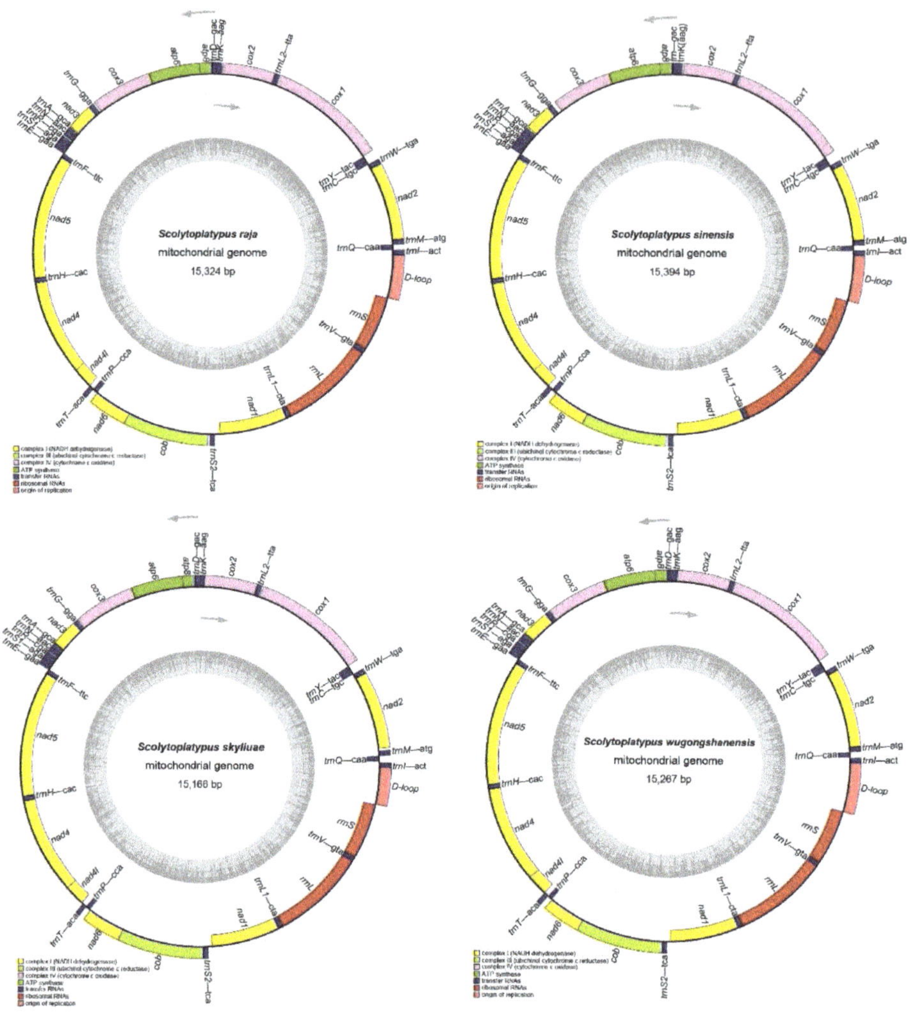

Figure 1. Circular map of the complete mitogenome of *S. raja*, *S. sinensis*, *S. skyliuae, and S. wugongshanensis*. Different colors indicate different types of genes and regions. Genes in the outer circle are located on the J-strand, and genes in the inner circle are located on the N-strand.

3.3. Nucleotide Diversity (Pi) and Nonsynonymous (Ka)/Synonymous (Ks) Mutation Rate Ratios

The Pi of the four Scolytoplatypodini species based on 13 PCGs was computed (Figure 3), and it ranged from 0.15 to 0.32. Among the PCGs, *atp8* (0.32) had the highest Pi values, followed by *nad6* (0.26) and *nad2* (0.25). The *cox1* (0.15) had the lowest Pi values, which implies that *cox1* is the most conserved gene in *Scolytoplatypus*.

The ratios of Ka/Ks for every gene of the 13 PCGs were also computed (Figure 3). The values of Ka/Ks ranged from 0.05 to 0.52; 13 PCGs displayed low evolutionary rates ($0 < \omega < 1$), indicating evolutions of 13 PCGs under the purification option [35]. *Cox1* (0.05) exhibits the lowest evolutionary rate, suggesting that it experienced the strongest purifying selection. The *atp8* (0.52) and *nad6* (0.38) exhibited a faster rate of evolution.

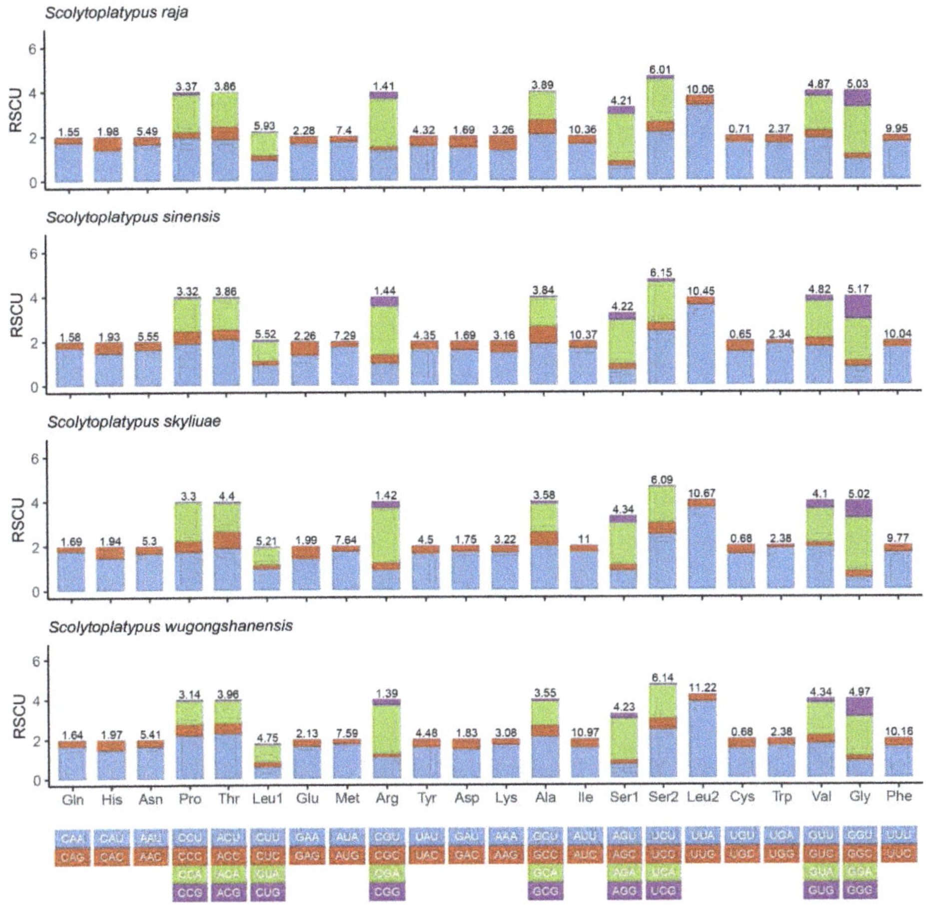

Figure 2. Relative synonymous codon usage (RSCU) of four *Scolytoplatypus* species. The ordinate represents the RSCU (the number of times a certain synonymous codon is used/the average number of times that all codons encoding the amino acid are used). The abscissa represents different amino acids. The number above the bar graph represents the ratio of amino acids (number of certain amino acids/total number of all amino acids).

3.4. Gene Overlaps and Intergenic Spacers

Gene overlaps were found in all four mitogenomes, and every overlap ranged from 1 bp to 7 bp (*S. raja*, seven gene junctions, 27 bp overlaps; *S. sinensis*, seven gene junctions, 21 bp overlaps; *S. skyliuae*, seven gene junctions, 23 bp overlaps; *S. wugongshanensis*, 10 gene junctions, 26 bp overlaps;). All mitogenomes shared the same two types of gene overlaps: *atp8-atp6* (7 bp) and *nad4-nad4l* (7 bp). Gene overlaps were also found in other known Scolytinae mitogenomes [15,27–29].

Intergenic spacers were identified in the four mitogenomes, including 14 intergenic spacers in *S. raja*, 15 in *S. sinensis*, 17 in *S. skyliuae*, and 13 in *S. wugongshanensis*. The length of intergenic spacers ranged from 1 bp to 68 bp (Table S3). The longest intergenic spacer was located between *trnS2* and *nad1* in *S. raja*.

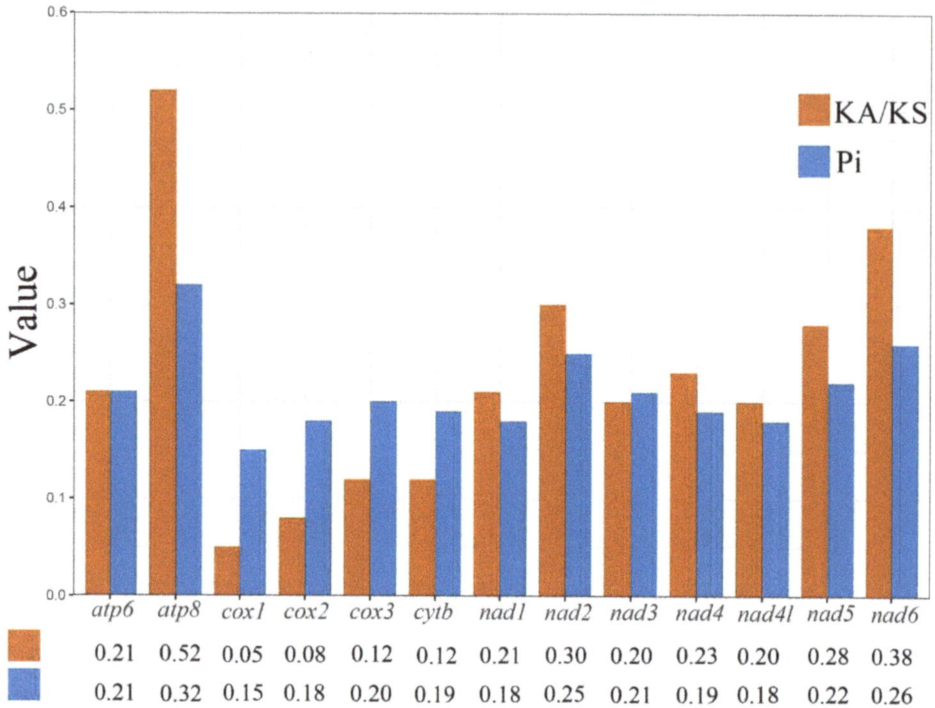

Figure 3. Nucleotide diversity (Pi) and nonsynonymous (Ka)/synonymous (Ks) mutation rate ratios of 13PCGs of *Scolytoplatypus* species (the Pi and Ka/Ks values of each PCG are shown under the gene name).

3.5. Transfer RNA, Ribosomal RNA Genes, and Non-Coding Regions

The position and secondary structures of tRNA genes were identified by the MITOS server. The mitogenomes of the four species each contained 22 typical tRNAs, and 14 tRNAs were encoded by the J-strand; the others were encoded by the N-strand (Figure 1 and Table S3). The length of the four Scolytoplatypodini mitogenomes ranged from 59 bp (*trnS1*, *trnY*) to 71 bp (*trnQ*) in *S. raja*, from 59 bp (*trnY*) to 71 bp (*trnQ*) in *S. sinensis*, from 55 bp (*trnS1*) to 68 bp (*trnQ*, *trnM*, *trnK*) in *S. skyliuae*, and from 57 bp (*trnS1*) to 70 bp (*trnK*) in *S. wugongshanensis* (Table S3).

A typical tRNA consists of a discriminator nucleotide, amino-acyl (AA) arm, TΨC (T) arm, variable (V) arm, anticodon (AC) arm, and dihydrouridine (DHU) arm. All tRNAs exhibit a canonical cloverleaf structure, except for *trnS1* lacking the DHU arm, it is common in most metazoan mitogenomes for *trnS1* to lack the DHU arm [8,11,36]. Such abnormal tRNAs may sustain their function through a posttranscriptional RNA editing mechanism [37,38]. Nine mismatched base pairs (UU, GG, AA, UG, UC, AG, AC, two A, and single U) of tRNAs were found in the four Scolytoplatypodini mitogenomes. The *trnY* not only lacks the discriminator nucleotide but also lost three bases, so the three bases (GGU) of the 5′ end are exposed (Figure 4). The *trnA* (*S. raja*) and *trnK* (*S. sinensis* and *S. skyliuae*) also lack the discriminator nucleotide.

The *rrnL* genes of the four Scolytoplatypodini mitogenomes are located at the intergenic region between *trnL1* and *trnV*, with lengths that range from 1290 bp to 1313 bp. The *rrnS* genes were located between *trnV* and the D-loop, with sizes ranging from 753 bp to 760 bp. The two rRNAs are all encoded on the N-strand and have a high A/T bias that reached 81.5% in *S. raja*, 80.9% in *S. sinensis*, 80.3% in *S. skyliuae*, and 80.5% in

S. wugongshanensis. Since rRNAs do not have functional annotation features like PCGs, it is so hard to establish their boundaries [11,39].

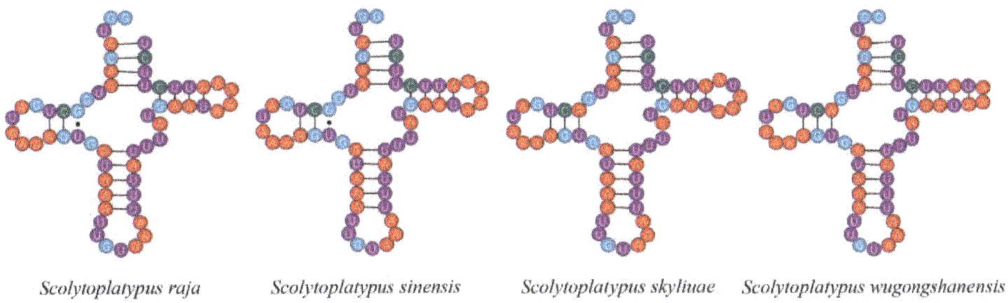

Figure 4. Secondary structures of the *trnY* in *Scolytoplatypus* mitogenomes.

The D-loop acts in the initiation and regulation of replication and transcription in insects [12,40]. The D-loops of the four Scolytoplatypodini mitogenomes are all located between the *rrnS* and *trnI*. The full lengths of the D-loops are 586 bp in *S. raja*, 621 bp in *S. sinensis*, 495 bp in *S. skyliuae*, and 670 bp in *S. wugongshanensis*. The AT content ranged from 81.2% (*S. raja*) to 85.3% (*S. sinensis*). There were different lengths of repeat sequences in the D-loop.

3.6. Phylogenetic Analysis

The phylogeny of Scolytinae, based on the 13 PCGs, was constructed (Figures 5 and S5) using 32 Scolytinae species and two out-groups (*Sitophilus oryzae* and *Sitophilus zeamais*). The ML tree has a similar topology to the BI tree, and their support values are reported above and below the nodes. Compared to the ML tree, the BI tree had higher confidence values, and the monophyly of most tribes and genera of the species studied was well supported. The node support values of BI trees were always higher than ML trees, especially in low-valued branches in the ML tree.

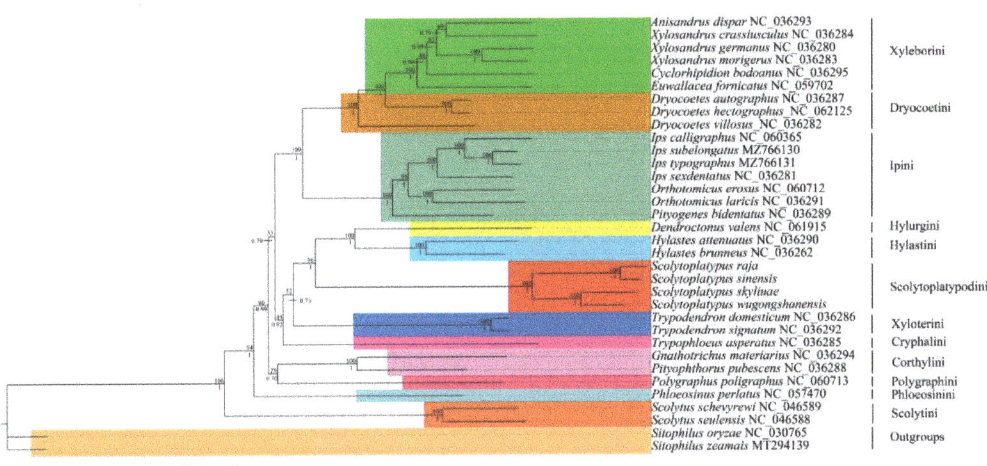

Figure 5. Phylogenetic tree of Scolytinae inferred from ML and BI methods based on 13 PCGs. The ML tree has the same topology as the BI tree, and their support values are reported above and below the nodes, respectively.

The clades consisting of Hylurgini, Hylastini, and Scolytoplatypodini had a high support value. However, within each of the close clades, Xyloterini and Cryphalini, the internal relationship among the included group was not strongly supported in the ML results. The monophyly of Scolytoplatypodini was confirmed. All *Scolytoplatypus* were clustered on one branch, and the nodes received high support. The *Scolytoplatypus* split into two branches; one branch was composed of *S. skyliuae* and *S. wugongshanensis*, and, despite their morphological similarities, the molecular data were diagnostic. The other branch was composed of *S. raja* and *S. sinensis*. The branch of *S. skyliuae* and *S. wugongshanensis* formed a sister lineage with *S. raja* and *S. sinensis*. These results support those of previous morphological and phylogenetic studies [1]. The Scolytini was the first diverging lineage in the subfamily Scolytinae lineage and was highly supported. This result was consistent with previous studies [41–43]. The Ipini, Dryocoetini, and Xyleborini were clustered in a branch and had high support values. This was consistent with the results of Du et al. [15]. The Corthylini and Polygraphini had low support values, although they were clustered together.

This is the first analysis of the relationship between Scolytoplatypodini in Scolytinae based on the mitogenome. However, the relationship between the Scolytoplatypodini and the other partial groups remains unclear. This is primarily due to the limited number of published mitogenomes within the Scolytinae. This issue could be resolved when the mitogenomes of additional Scolytinae species are obtained.

4. Conclusions

In this study, we sequenced and analyzed the mitogenomes of *S. raja*, *S. sinensis*, *S. skyliuae*, and *S. wugongshanensis*, in which mitogenomes had common and similar structures. Gene rearrangements are unusually rare in the mitogenomes in Coleoptera, especially PCGs. [44]. The complete mitogenomes of the four Scolytoplatypodini species, like other Scolytinae, revealed high A/T bias [15,45,46]. Most PCGs start with ATN except for *nad1* in *S. raja* and *S. sinensis*, which begin with TTG, and for *atp6* in *S. raja*, which begins with GTG. Only *atp6* and *cox2* (*S. skyliuae*) are terminated with an incomplete T residue, other PCGs terminated with TAA/TAG. By analysis of KA/KS, *cox1* was determined to be the most conserved gene. In contrast, *atp8* and *nad6* had relatively higher evolutionary rates that differed from several bark beetles [15]. Most tRNA molecules have a typical cloverleaf structure, but *trnS1* lacks the DHU and *trnA* (*S. raja*) and *trnK* (*S. sinensis* and *S. skyliuae*) lack the discriminator nucleotide. The *trnY* lacks not only the discriminator nucleotide but also the three bases. While the *cox1* gene is the most conserved, the *atp8* and *nad6* have relatively higher evolutionary rates, and this differs from other bark beetles [15]. The phylogenetic tree using 13PCGs showed the monophyly of Scolytoplatypodini. Compared to the results of Du et al. [15], our use of 13 PCGs to construct the ML tree yielded the same results as obtained with the BI tree. This implies that the phylogenetic tree that was constructed using 13 PCGs obtained more stable results.

On the issue of whether Scolytoplatypodini belongs to Scolytinae or Platypodinae, there is no doubt that Scolytoplatypodini belongs to Scolytinae. However, Scolytinae and Platypodinae have the same convergent evolution as resemblant wood-boring. Nevertheless, current molecular data show that Dryophthorinae and Platypodinae are indeed sister groups, refuting the close relationship between Scolytinae and Platypodinae [7,14,47]. *Remansus* was the second genus in Scolytoplatypodini. Although no samples of *Remansus* were obtained in this study, Pistone et al. [41] showed *Scolytoplatypus* as a sister group to *Remansus* with high node support, then both clustering with other Scolytinae, for example, genera *Scolytodes* and *Gymnochilus*. Jordal [6] suggested that the African *Scolytoplatypus* and the Asian *Scolytoplatypus* might represent two unique genera. In males, the Asian *Scolytoplatypus* differ from the African *Scolytoplatypus* by having a strongly modified prosternum with nodules or hooked projections, by the longer and more triangular antennal club, and in all but two species by the large fovea on the anterolateral angle of the male pronotum [6]. African *Scolytoplatypus* and Asian *Scolytoplatypus* have previously been divided into two major phylogenetic branches [1,6]. However, few representative specimens were sequenced in this study, making it difficult to provide a more detailed discussion. The four new mitogenomes were from species collected in Asia. It is necessary to sequence more *Scolytoplatypus* species to explore the relationship of African *Scolytoplatypus* with the Asian *Scolytoplatypus*.

Supplementary Materials: The following supporting information can be downloaded at: https://www.mdpi.com/article/10.3390/genes14010162/s1, Figure S1: Secondary structures for tRNA genes from the mtDNA of *S. raja*; Figure S2: Secondary structures for tRNA genes from the mtDNA of *S. sinensis*; Figure S3: Secondary structures for tRNA genes from the mtDNA of *S. skyliuae*; Figure S4: Secondary structures for tRNA genes from the mtDNA of *S. wugongshanensis*; Figure S5: Phylogenetic tree of Scolytinae using the BI method based on 13 PCGs. Table S1: The best-fit partitioning scheme and corresponding models used on the phylogenetic tree; Table S2: Nucleotide compositions in regions of the *S. raja*, *S. sinensis*, *S. skyliuae*, and *S. wugongshanensis* mitochondrial genomes; Table S3: Mitogenomic characteristics of *S. raja*/*S. sinensis*/*S. skyliuae*/*S. wugongshanensis*.

Author Contributions: Conceptualization, G.Y., S.L. (Shengchang Lai), H.T. and J.W.; data curation and methodology, G.Y., S.L. (Shengchang Lai) and W.L.; software, G.Y., S.L. (Shengchang Lai), Y.C., C.L. and W.L.; resources, S.L. (Shengchang Lai)., S.L. (Song Liao) and Y.C.; writing—original draft, G.Y. and S.L. (Shengchang Lai); writing—review and editing, G.Y., S.L. (Shengchang Lai), W.L. and J.W.; funding acquisition, H.T. and J.W. All authors have read and agreed to the published version of the manuscript.

Funding: This study was supported by the National Natural Science Foundation of China (No. 31160380, 31360457, 31760543) and the National Foreign Experts Program of Jiangxi Province (G20200222010, G2021022002, DL2022022001L).

Institutional Review Board Statement: Not applicable.

Informed Consent Statement: Not applicable.

Data Availability Statement: The mitogenomes were deposited at NCBI, with accession numbers OP719285, OP719283, OP719284, and OP712675.

Acknowledgments: We thank LetPub (www.letpub.com, accessed on 15 November 2022) for its linguistic assistance during the preparation of this manuscript.

Conflicts of Interest: The authors declare no conflict of interest.

References

1. Liao, S.; Lai, S.; Beaver, R.A.; Gebhardt, H.; Wang, J. New species and new records of *Scolytoplatypus* Schaufuss (Curculionidae, Scolytinae) from China, and resurrection of *Scolytoplatypus sinensis* (Tsai & Huang, 1965) as a distinct species. *ZooKeys* **2022**, *1082*, 27–50. [PubMed]
2. Gebhardt, H.; Beaver, R.A.; Allgaier, C. Three new species of *Scolytoplatypus* Schaufuss from China, and notes on the movement and functions of the prosternal processes (Coleoptera: Curculionidae: Scolytinae). *Zootaxa* **2021**, *5082*, 485–493. [CrossRef] [PubMed]
3. Mayers, C.G.; Harrington, T.C.; Masuya, H.; Jordal, B.H.; McNew, D.L.; Shih, H.H.; Roets, F.; Kietzka, G.J. Patterns of coevolution between ambrosia beetle mycangia and the *Ceratocystidaceae*, with five new fungal genera and seven new species. *Persoonia* **2020**, *44*, 41–66. [CrossRef]
4. Beaver, R.A.; Gebhardt, H. A review of the Oriental species of *Scolytoplatypus* Schaufuss (Coleoptera, Curculionidae, Scolytinae). *Dtsch. Entomol. Z.* **2006**, *53*, 155–178. [CrossRef]
5. Bhalla, O.; Sharma, P. A new record of shot-holeborer, *scolytoplatypus raja* blandaf. (scolytidae: Col.) of apple trees in H.P. *Curr. Sci.* **1963**, *32*, 86.
6. Jordal, B.H. Deep phylogenetic divergence between *Scolytoplatypus* and *Remansus*, a new genus of Scolytoplatypodini from Madagascar (Coleoptera, Curculionidae, Scolytinae). *ZooKeys* **2013**, *352*, 9–33. [CrossRef]
7. Gillett, C.P.; Crampton-Platt, A.; Timmermans, M.J.; Jordal, B.H.; Emerson, B.C.; Vogler, A.P. Bulk de novo mitogenome assembly from pooled total DNA elucidates the phylogeny of weevils (Coleoptera: Curculionoidea). *Mol. Biol. Evol.* **2014**, *31*, 2223–2237. [CrossRef]
8. Cameron, S.L. Insect mitochondrial genomics: Implications for evolution and phylogeny. *Annu. Rev. Entomol.* **2014**, *59*, 95–117. [CrossRef]
9. Li, R.; Zhang, W.; Ma, Z.; Zhou, C. Novel gene rearrangement pattern in the mitochondrial genomes of *Torleya mikhaili* and *Cincticostella fusca* (Ephemeroptera: Ephemerellidae). *Int. J. Biol. Macromol.* **2020**, *165*, 3106–3114. [CrossRef]

10. Sharma, A.; Siva, C.; Ali, S.; Sahoo, P.K.; Nath, R.; Laskar, M.A.; Sarma, D. The complete mitochondrial genome of the medicinal fish, *Cyprinion semiplotum*: Insight into its structural features and phylogenetic implications. *Int. J. Biol. Macromol.* **2020**, *164*, 939–948. [CrossRef] [PubMed]
11. Boore, J.L. Animal mitochondrial genomes. *Nucleic Acids Res.* **1999**, *27*, 1767–1780. [CrossRef] [PubMed]
12. Zhang, D.-X.; Hewitt, G.M. Insect mitochondrial control region: A review of its structure, evolution and usefulness in evolutionary studies. *Biochem. Syst. Ecol.* **1997**, *25*, 99–120. [CrossRef]
13. Barr, C.M.; Neiman, M.; Taylor, D.R. Inheritance and recombination of mitochondrial genomes in plants, fungi and animals. *New Phytol.* **2005**, *168*, 39–50. [CrossRef]
14. Haran, J.; Timmermans, M.J.; Vogler, A.P. Mitogenome sequences stabilize the phylogenetics of weevils (Curculionoidea) and establish the monophyly of larval ectophagy. *Mol. Phylogenet. Evol.* **2013**, *67*, 156–166. [CrossRef] [PubMed]
15. Du, H.; Fang, J.; Shi, X.; Zhang, S.; Liu, F.; Yu, C.; Zhang, Z.; Kong, X. Comparative Analysis of Eight Mitogenomes of Bark Beetles and Their Phylogenetic Implications. *Insects* **2021**, *12*, 949. [CrossRef]
16. Li, J.Y.; Li, W.X.; Wang, A.T.; Yu, Z. MitoFlex: An efficient, high-performance toolkit for animal mitogenome assembly, annotation, and visualization. *Bioinformatics* **2021**, *37*, 3001–3003. [CrossRef]
17. Bernt, M.; Donath, A.; Jühling, F.; Externbrink, F.; Florentz, C.; Fritzsch, G.; Pütz, J.; Middendorf, M.; Stadler, P.F. MITOS: Improved de novo metazoan mitochondrial genome annotation. *Mol. Phylogenet. Evol.* **2013**, *69*, 313–319. [CrossRef]
18. Kearse, M.; Moir, R.; Wilson, A.; Stones-Havas, S.; Cheung, M.; Sturrock, S.; Buxton, S.; Cooper, A.; Markowitz, S.; Duran, C.; et al. Geneious Basic: An integrated and extendable desktop software platform for the organization and analysis of sequence data. *Bioinformatics* **2012**, *28*, 1647–1649. [CrossRef]
19. Greiner, S.; Lehwark, P.; Bock, R. OrganellarGenomeDRAW (OGDRAW) version 1.3.1: Expanded toolkit for the graphical visualization of organellar genomes. *Nucleic Acids Res.* **2019**, *47*, W59–W64. [CrossRef]
20. Kumar, S.; Stecher, G.; Li, M.; Knyaz, C.; Tamura, K. MEGA X: Molecular Evolutionary Genetics Analysis across Computing Platforms. *Mol. Biol. Evol.* **2018**, *35*, 1547–1549. [CrossRef]
21. Zhang, D.; Gao, F.; Jakovlic, I.; Zou, H.; Zhang, J.; Li, W.X.; Wang, G.T. PhyloSuite: An integrated and scalable desktop platform for streamlined molecular sequence data management and evolutionary phylogenetics studies. *Mol. Ecol. Resour.* **2020**, *20*, 348–355. [CrossRef] [PubMed]
22. Perna, N.T.; Kocher, T.D. Patterns of nucleotide composition at fourfold degenerate sites of animal mitochondrial genomes. *J. Mol. Evol.* **1995**, *41*, 353–358. [CrossRef] [PubMed]
23. Rozas, J.; Ferrer-Mata, A.; Sanchez-DelBarrio, J.C.; Guirao-Rico, S.; Librado, P.; Ramos-Onsins, S.E.; Sanchez-Gracia, A. DnaSP 6: DNA Sequence Polymorphism Analysis of Large Data Sets. *Mol. Biol. Evol.* **2017**, *34*, 3299–3302. [CrossRef]
24. Katoh, K.; Standley, D.M. MAFFT Multiple Sequence Alignment Software Version 7: Improvements in Performance and Usability. *Mol. Biol. Evol.* **2013**, *30*, 772–780. [CrossRef] [PubMed]
25. Talavera, G.; Castresana, J. Improvement of Phylogenies after Removing Divergent and Ambiguously Aligned Blocks from Protein Sequence Alignments. *Syst. Biol.* **2007**, *56*, 564–577. [CrossRef]
26. Lanfear, R.; Frandsen, P.B.; Wright, A.M.; Senfeld, T.; Calcott, B. PartitionFinder 2: New Methods for Selecting Partitioned Models of Evolution for Molecular and Morphological Phylogenetic Analyses. *Mol. Biol. Evol.* **2016**, *34*, 772–773. [CrossRef]
27. Wang, L.J.; Hsu, M.H.; Liu, T.Y.; Lin, M.Y.; Sung, C.H. Characterization of the complete mitochondrial genome of *Euwallacea fornicatus* (Eichhoff, 1868) (Coleoptera: Curculionidae: Scolytinae) and its phylogenetic implications. *Mitochondrial DNA B* **2020**, *5*, 3502–3504. [CrossRef]
28. Xu, M.F.; Meng, R.; Shui, K.J.; Cai, B.; Lin, W. The complete mitochondrial genome of *Ips calligraphus* (Germar 1824) (Coleoptera: Curculionidae: Scolytinae). *Mitochondrial DNA B* **2021**, *6*, 2494–2495. [CrossRef]
29. Zhang, L.; Wen, Y.; Lai, S.; He, P.; Li, T.; Zhou, Q.; Wang, J. The complete mitochondrial genome of *Scolytus schevyrewi* Semenov (Coleoptera: Curculionidae). *Mitochondrial DNA Part B* **2020**, *5*, 1841–1842. [CrossRef]
30. Ojo, J.A.; Valero, M.C.; Sun, W.; Coates, B.S.; Omoloye, A.A.; Pittendrigh, B.R. Comparison of full mitochondrial genomes for the rice weevil, *Sitophilus oryzae* and the maize weevil, *Sitophilus zeamais* (Coleoptera: Curculionidae). *Agri Gene* **2016**, *2*, 29–37. [CrossRef]
31. Ronquist, F.; Teslenko, M.; van der Mark, P.; Ayres, D.L.; Darling, A.; Höhna, S.; Larget, B.; Liu, L.; Suchard, M.A.; Huelsenbeck, J.P. MrBayes 3.2: Efficient Bayesian Phylogenetic Inference and Model Choice Across a Large Model Space. *Syst. Biol.* **2012**, *61*, 539–542. [CrossRef] [PubMed]
32. Nguyen, L.-T.; Schmidt, H.A.; von Haeseler, A.; Minh, B.Q. IQ-TREE: A Fast and Effective Stochastic Algorithm for Estimating Maximum-Likelihood Phylogenies. *Mol. Biol. Evol.* **2014**, *32*, 268–274. [CrossRef] [PubMed]
33. Minh, B.Q.; Nguyen, M.A.T.; von Haeseler, A. Ultrafast Approximation for Phylogenetic Bootstrap. *Mol. Biol. Evol.* **2013**, *30*, 1188–1195. [CrossRef] [PubMed]
34. Ojala, D.; Montoya, J.; Attardi, G. tRNA punctuation model of RNA processing in human mitochondria. *Nature* **1981**, *290*, 470–474. [CrossRef]
35. Hurst, L.D. The Ka/Ks ratio: Diagnosing the form of sequence evolution. *Trends Genet.* **2002**, *18*, 486. [CrossRef]
36. Garey, J.R.; Wolstenholme, D.R. Platyhelminth mitochondrial DNA: Evidence for early evolutionary origin of a tRNA(serAGN) that contains a dihydrouridine arm replacement loop, and of serine-specifying AGA and AGG codons. *J. Mol. Evol.* **1989**, *28*, 374–387. [CrossRef]

37. Lavrov, D.V.; Brown, W.M.; Boore, J.L. A novel type of RNA editing occurs in the mitochondrial tRNAs of the centipede Lithobius forficatus. *Proc. Natl. Acad. Sci. USA* **2000**, *97*, 13738–13742. [CrossRef]
38. Masta, S.E.; Boore, J.L. The Complete Mitochondrial Genome Sequence of the Spider *Habronattus oregonensis* Reveals Rearranged and Extremely Truncated tRNAs. *Mol. Biol. Evol.* **2004**, *21*, 893–902. [CrossRef]
39. Ballard, J.W. Comparative genomics of mitochondrial DNA in members of the *Drosophila melanogaster* subgroup. *J. Mol. Evol.* **2000**, *51*, 48–63. [CrossRef]
40. Zhang, D.-X.; Szymura, J.M.; Hewitt, G.M. Evolution and structural conservation of the control region of insect mitochondrial DNA. *J. Mol. Evol.* **1995**, *40*, 382–391. [CrossRef]
41. Pistone, D.; Gohli, J.; Jordal, B.H. Molecular phylogeny of bark and ambrosia beetles (Curculionidae: Scolytinae) based on 18 molecular markers. *Syst. Entomol.* **2017**, *43*, 387–406. [CrossRef]
42. Jordal, B.H.; Cognato, A.I. Molecular phylogeny of bark and ambrosia beetles reveals multiple origins of fungus farming during periods of global warming. *BMC Evol. Biol.* **2012**, *12*, 133. [CrossRef] [PubMed]
43. Smith, S.M.; Cognato, A.I. A taxonomic monograph of Nearctic Scolytus Geoffroy (Coleoptera, Curculionidae, Scolytinae). *ZooKeys* **2014**, *450*, 1–182. [CrossRef] [PubMed]
44. Timmermans, M.J.; Vogler, A.P. Phylogenetically informative rearrangements in mitochondrial genomes of Coleoptera, and monophyly of aquatic elateriform beetles (Dryopoidea). *Mol. Phylogenet. Evol.* **2012**, *63*, 299–304. [CrossRef]
45. Liu, N.; Fang, L.; Zhang, Y. The Complete Mitochondrial Genomes of Four Species in the Subfamily Limenitidinae (Lepidoptera, Nymphalidae) and a Phylogenetic Analysis. *Insects* **2021**, *13*, 16. [CrossRef]
46. Zou, Z.; Min, Q.; Cheng, S.; Xin, T.; Xia, B. The complete mitochondrial genome of *Thitarodes sejilaensis* (Lepidoptera: Hepialidae), a host insect of Ophiocordyceps sinensis and its implication in taxonomic revision of Hepialus adopted in China. *Gene* **2017**, *601*, 44–55. [CrossRef]
47. Mugu, S.; Pistone, D.; Jordal, B. New molecular markers resolve the phylogenetic position of the enigmatic wood-boring weevils Platypodinae (Coleoptera: Curculionidae). *Arthropod Syst. Phylogeny* **2018**, *76*, 45–58.

genes

Article

The Adaptive Evolution in the Fall Armyworm *Spodoptera frugiperda* (Lepidoptera: Noctuidae) Revealed by the Diversity of Larval Gut Bacteria

Yan-Ping Wang [1], Xu Liu [1], Chun-Yan Yi [1], Xing-Yu Chen [2], Chang-Hua Liu [1], Cui-Cui Zhang [1], Qing-Dong Chen [1], Song Chen [1], Hong-Ling Liu [1] and De-Qiang Pu [1,*]

[1] Key Laboratory of Integrated Pest Management of Southwest Crops, Institute of Plant Protection, Sichuan Academy of Agricultural Sciences, Chengdu 610066, China
[2] Service Center of Sichuan Academy of Agricultural Sciences, Chengdu 610066, China
* Correspondence: pudeqiang@ippsaas.org.cn

Abstract: Insect gut microbes have important roles in host feeding, digestion, immunity, development, and coevolution with pests. The fall armyworm, *Spodoptera frugiperda* (Smith, 1797), is a major migratory agricultural pest worldwide. The effects of host plant on the pest's gut bacteria remain to be investigated to better understand their coevolution. In this study, differences in the gut bacterial communities were examined for the fifth and sixth instar larvae of *S. frugiperda* fed on leaves of different host plants (corn, sorghum, highland barley, and citrus). The 16S rDNA full-length amplification and sequencing method was used to determine the abundance and diversity of gut bacteria in larval intestines. The highest richness and diversity of gut bacteria were in corn-fed fifth instar larvae, whereas in sixth instar larvae, the richness and diversity were higher when larvae were fed by other crops. Firmicutes and Proteobacteria were dominant phyla in gut bacterial communities of fifth and sixth instar larvae. According to the LDA Effect Size (LEfSe) analysis, the host plants had important effects on the structure of gut bacterial communities in *S. frugiperda*. In the PICRUSt2 analysis, most predicted functional categories were associated with metabolism. Thus, the host plant species attacked by *S. frugiperda* larvae can affect their gut bacterial communities, and such changes are likely important in the adaptive evolution of *S. frugiperda* to host plants.

Keywords: *Spodoptera frugiperda*; coevolution; 16S rDNA; gut bacteria diversity; bacterial functional analysis

Citation: Wang, Y.-P.; Liu, X.; Yi, C.-Y.; Chen, X.-Y.; Liu, C.-H.; Zhang, C.-C.; Chen, Q.-D.; Chen, S.; Liu, H.-L.; Pu, D.-Q. The Adaptive Evolution in the Fall Armyworm *Spodoptera frugiperda* (Lepidoptera: Noctuidae) Revealed by the Diversity of Larval Gut Bacteria. *Genes* **2023**, *14*, 321. https://doi.org/10.3390/genes14020321

Academic Editor: Alfred M. Handler

Received: 16 November 2022
Revised: 19 January 2023
Accepted: 24 January 2023
Published: 26 January 2023

1. Introduction

The fall armyworm, *Spodoptera frugiperda* (Smith, 1797) [1] (Lepidoptera: Noctuidae), is a major migratory agricultural pest of global concern. It originated in tropical and subtropical regions of America, where it is widely distributed [2–4]. *S. frugiperda* invaded Ghana and Nigeria in southwest Africa in 2016 [5,6] and was later reported in India in 2018 [7]. The pest was subsequently recorded in Sri Lanka, Thailand, Myanmar, and other Asian countries [8]. The armyworm invaded Yunnan in southern China in 2018 and then rapidly expanded to most areas of the country [9–11]. *S. frugiperda* is an omnivorous pest with strong migratory ability, a high rate of reproduction, and a short life cycle. It damages 353 species of plants in 76 families, including the major crops corn, sorghum, sugarcane, barley, rice, pepper, wild oat, and potato [12]. Maize crops can suffer serious economic loss when attacked [13]. However, information on the effective prevention and control strategies for *S. frugiperda* is lacking. *S. frugiperda* has developed resistance to a variety of insecticides, including diamide and neonicotinoids [14]. Because *S. frugiperda* is a newly invaded major agricultural pest in China, native natural enemy insects are being investigated to provide biological control [15]. Nevertheless, these control measures might be harmful to the environment and not economically viable. Therefore, there is a great need for alternative green control methods for *S. frugiperda*.

Insect gut microbiomes are important for the digestion of food, absorption of nutrients, and general metabolism [16,17]. The gut and associated microbes resist invasion and colonization by external pathogens [18,19], degrade harmful substances, and produce drug resistance [20]. Communities of gut bacteria can promote host absorption and utilization of food [21], and different foods can also affect the composition and metabolic function of gut bacterial communities [22]. Although many plant tissues are low in nutrients, indigestible, or toxic, herbivorous insects are among the most numerous and diverse groups of organisms [23,24]. Herbivorous insects have numerous morphological, behavioral, and physiological characteristics that enable them to overcome dietary barriers [25–27]. Some insects can adapt to new host plants, and in that process, changes occur in the abundance and composition of gut enzymes that reduce the toxicity of plant allelochemicals [28–30]. Therefore, the composition and diversity of gut bacterial communities have been one of the recent hotspots of entomological research. Enriching the understanding of the coevolution between insects and their gut bacteria can provide a theoretical basis for pest control [31–33].

The host plant is an important factor affecting insect gut bacteria. The feeding habits of insects impact the composition and structure as well as the diversity and function of gut bacterial communities [34]. Gut bacterial communities have been investigated in an increasing number of insects, including bees [35,36], fruit flies [37,38], beetles [39], termites [40], and other common pests [41–44]. In ground beetles, their food habits and habitats affect their gut bacterial and fungal communities [39]. In cluster analysis of relative abundances of orthologous gene clusters, high similarities were observed among wood- and litter-feeding termites, but those groups had strong differences with humivorous species [40]. In a cockroach pest, the gut bacteria are highly dynamic, and bacterial communities reassemble relatively rapidly and with different compositions in a diet-specific manner (the highest diversity was associated with a no-protein diet) [41]. The flexibility of the gut bacteria is most likely due to the fact that cockroaches are omnivorous with variable diets [41]. In a comparative analysis of the moth pests' midgut bacterial diversity, the plant species influenced the composition of the gut bacterial community; the moth larvae reared on an artificial diet and different host plants revealed significantly different compositions and diversity of gut bacterial communities [42]. Thus, host plants can greatly influence the composition and structure of gut bacterial communities in pests, which may be essential in long-term adaptation to host plants [43]. Similarly, gut bacterial communities of another moth pest are influenced by the host diet and therefore may also be important in adaptation to the hosts [44].

Although the gut bacteria of *S. frugiperda* have been examined previously [45–48], how feeding on different host plants affects the composition and functions of gut bacterial communities is not fully clear. In this study, *S. frugiperda* larvae were reared on leaves of corn, sorghum, highland barley, and citrus. Then, 16S rDNA sequence amplification was used to compare the effects of different host plants on the structure, diversity, and functions of gut bacterial communities in *S. frugiperda*. The results will provide a foundation to generate new ideas for further study of the effects of host plants on gut bacterial communities of *S. frugiperda* and the adaptive evolution of this important pest. In addition, new insights may lead to the manipulation of gut bacterial communities for pest control of *S. frugiperda*.

2. Materials and Methods

2.1. Insect Collection and Laboratory Feeding

The samples of *S. frugiperda* were collected from a corn field in the Base of Xindu, Sichuan Academy of Agricultural Sciences, Chengdu, China. The larvae were fed on an artificial diet containing the following contents (g/L): soybean powder (225), wheat powder (125), yeast (40), casein (20), cholesterol (0.6), and agar (30) [49]. Insects were reared at our laboratory for three generations at 27 ± 1 °C with 70 ± 5% relative humidity, and a light:dark = 16:8 h photoperiod. Larvae were separately reared on leaves of corn, sorghum, highland barley, and citrus in the laboratory. Host plants included corn (*Zea mays* L. var. Chengdan 11, ZmL), sorghum (*Sorghum bicolor* L. Moench. Chuannuo 15, SbL), highland barley (*Hordeum vulgare* L. var. Kangqing 9, HvL), and citrus (*Citrus reticulata* Blanco. Chunjian, CrB). Plants were cultivated to the 3–4-true-leaf stages. Newly hatched larvae were reared to 5th and 6th instars on fresh young leaves of the four different host plants, respectively. All experiments and procedures for this study complied with the current animal ethics guidelines and did not involve any protected animals.

A total of 32 gut samples of 5th and 6th instar larvae of *S. frugiperda* fed on different host plants were collected and profiled. Thus, the experiment had two treatment factors: host plant and larval stage. Host plants were corn (Zm), sorghum (Sb), highland barley (Hv), and citrus (Cr), and larval stages were 5th instar (L1 or B1) and 6th instar (L2 or B2). Therefore, there were eight treatment combinations: ZmL1, ZmL2, SbL1, SbL2, HvL1, HvL2, CrB1, CrB2.

2.2. Processing of *S. frugiperda* Larvae

To ensure the gut bacteria were in a relatively stable state, the *S. frugiperda* larvae were transferred to new centrifuge tubes and starved for 24 h in a natural environment. After all materials were prepared, dissections were performed on an ultra-clean bench. First, beakers were prepared with sterile water and absolute ethanol. Larvae were removed from centrifuge tubes, soaked in absolute ethanol for 90 s, and then blotted on filter paper. Larvae were then washed three times with sterile water, blotted dry, and placed in petri dishes. Under a stereomicroscope, the head of a larva was held with pointed tweezers, and medical scissors were used to cut along the abdomen below the mouth. Ganglion, salivary glands, martensitic ducts, fat bodies, and other organs were carefully removed. Then, the intestine was completely removed, placed in a sterile centrifuge tube, quickly frozen with liquid nitrogen, and stored at −80 °C.

2.3. DNA Extraction and 16S rDNA Sequencing

To extract the total DNA from gut contents, a PowerSoil DNA Isolation Kit (MOBIO Laboratories Inc., Carlsbad, CA, USA) was used according to the protocol provided by the manufacturer. The integrity of extracted DNA was confirmed by agarose gel electrophoresis. Extracted DNA was quantified in a Qubit 2.0 (Invitrogen, Carlsbad, CA, USA), and 10 ng/uL was used for amplification and sequencing of the 16S rDNA genes from 32 samples. PCR full-length amplification was performed using the 16S primers F (5′AGAGTTTGA TCCTGGCTCAG3′) and R (5′GNTACCTTGTTACGACTT3′) with a Phusion® High-Fidelity PCR Master Mix (New England Biolabs Inc., Ipswich, MA, USA) under the following conditions: 94 °C for 5 min; 35 cycles of 94 °C for 30 s, 56 °C for 30 s, 72 °C for 30 s; 72 °C for 5 min. This was followed by product purification, construction of a SMART bel library, and sequencing on PacBio [50]. Total DNA was sent to Beijing Novogene Bioinformatics Technology Co., Ltd. (Beijing, China) for sequencing.

2.4. Statistical Analyses

PacBio offline data were exported to a bam file through PacBio's SMRT analysis software (version 7.0). After samples were distinguished according to barcodes, operational taxonomic unit (OTU) clustering and classification analysis were performed. Sequences that were less than 1340 bp or greater than 1640 bp were removed. Uparse software (http://drive5.com/uparse, accessed on 15 January 2022) was used to cluster the clean reads. The sequences were clustered into OTUs (operational taxonomic units) with 97% identity. Species annotation analysis was performed using the Mothur method with the SSUrRNA database of SILVA (http://www.arb-silva.de, accessed on 15 January 2022). We used MUSCLE (http://www.drive5.com/muscle, accessed on 15 January 2022) to perform the rapid multiple sequence alignment and then obtained all the OTUs' representative sequences. The subsequent analysis of α diversity and β diversity was based on the standardized data.

The abundances of OTUs were analyzed according to the results of OTU clustering, and a petal map was prepared. α diversity reflects the abundance and species diversity of sample species. QIIME software v1.9.1 was used to calculate α diversity indices, including Chao1, Simpson, and Shannon. The raw data were tested for normality and homogeneity of variance using the Shapiro–Wilk and Levene's test, respectively. After log-transforming the data, normality was confirmed ($p > 0.05$) and the data were suitable for parametric analysis. ANOVA (one-way analysis of variance) followed by Tukey's tests were performed to test the difference between host plants, where the diversity of gut bacteria was the response variable. Differences were considered significant when $p < 0.05$. R software v2.15.3 was used to analyze the differences between groups in the β diversity index, and the LDA Effect Size (LEfSe) analysis was used to test the significance of differences in the composition and structure of bacterial communities in samples from different treatments. Last, PICRUSt2 (https://github.com/picrust/picrust2, accessed on 15 January 2022) was used to predict the metabolic functions of bacterial communities based on the KEGG database (https://www.kegg.jp, accessed on 15 January 2022).

3. Results

3.1. Sequence Analysis

Thirty-two *S. frugiperda* gut samples were examined. A total of 568,300 original reads and 748,360,295 bp of original bases were obtained (Table 1). After filtration, 15,802 high-quality average reads and 566 unique average reads were obtained (Figure S1). From the fifth and sixth instar larvae raised on different host plants, 498 and 562 OTUs, respectively, were obtained from sequencing data (Tables S1 and S2). Gut bacteria were classified into nine phyla, 14 classes, 32 orders, 56 families, 93 genera, and 66 species. Differences in the OTUs of gut bacteria in different larval instars fed on different host plants were compared in a flower plot (Figure 1). Although only seven OTUs of gut bacteria were shared among different *S. frugiperda* instars fed on different host plants, they indicated there were similarities in the composition of bacterial communities. In fifth instar larvae of *S. frugiperda*, the number of unique OTUs was 68 in those fed on corn (ZmL1), 14 in those fed on citrus (CrB1), 12 in those fed on sorghum (SbL1), and 2 in those fed on highland barley (HvL1). In sixth instar larvae, the number of unique OTUs was 54 in those fed on sorghum (SbL2), 26 in those fed on highland barley (HvL2), 15 in those fed on corn (ZmL2), and 2 in those fed on citrus (CrB2). Thus, the composition of gut bacterial communities was different in *S. frugiperda* fed on different host plants.

Table 1. Effective reads data for subsequent analysis after quality control.

Sample Name	Raw Reads *	Clean Reads *	Base (nt)	AvgLen (nt) *	Effective (%) *
HvL1.1	18,434	16,712	24,823,416	1485	90.66
HvL1.2	13,733	12,297	18,259,102	1484	89.54
HvL1.3	12,142	10,472	15,560,783	1485	86.25
HvL1.4	16,263	14,725	21,871,158	1485	90.54
SbL1.1	16,291	14,668	21,773,693	1484	90.04
SbL1.2	20,232	18,538	27,167,447	1465	91.63
SbL1.3	23,156	19,666	29,188,401	1484	84.93
SbL1.4	23,467	21,067	31,264,901	1484	89.77
ZmL1.1	17,639	15,895	23,578,489	1483	90.11
ZmL1.2	10,148	9020	13,191,119	1462	88.88
ZmL1.3	12,649	10,831	15,562,182	1436	85.63
ZmL1.4	13,328	12,007	17,818,003	1483	90.09
CrB1.1	26,445	23,292	34,155,876	1466	88.08
CrB1.2	22,652	20,542	30,485,052	1484	90.69
CrB1.3	18,589	16,033	23,804,488	1484	86.25
CrB1.4	14,176	12,932	18,909,984	1462	91.22

Table 1. *Cont.*

Sample Name	Raw Reads *	Clean Reads *	Base (nt)	AvgLen (nt) *	Effective (%) *
HvL2.1	12,065	11,046	16,295,806	1475	91.55
HvL2.2	15,120	13,924	20,641,745	1482	92.09
HvL2.3	25,269	21,263	31,586,883	1485	84.15
HvL2.4	21,157	19,133	28,414,494	1485	90.43
SbL2.1	15,491	13,452	19,895,164	1478	86.84
SbL2.2	15,980	14,028	20,809,492	1483	87.78
SbL2.3	18,850	16,365	24,214,434	1479	86.82
SbL2.4	25,471	22,426	33,299,817	1484	88.05
ZmL2.1	12,346	11,205	16,617,495	1483	90.76
ZmL2.2	19,422	17,006	25,168,430	1479	87.56
ZmL2.3	17,704	16,464	24,431,320	1483	93
ZmL2.4	13,393	12,090	17,803,176	1472	90.27
CrB2.1	25,248	21,817	32,396,921	1484	86.41
CrB2.2	13,162	11,940	17,728,360	1484	90.72
CrB2.3	16,103	14,615	21,684,846	1483	90.76
CrB2.4	22,175	20,180	29,957,818	1484	91
Total	568,300	505,651	748,360,295	47319	2852.5

* Raw reads represent the number of original reads sequenced by PacBio. Clean reads are the number of high-quality reads obtained after quality control and splicing. AvgLen (nt) is the average sequence length of all samples. Effective (%) is the percentage of effective reads in raw reads.

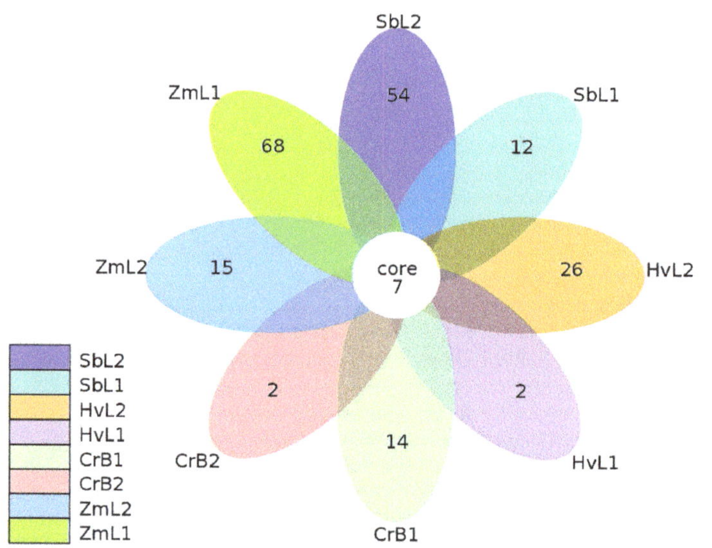

Figure 1. Flower plot of bacterial operational taxonomic units (OTUs) in guts of *Spodoptera frugiperda* larvae fed on leaves of different host plants. Each petal in the flower represents a treatment, and the core number in the overlapped parts of the petals represents the number of OTUs shared among treatments. The numbers at petal edges represent the number of unique OTUs in a treatment. Treatments: letters represent different host plants (Zm, corn; Sb, sorghum; Hv, highland barley; Cr, citrus), and numbers represent different larval instars (L1 or B1, fifth instar; L2 or B2, sixth instar).

3.2. *Taxa Annotation and Relative Abundance*

The relative abundance of gut bacteria was determined at different taxonomic levels. All samples typically included nine main phyla (Figure 2A and Table S3). In the fifth instar *S. frugiperda*, Firmicutes (avg. 78.48%) was the most abundant phylum among gut bacteria, followed by Proteobacteria (avg. 20.27%), with other phyla at much lower relative abundance, including Bacteroidetes, Cyanobacteria, Actinobacteria, Verrucomicrobia, Chloroflexi, and unidentified Bacteria. In sixth instar larvae, Firmicutes (avg. 90.76%) was also the most abundant phylum, followed by Proteobacteria (avg. 7.57%). Phyla of bacteria were highly consistent between the two instars, with Firmicutes and Proteobacteria dominant in both instars. There were no significant effects of host plants on the phyla of gut bacteria ($p > 0.05$). Although the same phyla were dominant in the guts of the two larval instars, their relative abundances were different.

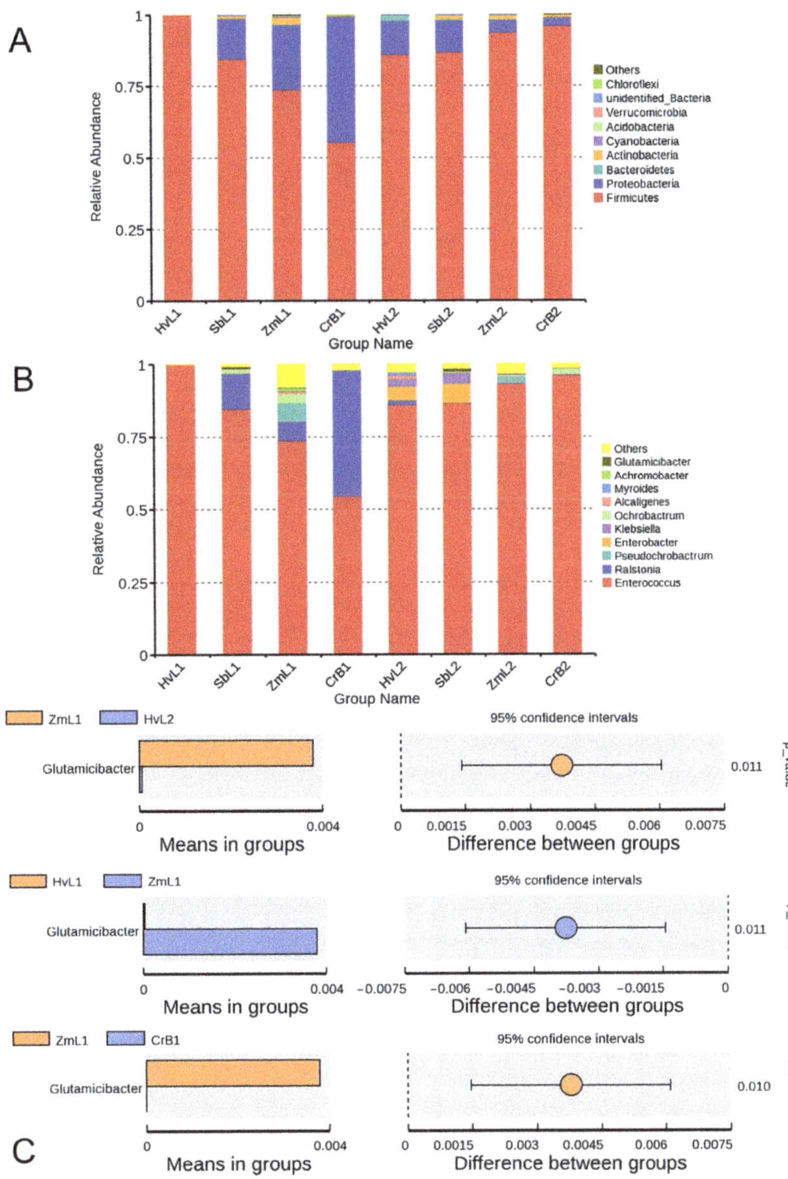

Figure 2. Relative abundance of the most predominant taxa of gut bacteria in *S. frugiperda* larvae fed on leaves of different host plants at the (**A**) phyla and (**B**) genera levels. (**C**) *T*-test analysis of species differences between groups. Others represents the sum of the relative abundances of all phyla (genera) other than the phyla (genera) in the figure. Each bar in the figure represents the mean value of species with significant differences in abundance between groups. It is the *p*-value for the between-group significance test for the corresponding differing species.

The dominant genera in fifth instar *S. frugiperda* were primarily *Enterococcus* (avg. 78.26%) and *Ralstonia* (avg. 15.54%), with other genera at a lower relative abundance, including *Pseudochrobactrum*, *Enterobacter*, *Klebsiella*, *Ochrobactrum*, *Alcaligenes*, *Myroides*, *Achromobacter*, and *Glutamicibacter* (Figure 2B and Table S4). In sixth instar larvae, *Enterococcus* (avg. 90.54%) was also the dominant genus, but *Ralstonia* (avg. 0.43%) was less abundant. Whereas *Ralstonia* composed 43.33% of the gut community in the fifth instar larvae fed on citrus (CrB1), the genus composed only 0.03% in the sixth instar larvae fed on corn (ZmL2). The relative abundance of *Glutamicibacter* in the fifth instar larvae fed on corn (ZmL1) was 0.38%, which was significantly different from that on other host treatments, especially in fifth and sixth instar larvae fed on highland barley (HvL1 and HvL2) and fifth instar larvae fed on citrus (CrB1) ($p < 0.05$) (Figure 2C). Thus, the phyla and genera of gut bacteria in *S. frugiperda* reared on different hosts were the same, but relative abundances at each taxonomic level were different.

3.3. Diversity of Gut Bacteria

α diversity of bacterial communities in different treatments was analyzed (Figure 3 and Table S5). The highest Chao index of gut bacteria was 108.34 in ZmL1, followed by 60.073 in SbL2. Thus, the richness of gut bacterial communities was highest in the larvae fed on corn and sorghum. Fifth instar larvae fed on corn (ZmL1) had the highest Shannon and Simpson diversity values (2.153 and 0.654, respectively). Shannon and Simpson indices between ZmL1 and CrB1 were significantly different ($p < 0.05$). Compared with fifth instar larvae fed on corn, the richness and diversity of gut bacteria decreased when larvae fed on leaves of other hosts. Compared with the fifth instar larvae, diversity indices of gut bacterial communities in the sixth instar larvae increased when fed on citrus, sorghum, and highland barley. Thus, there were differences in gut bacterial communities between larval stages in *S. frugiperda*.

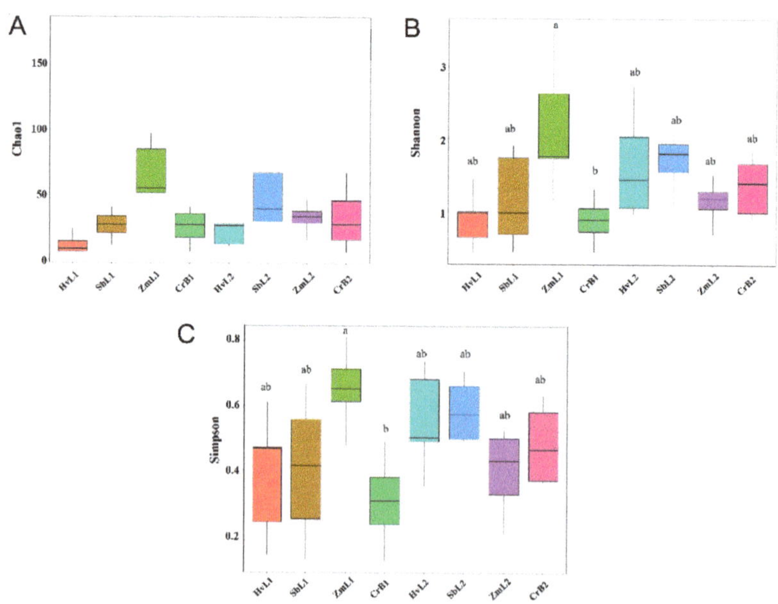

Figure 3. Box plots of (**A**) Chao1, (**B**) Shannon, and (**C**) Simpson α diversity indices of gut bacterial communities in *S. frugiperda* larvae fed on leaves of different host plants. The a and b indicate the significant differences in relative abundance in the same column in the mean values. Different letters above boxes indicate significant differences among treatments (one-way ANOVA, Tukey's post hoc test) in the mean values.

To better reflect the nonlinear structure of data on gut bacteria in *S. frugiperda* fed on different hosts, nonmetric multidimensional calibration (NMDS) was performed on sequencing data based on Bray–Curtis distances (Figure 4). The distance between gut bacteria in the fifth instars fed on corn and those in other host plant treatments was relatively large, indicating there were differences in gut bacteria among the different treatments. Differences in gut bacterial communities in fifth and sixth instars fed on different plants were analyzed (Figure 5). In fifth instars fed on corn, the gut bacterial community was significantly enriched from genus to phylum levels. According to linear discriminant analysis effect size (LEfSe), nine bacterial clades were consistently significantly enriched in ZmL1 samples (Figure 5C). Each larval stage had a unique, significantly enriched set of bacteria at taxonomic levels ranging from phylum to species. For example, the genera *Pseudochrobactrum*, *Paenochrobactrum*, and *Ochrobactrum* were notably enriched in ZmL1 when compared with other hosts, whereas *Enterobacter* was notably enriched in HvL1, and *Providencia* was enriched in HvL2. Thus, different bacterial groups were enriched in different larval stages fed on different host plants.

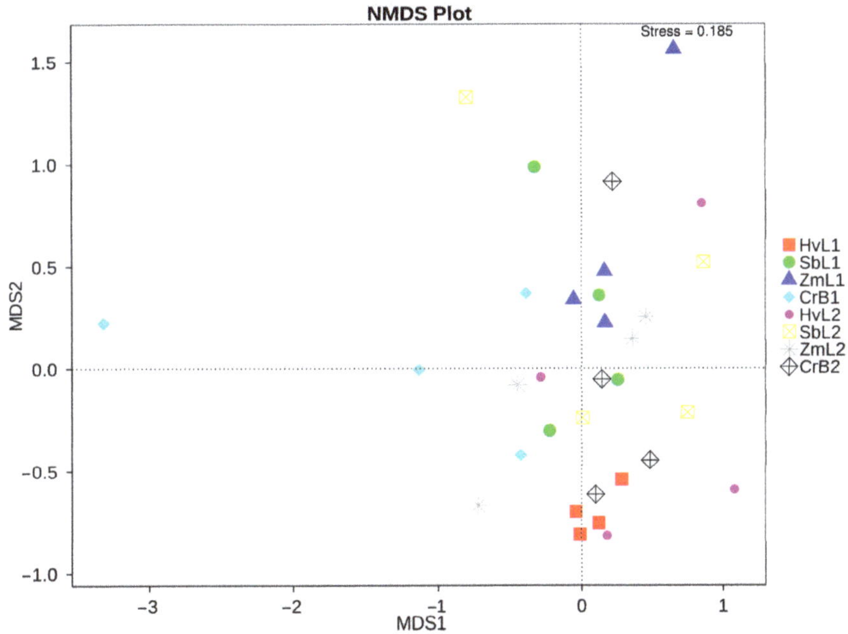

Figure 4. Nonmetric multi-dimensional scaling (NMDS) analysis of gut bacterial communities in *S. frugiperda* larvae fed on leaves of different host plants. Each point in the figure represents a sample; distance between points represents degree of difference, and samples in the same treatment are the same color. When stress is less than 0.2, the NMDS accurately reflects degree of difference between samples.

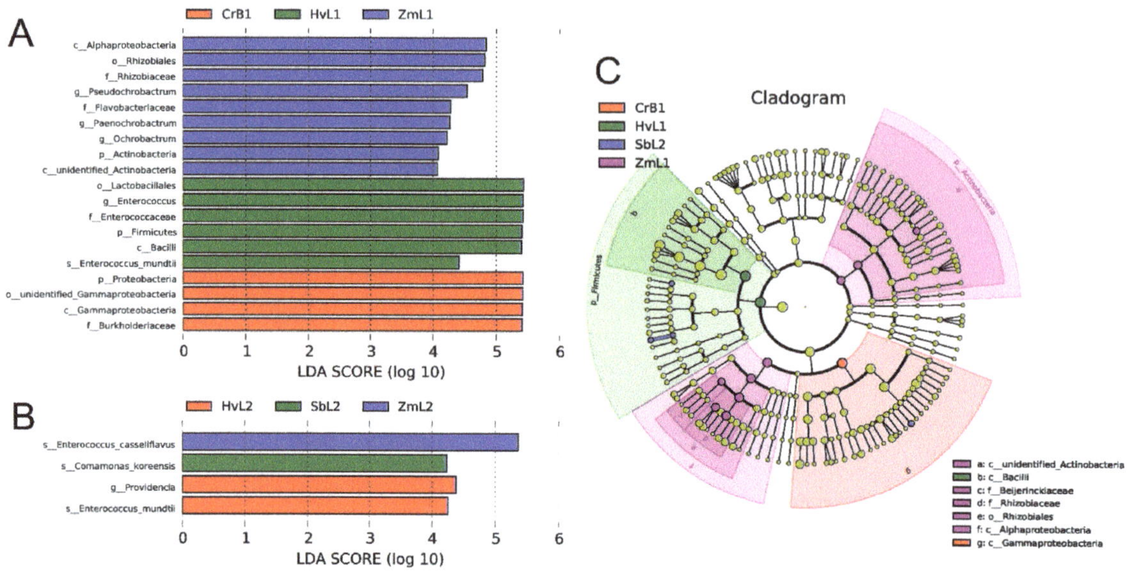

Figure 5. Linear discriminant analysis (LDA) of gut bacterial taxa in *S. frugiperda* larvae fed on leaves of different host plants. Taxa with LDA score (log 10) greater than four in (**A**) fifth and (**B**) sixth instar larvae. (**C**) LEfSe (LDA Effect Size) analysis showing significant differences in bacterial taxa at the level of phylum, class, order, family, and genus, from inside to outside. Small circles at different classification levels represent classifications at a particular level, and their diameters represent relative abundances. Nodes of different colors represent bacteria that were significantly enriched with the corresponding host. Significantly different biomarkers follow the group for coloring. Small yellow nodes indicate bacterial taxa that were not significantly different in guts of larvae fed on different hosts.

3.4. Cluster Analysis of Predominant Bacteria

The cluster heat map in Figure 6 shows annotation and abundance information for the top 35 genera based on relative abundance. The genera of gut bacteria in *S. frugiperda* fed on corn (ZmL1, ZmL2), sorghum (SbL1, SbL2), barley (HvL1, HvL2), and citrus (CrB1, CrB2) were clustered in different branches. As shown in the horizontal direction, the abundance of each genus was different in different larval stages fed on different hosts. For both the fifth and sixth instar larvae of *S. frugiperda*, the dominant genera of gut bacteria were also different when reared on different plants.

3.5. Prediction of Bacterial Functions

To better understand the important functions of gut bacteria in *S. frugiperda*, relative abundances of Kyoto Encyclopedia of Genes and Genomes (KEGG) pathways were predicted based on the 16S rDNA gene sequences using PICRUSt2. Functions of gut bacteria primarily involved six types of metabolic pathways: metabolism, genetic information processing, environmental information processing, cellular processes, organismal systems, and human diseases (Figure 7A). Gut bacteria primarily functioned in metabolism-associated pathways, which accounted for $45.39 \pm 1.07\%$. In the analysis of the second functional layer of predicted genes (Figure 7B), functions included membrane transport, signal transduction, carbohydrate metabolism, amino acid metabolism, energy metabolism, cell motility, and xenobiotic biodegradation and metabolism among other pathways.

The guts of the fifth and sixth instar larvae of *S. frugiperda* reared on different plants were enriched in different functional proteins (Figure 7C). For example, cold shock protein (K03704) and chitin-binding protein (K03933) were significantly enriched in the fifth instar larvae reared on highland barley (HvL1). Gut microbiomes were enriched in several ABC transporter-related KOs (KEGG Orthogroups), including phosphate and amino acid transporters (K01999), permease protein (K02029, K01997, K01998), ATP-binding protein (K01996), periplasmic binding protein (K01999), hypothetical protein (K02030), and peptide and nickel transporters (K02035). All of the predicted pathways perform the most important functions in the gut and therefore are important in the overall growth and development of *S. frugiperda* larvae.

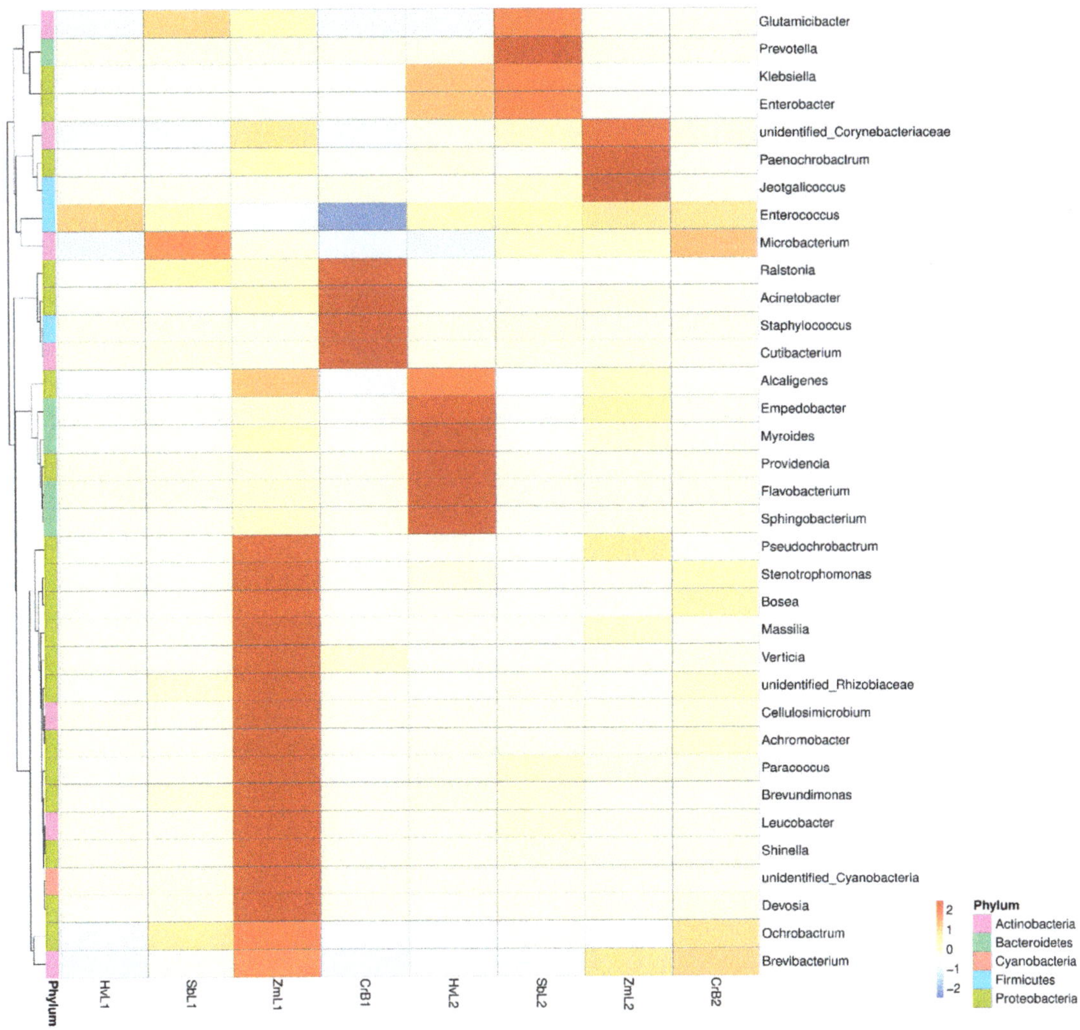

Figure 6. Heat map of relative abundances of the top thirty-five predominant genera of gut bacteria separated by phylum in *S. frugiperda* larvae fed on leaves of different host plants. Treatment names are on the *x*-axis, and genus annotation is on the *y*-axis. The clustering tree for genera is on the left, and heat map values are Z-values obtained after relative abundances of each genus were standardized.

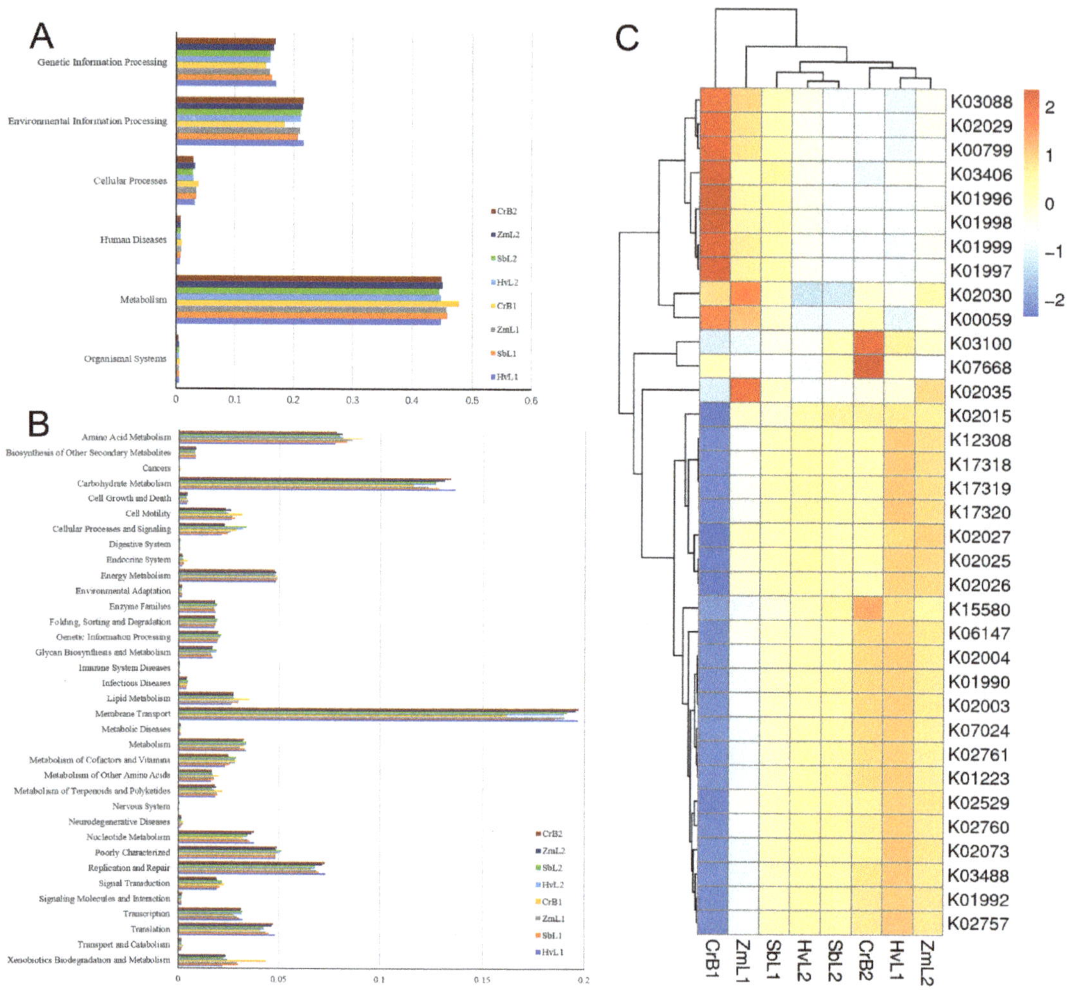

Figure 7. Annotations of KEGG-predicted functions of gut bacterial communities in *S. frugiperda* larvae fed on leaves of different host plants. (**A**) Level 1 and (**B**) level 2. (**C**) PICRUSt2 prediction of proteins based on functions in the KEGG database.

4. Discussion

Multiple external and internal factors could influence the gut microbial community of insect herbivores, such as weather, temperature, usage of antibiotics, host phylogeny, host treatments, diet of insect, host immune recognition, bacterial sources, immigration and competition of microbes, and different developmental stages of insects [51–54]. Meanwhile, the natural populations of *S. frugiperda* were found to have more diverse gut microbiota and significantly higher diversity of bacteria functional in metabolizing insecticides than laboratory-reared populations [55]. Among these factors, diet or the host plant is the most important factor shaping gut microbiota in *S. frugiperda* [56]. Conversely, the plasticity of the gut microbes can help the insects utilize different foods and enhance the fitness of *S. frugiperda* as a pest.

The omnivorous pest *S. frugiperda* feeds on a wide range of crops, and the analysis of the metagenomic DNA of its gut bacteria can provide the basis for pest control research. However, little is known about how host plants affect the diversity of gut bacteria in

S. frugiperda. It is essential to identify differences in gut bacteria in *S. frugiperda* feeding on different plants because of their potentially significant effects on larval growth and development. In this study, the total number of OTUs was different in gut bacterial communities of *S. frugiperda* reared on leaves of different hosts, with numbers of OTUs in the larvae fed on corn and sorghum higher than those in larvae fed on highland barley and citrus. This result indicates that different host plants can strongly influence the microbial communities of *S. frugiperda*, similarly to other lepidopterans [57]. Different host plants have variable nutritional content, palatability, and secondary metabolites, which could affect the growth and development of *S. frugiperda* [58]. The significant microbial differences in larvae of *S. frugiperda* fed on different hosts might have resulted from the different nutritional content and inhibitory secondary metabolites in the four hosts used herein.

In this study, the abundance and diversity of gut bacteria were analyzed in fifth and sixth instar larvae of *S. frugiperda* reared on leaves of corn, sorghum, highland barley, and citrus. We found that the dominant microbes were the same at different larval stages of *S. frugiperda*, but their proportions and compositions were variable. In the α diversity analysis, the abundance and diversity of gut bacteria in HvL1, CrB1, and SbL1 showed an upward trend when compared with ZmL1. There were also differences in the abundance and the diversity of gut bacteria in larvae fed on different plants, and the diversity and abundance of gut bacteria in ZmL1 decreased from the fifth to sixth instars, which may be a result of the presence of one or more inhibitory components in the corn diet provided for the larvae. Such a trend of decrease in α diversity has also been recorded for the larvae of *S. frugiperda* fed on maize leaves in a recent study [52]. Different host plants can cause differences in insect gut microenvironments, which, in turn, lead to differences in gut microbial diversity [59]. The results of this study are consistent with those on other insects. A previous study [60] found differences in dominant flora and their abundances in the guts of a moth pest feeding on three different types of pine trees. Another study [61] found that dietary substrate affects the gut bacteria in cockroaches, with changes in food leading to changes in the dietary matrix available for gut bacteria and, ultimately, changes in gut flora. The inconsistency in α diversity analysis between different studies may be caused by the different sources and treatment of plants, as well as the different sources and rearing conditions of insect samples.

At the phylum level, the composition of gut bacterial communities in fifth and sixth larvae was different among larvae fed on different plants, but Firmicutes was the dominant phylum in both larval stages, followed by Proteobacteria. Previous studies found similar results in *S. frugiperda* [62], and gut bacterial communities of most samples were dominated by Firmicutes. The high abundance of Firmicutes in wild oats is due to better absorption of different nutrients [63]. However, in larvae and adults of *S. frugiperda* from different maize-growing areas in Kenya, Firmicutes was only dominant in one Ngeria (Ngeria-l2) larva and two Kitale (Kitale-m2 and Kitale-m3) adult males [57]. In a closely related lepidopteran pest, *Spodoptera exigua* (Hübner, 1808) [64], the larval gut microbes were found to be 97.9% Proteobacteria and 2.1% Firmicutes [65]. The difference in bacterial proportion between closely related species might have resulted from the different origins of the insect colonies and the remarkable differences in bacterial identification methods [65].

At the genus level, *Enterococcus* was the dominant genus of gut bacteria in all eight treatment combinations of *S. frugiperda*. *Enterococcus* is also known as the most common intestinal bacterium in the order Lepidoptera [42–44,65]. *Enterococcus* can degrade alkaloids and latex and, therefore, has a stabilizing role in insect tolerance to their toxic diet [63]. In this study, the relative abundance of *Enterococcus* increased significantly with the increase in food intake from the fifth to sixth instars of *S. frugiperda*. The higher abundance of *Enterococcus* could assist digestion and accelerate the development of resistance to insecticides [52]. The dramatic loss of *Ralstonia* from the fifth to sixth instars was also detected in this study. *Ralstonia* has been commonly found in the guts of many insects; it originates from leaf surfaces and comprises important plant pathogens that cause serious damage on a global scale [66,67]. Some studies also speculated that *Ralstonia* may play a vital role in the nitrogen cycle of host insects and the degradation of bactericides [68,69]. However, it is impossible to identify all the *Ralstonia* species accurately, which prevents us from understanding their exact biological function in *S. frugiperda* larvae. The loss of *Ralstonia* from the fifth to sixth instars might be helpful to protect the host plant from being a stable food source for the larvae of *S. frugiperda*. The larval stage of lepidopteran insects suffers the most damage, and changes in larval intestinal microbes can influence the feeding, digestion, growth, and development of these insects [52]. The late larval instars of *S. frugiperda* significantly increase the food intake and their body size grew faster; therefore, the changes in the gut microbiota are associated with the growth and development of *S. frugiperda* [70]. The differences in gut microbes between the fifth to sixth instars of *S. frugiperda* may also be related to the development of their immune systems, gut morphology, and gut physicochemical conditions [71,72]. The significant differences in the gut microbes of different developmental stages have also been reported in previous studies on *S. frugiperda* [70]. The dynamic change in gut bacterial communities might help herbivorous insects adapt to the host plants and play an important role in the physiological metabolism of the insects.

To further study the effects of host plants on the diversity of gut bacterial communities, NMDS was used to examine the β diversity. Among samples from four different hosts, all samples of CrB1 were closely clustered together and separately from samples in other treatments, whereas there was an overlap among samples in the other treatments. Therefore, components of CrB1 gut bacterial communities were apparently different from those in larvae fed on other hosts. There were significant differences in the structure of gut bacterial communities in larvae fed on different hosts. The LEfSe analysis effectively detected differentially abundant bacterial taxa in gut microbiomes. A comparison with existing statistical methods and metagenomic analyses of the environmental, gut microbiome, and synthetic data shows that LEfSe analysis consistently provides lower false positive rates and can effectively aid in explaining the biology underlying differences in microbial communities [73]. In general, the results in this study confirm that feeding on different host plants alters the structure of gut bacterial communities in *S. frugiperda* larvae, which is similar to the results for other lepidopterans [56].

PICRUSt2 software was used to analyze the functions of gut bacteria [74]. There were 35 predicted functions of gut bacteria in larvae feeding on different hosts, with most related to metabolic functions. In the analysis of differences in KEGG metabolic pathways, guts of the fifth and sixth instar larvae fed on different hosts were obviously enriched with different functional proteins in most metabolic pathways. In ZmL1 and ZmL2 treatments (larvae fed on corn), samples were enriched with different functional proteins, with genes associated with ABC transport function accounting for the largest proportion. Notably, the bacterial detoxification pump is based on ABC transporters in several main categories: the ABC superfamily [75], the major promoter superfamily [76], and the small multidrug resistance family [77]. The composition of gut bacterial communities in *S. frugiperda* and the ability of members of those communities to metabolize insecticides differ depending on the diversity of chemicals used to treat the host [55]. Accumulation of detoxification and defense genes in the gut of *S. frugiperda* may be related to the diversity of food intake or the variable host environment. The specific factors of influence still need to be verified with further experiments.

This study showed that different host plants had important effects on the structure and diversity of gut bacterial communities in *S. frugiperda*. Host-induced changes in the structure and metabolic functions of gut bacterial communities likely assist *S. frugiperda* larvae in adapting to different food sources. This work provides a good foundation for further exploration of interactions between gut bacteria and hosts for *S. frugiperda*. The results also provide insights into the selection of dominant gut microbial members as potential targets for biological control of the pest. Further research on gut microbes should include more life stages and more host plants, which could provide more perspectives and directions for the adaptive evolution and integrated management of *S. frugiperda*.

Supplementary Materials: The following supporting information can be downloaded at: https://www.mdpi.com/article/10.3390/genes14020321/s1, Figure S1: Statistics of valid tags and operational taxonomic unit (OTU) clustering of each sample; Table S1: Statistics of valid tags and operational taxonomic unit (OTU) clustering of fifth instar larvae; Table S2: Statistics of valid tags and operational taxonomic unit (OTU) clustering of sixth instar larvae; Table S3: Relative abundance of the most predominant taxa of gut bacteria in different treatment at the phylum level; Table S4: Relative abundance of the most predominant taxa of gut bacteria in different treatment at the genus level; Table S5: α diversity indices of gut bacterial communities in different treatment.

Author Contributions: Conceptualization, H.-L.L. and X.L.; methodology, C.-Y.Y. and C.-H.L.; software, Y.-P.W. and X.-Y.C.; writing—original draft preparation, Y.-P.W.; writing—review and editing, Y.-P.W. and D.-Q.P.; visualization, Y.-P.W.; investigation, Q.-D.C., S.C. and C.-C.Z.; funding acquisition, X.L. and H.-L.L.; project administration, D.-Q.P. and X.L. All authors have read and agreed to the published version of the manuscript.

Funding: This research was funded by the Research on Feeding Mechanism of the Fall Armyworm larvae, grant number 2021YJ0261.

Institutional Review Board Statement: No special permits were required to retrieve and process the samples because the study did not involve any live vertebrates, nor regulated invertebrates.

Informed Consent Statement: Not applicable.

Data Availability Statement: The data presented in this study are available in the Supplementary Materials.

Acknowledgments: We thank LetPub (www.letpub.com, accessed on 4 February 2022) for the linguistic assistance during the preparation of this manuscript.

Conflicts of Interest: The authors declare no conflict of interest.

References

1. Smith, J.E.; Abbott, J. *The Natural History of the Rarer Lepidopterous Insects of Georgia: Including their Systematic Characters, the Particulars of Their Several Metamorphoses, and the Plants on Which They Feed*; Missouri Botanical Garden Press: London, UK, 1797. [CrossRef]
2. Sparks, A.N. A review of the biology of the fall armyworm. *Fla. Entomol.* **1979**, *62*, 82–87. [CrossRef]
3. Johnson, S.J. Migration and the life history strategy of the fall armyworm, *Spodoptera frugiperda* in the western hemisphere. *Int. J. Trop. Insect Sci.* **1987**, *8*, 543–549. [CrossRef]
4. Todd, E.L.; Poole, R.W. Keys and illustrations for the armyworm moths of the noctuid genus Spodoptera Guenée from the western hemisphere. *Ann. Entomol. Soc. Am.* **1980**, *73*, 722–738. [CrossRef]
5. Goergen, G.; Kumar, P.L.; Sankung, S.B.; Togola, A.; Tamò, M. First report of outbreaks of the fall armyworm *Spodoptera frugiperda* (J E Smith) (Lepidoptera, Noctuidae), a new alien invasive pest in west and central Africa. *PLoS ONE* **2016**, *11*, e165632. [CrossRef] [PubMed]
6. Cock, M.; Beseh, P.K.; Buddie, A.G.; Cafá, G.; Crozier, J. Molecular methods to detect *Spodoptera frugiperda* in Ghana, and implications for monitoring the spread of invasive species in developing countries. *Sci. Rep.* **2017**, *7*, 4103. [CrossRef]
7. Sharanabasappa, C.; Kalleshwaraswamy, C.M.; Asokan, R.; Swamy, H.M.M.; Maruthi, M.S.; Pavithra, H.B.; Hegbe, K.; Navi, S.; Prabhu, S.T.; Goergen, G.E. First report of the fall armyworm, *Spodoptera frugiperda* (J E Smith) (Lepidoptera: Noctuidae), an alien invasive pest on maize in India. *Pest Manag. Hortic. Ecosyst.* **2018**, *24*, 23–29.
8. Yang, P.; Chang, X. The occurrence and impact of *Spodoptera frugiperda* in Asia and Africa and its prevention and control strategies. *China Plant Prot. Guide* **2019**, *39*, 88–90.
9. Jiang, Y.; Liu, J.; Zhu, X. Analysis of the occurrence and future trends of the invasion of *Spodoptera frugiperda* in my country. *China Plant Prot.* **2019**, *39*, 33–35.
10. Jiang, Y.; Liu, J.; Xie, M.; Li, Y.; Yang, J.; Zhang, M.; Qiu, K. Observation on law of diffusion damage of *Spodoptera frugiperda* in China in 2019. *Plant Prot.* **2019**, *45*, 10–19.
11. Sun, X.; Hu, C.; Jia, H.; Wu, Q.; Shen, X.; Zhao, S.; Jiang, Y.; Wu, K. Case study on the first immigration of fall armyworm, *Spodoptera frugiperda* invading into China. *J. Integr. Agric.* **2021**, *20*, 664–672. [CrossRef]
12. Montezano, D.G.; Specht, A.; Sosa-Gómez, D.R.; Specht, A.; Roque-Specht, V.F.; Sousa-Silva, J.C.; Paula-Moraes, S.V.; Peterson, J.A.; Hunt, T.E. Host plants of *Spodoptera frugiperda* (Lepidoptera: Noctuidae) in the Americas. *Afr. Entomol.* **2018**, *26*, 286–300. [CrossRef]
13. Yang, X.; Wyckhuys, K.; Jia, X.; Nie, F.; Wu, K. Fall armyworm invasion heightens pesticide expenditure among Chinese smallholder farmers. *J. Environ. Manag.* **2021**, *282*, 111949. [CrossRef]
14. Silver, K.S.; Soderlund, D.M. Action of pyrazoline-type insecticides at neuronal target sites. *Pestic. Biochem. Phys.* **2005**, *81*, 136–143. [CrossRef]

15. Zhang, Y. Natural enemy insect resources of *Spodoptera frugiperda* and domestic research progress. *Agric. Technol. Serv.* **2020**, *37*, 38–39.
16. Sharon, G.; Segal, D.; Ringo, J.M.; Hefetz, A.; Zilber-Rosenberg, I.; Rosenberg, E. Commensal bacteria play a role in mating preference of Drosophila melanogaster. *Proc. Natl. Acad. Sci. USA* **2010**, *107*, 20051–20056. [CrossRef]
17. Sharon, G.; Segal, D.; Zilber-Rosenberg, I.; Rosenberg, E. Symbiotic bacteria are responsible for diet-induced mating preference in Drosophila melanogaster, providing support for the hologenome concept of evolution. *Gut Microbes* **2011**, *2*, 190–192. [CrossRef]
18. Vollaard, E.J.; Clasener, H.A. Colonization resistance. *Antimicrob. Agents Chemother.* **1994**, *38*, 409–414. [CrossRef]
19. Takatsuka, J.; Kunimi, Y. Intestinal bacteria affect growth of Bacillus thuringiensis in larvae of the oriental tea tortrix, Homona magnanima diakonoff (Lepidoptera: Tortricidae). *J. Invertebr. Pathol.* **2000**, *76*, 222–226. [CrossRef]
20. Paramasiva, I.; Sharma, H.C.; Krishnayya, P.V. Antibiotics influence the toxicity of the delta endotoxins of Bacillus thuringiensis towards the cotton bollworm, Helicoverpa armigera. *BMC Microbiol.* **2014**, *14*, 200. [CrossRef]
21. Song, Y.; Li, N.; Yue, Y.X.; Yan, F.F.; Wang, N.N.; Huo, G.C. Research progress on the effects of dietary main components in diet on intestinal microbial composition and metabolism. *Food Ind. Sci. Technol.* **2019**, *40*, 354–360.
22. Liu, C. *The Effects of Different Sugar Source Feeds on The Gut Health and Gut Microflora of Overwintering Honeybees*; Shandong Agricultural University: Taian, China, 2017.
23. Acevedo, F.E.; Peiffer, M.; Tan, C.W.; Stanley, B.A.; Stanley, A.; Wang, J.; Jones, A.G.; Hoover, K.; Rosa, C.; Luthe, D.; et al. Fall armyworm-associated gut bacteria modulate plant defense responses. *Mol. Plant Microbe Interact.* **2017**, *30*, 127–137. [CrossRef] [PubMed]
24. Broderick, N.A.; Buchon, N.; Lemaitre, B. Microbiota-induced changes in Drosophila melanogaster host gene expression and gut morphology. *mBio* **2014**, *5*, e01117-14. [CrossRef] [PubMed]
25. Chaturvedi, S.; Rego, A.; Lucas, L.K.; Gompert, Z. Sources of variation in the gut microbial community of Lycaeides melissa Caterpillars. *Sci. Rep.* **2017**, *7*, 11335. [CrossRef] [PubMed]
26. Chen, Y.; Zhou, H.; Lai, Y.; Chen, Q.; Yu, X.Q.; Wang, X. Gut microbiota dysbiosis influences metabolic homeostasis in *Spodoptera frugiperda*. *Front. Microbiol.* **2021**, *12*, 727434. [CrossRef] [PubMed]
27. Hood, G.R.; Forbes, A.A.; Powell, T.H.; Egan, S.P.; Hamerlinck, G.; Smith, J.J.; Feder, J.L. Sequential divergence and the multiplicative origin of community diversity. *Proc. Natl. Acad. Sci. USA* **2015**, *112*, E5980–E5989. [CrossRef]
28. Ceja-Navarro, J.A.; Vega, F.E.; Karaoz, U.; Hao, Z.; Jenkins, S.; Lim, H.C.; Kosina, P.; Infante, F.; Northen, T.R.; Brodie, E.L. Gut microbiota mediate caffeine detoxification in the primary insect pest of coffee. *Nat. Commun.* **2015**, *6*, 7618. [CrossRef]
29. Zhu-Salzman, K.; Koiwa, H.; Salzman, R.A.; Shade, R.E.; Ahn, J.E. Cowpea bruchid Callosobruchus maculatus uses a three-component strategy to overcome a plant defensive cysteine protease inhibitor. *Insect Mol. Biol.* **2003**, *12*, 135–145. [CrossRef]
30. Després, L.; David, J.P.; Gallet, C. The evolutionary ecology of insect resistance to plant chemicals. *Trends Ecol. Evol.* **2007**, *22*, 298–307. [CrossRef]
31. Berenbaum, M.R. Allelochemicals in insect-microbe-plant Interactions; agents provocateurs in the coevolutionary arms race. In *Novel Aspects of Insect-Plant Interactions*; Barbosa, P., Letourneau, D.K., Eds.; Wiley-Interscience: New York, NY, USA, 1988; pp. 97–123.
32. Feldhaar, H. Bacterial symbionts as mediators of ecologically important traits of insect hosts. *Ecol. Entomol.* **2011**, *36*, 533–543. [CrossRef]
33. Hansen, A.K.; Moran, N.A. The impact of microbial symbionts on host plant utilization by herbivorous insects. *Mol. Ecol.* **2014**, *23*, 1473–1496. [CrossRef]
34. Semova, I.; Carten, J.D.; Stombaugh, J.; Mackey, L.C.; Knight, R.; Farber, S.A.; Rawls, J.F. Microbiota regulate intestinal absorption and metabolism of fatty acids in the zebrafish. *Cell Host Microbe* **2012**, *12*, 277–288. [CrossRef]
35. Ellegaard, K.M.; Engel, P. Genomic diversity landscape of the honey bee gut microbiota. *Nat. Commun.* **2019**, *10*, 446. [CrossRef]
36. Moran, N.A.; Hansen, A.K.; Powell, J.E.; Sabree, Z.L. Distinctive gut microbiota of honey bees assessed using deep sampling from individual worker bees. *PLoS ONE* **2012**, *7*, e36393. [CrossRef]
37. Bing, X.; Gerlach, J.; Loeb, G.; Buchon, N. Nutrient-Dependent Impact of Microbes on Drosophila suzukii Development. *mBio* **2018**, *9*, e02199-17. [CrossRef]
38. Erkosar, B.; Kolly, S.; van der Meer, J.R.; Kawecki, T.J. Adaptation to chronic nutritional stress leads to reduced dependence on microbiota in Drosophila melanogaster. *MBio* **2017**, *8*, e01496-17. [CrossRef]
39. Kudo, R.; Masuya, H.; Endoh, R.; Kikuchi, T.; Ikeda, H. Gut bacterial and fungal communities in ground-dwelling beetles are associated with host food habit and habitat. *ISME J.* **2019**, *13*, 676–685. [CrossRef]
40. Rossmassler, K.; Dietrich, C.; Thompson, C.; Mikaelyan, A.; Nonoh, J.O.; Scheffrahn, R.H.; Sillam-Dusses, D.; Brune, A. Metagenomic analysis of the microbiota in the highly compartmented hindguts of six wood- or soil-feeding higher termites. *Microbiome* **2015**, *3*, 56. [CrossRef]
41. Pérez-Cobas, A.E.; Maiques, E.; Angelova, A.; Carrasco, P.; Moya, A.; Latorre, A. Diet shapes the gut microbiota of the omnivorous cockroach Blattella germanica. *FEMS Microbiol. Ecol.* **2015**, *91*, fiv022. [CrossRef]
42. Broderick, N.A.; Raffa, K.F.; Goodman, R.M.; Handelsman, J. Census of the bacterial community of the gypsy moth larval midgut by using culturing and culture-independent methods. *Appl. Environ. Microbiol.* **2004**, *70*, 293–300. [CrossRef]
43. Wu, X.; Xia, X.; Chen, J.; Geoff, M.G.; You, M. Effects of different diets on the diversity of larval gut bacteria of the diamondback moth, Plutella xylostella (Lepidoptera: Plutellidae). *Acta Entomol.* **2019**, *62*, 1172–1185.

44. Yuan, X.; Zhang, X.; Liu, X.; Dong, Y.; Yan, Z.; Lv, D.; Wang, P.; Li, Y. Comparison of gut bacterial communities of Grapholita molesta (Lepidoptera: Tortricidae) reared on different host plants. *Int. J. Mol. Sci.* **2021**, *22*, 6843. [CrossRef] [PubMed]
45. Tang, Y.; Wu, Y.; Gu, R.; Zou, X.; Zhang, Z.; Niu, X.; Wang, Z.; Chen, J.; Wu, Y.; Li, T.; et al. Isolation and identification of gut bacteria of *Spodoptera frugiperda* that migrated to Chongqing area. *J. Southwest Univ. Nat. Sci. Ed.* **2019**, *41*, 8–14. [CrossRef]
46. Wei, J.; Tang, Y.; Gu, R.; Wu, Y.; Zou, X.; Zhang, Z.; Chen, J.; Li, C.; Pan, G.; Zhou, Z. Isolation and identification of endophytic bacteria of maize leaf in Chongqing area and comparison between maize microbes and gut bacteria of *Spodoptera frugiperda*. *J. Southwest Univ. Nat. Sci. Ed.* **2019**, *41*, 17–23.
47. Li, Q.; Tang, Y.; Jiang, R.; Zhang, Y.; Feng, Z.; Bai, X.; Gu, R.; Wu, Y.; Wu, Y.; Chen, J.; et al. Isolation and identification of gut bacteria of *Spodoptera frugiperda* feeding on maize in Yunnan, China. *J. Southwest Univ. Nat. Sci. Ed.* **2020**, *42*, 1–8.
48. Xu, T.M.; Fu, C.Y.; Su, Z.T.; Xiao, G.L.; Li, W.W.; Chen, B. Composition and diversity of gut bacteria community of *Spodoptera frugiperda* from the first invasion site and the dispersal area in Yunnan province. *Plant Prot.* **2020**, *46*, 116–125.
49. Wang, S.; Zhu, Q.; Tan, Y.; Ma, Q.; Wang, R.; Zhang, M.; Xu, H.; Zhang, Z. Artificial diets and rearing technique of *Spodoptera frugiperda* (J. E. Smith) in laboratory. *J. Environ. Entomol.* **2019**, *41*, 742–747.
50. Franzén, O.; Hu, J.; Bao, X.; Itzkowitz, S.H.; Peter, I.; Bashir, A. Improved OTU-picking using long-read 16S rRNA gene amplicon sequencing and generic hierarchical clustering. *Microbiome* **2015**, *3*, 43. [CrossRef]
51. Mason, C.J.; Hoover, K.; Felton, G.W. Effects of maize (Zea mays) genotypes and microbial sources in shaping fall armyworm (*Spodoptera frugiperda*) gut bacterial communities. *Sci. Rep.* **2021**, *11*, 4429. [CrossRef]
52. Lü, D.; Dong, Y.; Yan, Z.; Liu, X.; Zhang, Y.; Yang, D.; He, K.; Wang, Z.; Wang, P.; Yuan, X.; et al. Dynamics of gut microflora across the life cycle of *Spodoptera frugiperda* and its effects on the feeding and growth of larvae. *Pest Manag. Sci.* **2023**, *79*, 173–182. [CrossRef]
53. Chen, Y.P.; Li, Y.H.; Sun, Z.X.; Du, E.W.; Lu, Z.H.; Li, H.; Gui, F.R. Effects of host plants on bacterial community structure in larvae midgut of *Spodoptera frugiperda*. *Insects* **2022**, *13*, 373. [CrossRef]
54. Li, Y.; Liu, L.; Cai, X.; Yang, X.; Lin, J.; Shu, B. The bacterial and fungal communities of the larval midgut of *Spodoptera frugiperda* (Lepidoptera: Noctuidae) varied by feeding on two cruciferous vegetables. *Sci. Rep.* **2022**, *12*, 13063.
55. Gomes, A.F.F.; Omoto, C.; Cônsoli, F.L. Gut bacteria of field-collected larvae of *Spodoptera frugiperda* undergo selection and are more diverse and active in metabolizing multiple insecticides than laboratory-selected resistant strains. *J. Pest Sci.* **2020**, *93*, 833–851. [CrossRef]
56. Jones, A.G.; Mason, C.J.; Felton, G.W.; Hoover, K. Host plant and population source drive diversity of microbial gut communities in two polyphagous insects. *Sci. Rep.* **2019**, *9*, 2792. [CrossRef]
57. Gichuhi, J.; Sevgan, S.; Khamis, F.; Van den Berg, J.; du Plessis, H.; Ekesi, S.; Herren, J.K. Diversity of fall armyworm, *Spodoptera frugiperda* and their gut bacterial community in Kenya. *PeerJ* **2020**, *8*, e8701. [CrossRef]
58. He, L.; Wu, Q.; Gao, X.; Wu, K. Population life tables for the invasive fall armyworm, *Spodoptera frugiperda* fed on major oil crops planted in China. *J. Integr. Agric.* **2021**, *20*, 745–754. [CrossRef]
59. Lee, J.H.; Lee, K.A.; Lee, W.J. Microbiota, gut physiology, and insect immunity. *Adv. Insect Physiol.* **2017**, *52*, 111–138. [CrossRef]
60. Strano, C.P.; Malacrinò, A.; Campolo, O.; Palmeri, V. Influence of host plant on Thaumetopoea pityocampa gut bacterial community. *Microb. Ecol.* **2018**, *75*, 487–494. [CrossRef]
61. Tinker, K.A.; Ottesen, E.A. The core gut microbiome of the American cockroach, Periplaneta americana, is stable and resilient to dietary shifts. *Appl. Environ. Microbiol.* **2016**, *82*, 6603–6610. [CrossRef]
62. Lv, D.; Liu, X.; Dong, Y.; Yan, Z.; Zhang, X.; Wang, P.; Yuan, X.; Li, Y. Comparison of gut bacterial communities of fall armyworm (*Spodoptera frugiperda*) reared on different host plants. *Int. J. Mol. Sci.* **2021**, *22*, 11266. [CrossRef]
63. Liu, Y.J.; Shen, Z.J.; Yu, J.M.; Li, Z.; Liu, X.X.; Xu, H.L. Comparison of gut bacterial communities and their associations with host diets in four fruit borers. *Pest Manag. Sci.* **2020**, *76*, 1353–1362. [CrossRef]
64. Hübner, J. *Zuträge zur Sammlung Exotischer Schmettlinge [sic]*; Jacob Hübner: Augsburg, Germany, 1808.
65. Presa-Parra, E.; Lasa, R.; Reverchon, F.; Simón, O.; Williams, T. Use of biocides to minimize microbial contamination in Spodoptera exigua multiple nucleopolyhedrovirus preparations. *Biol. Control* **2020**, *151*, 104408. [CrossRef]
66. Xu, J.; Feng, J. Research progress on genetic diversity and pathogenic genomics of R. *solanacearum*. *Chin. Agric. Sci.* **2013**, *46*, 2902–2909.
67. Gomes, S.I.; Kielak, A.M.; Hannula, S.E.; Heinen, R.; Jongen, R.; Keesmaat, I.; De Long, J.R.; Bezemer, T.M. Microbiomes of a specialist caterpillar are consistent across different habitats but also resemble the local soil microbial communities. *Anim. Microbiome* **2020**, *2*, 37. [CrossRef] [PubMed]
68. Paulson, A.R.; von Aderkas, P.; Perlman, S.J. Bacterial associates of seed-parasitic wasps (Hymenoptera: Megastigmus). *BMC Microbiol.* **2014**, *14*, 224. [CrossRef]
69. Wang, G.; Liang, B.; Li, F.; Li, S. Recent advances in the biodegradation of chlorothalonil. *Curr. Microbiol.* **2011**, *63*, 450–457. [CrossRef]
70. Li, D.D.; Li, J.Y.; Hu, Z.Q.; Liu, T.X.; Zhang, S.Z. Fall armyworm gut bacterial diversity associated with different developmental stages, environmental habitats, and diets. *Insects* **2022**, *13*, 762. [CrossRef]
71. Engel, P.; Moran, N.A. The gut microbiota of insects—Diversity in structure and function. *FEMS Microbiol. Rev.* **2013**, *37*, 699–735. [CrossRef]
72. Dillon, R.J.; Dillon, V.M. The gut bacteria of insects: Nonpathogenic interactions. *Annu. Rev. Entomol.* **2004**, *49*, 71–92. [CrossRef]

73. Segata, N.; Izard, J.; Waldron, L.; Gevers, D.; Miropolsky, L.; Garrett, W.S.; Huttenhower, C. Metagenomic biomarker discovery and explanation. *Genome Biol.* **2011**, *12*, 60. [CrossRef]
74. Douglas, G.M.; Maffei, V.J.; Zaneveld, J.R.; Yurgel, S.N.; Brown, J.R.; Taylor, C.M.; Huttenhower, C.; Langille, M.G. PICRUSt2 for prediction of metagenome functions. *Nat. Biotechnol.* **2020**, *38*, 685–688. [CrossRef]
75. Higgins, C.F. ABC transporters: From microorganisms to man. *Annu. Rev. Cell Biol.* **1992**, *8*, 67–113. [CrossRef]
76. Marger, M.D.; Saier, M.H. A major superfamily of transmembrane facilitators that catalyze uniport, symport and antiport. *Trends Biochem Sci.* **1993**, *18*, 13–20. [CrossRef]
77. Paulsen, I.T.; Skurray, R.A.; Tam, R.; Saier, M.H., Jr.; Turner, R.J.; Weiner, J.H.; Goldberg, E.B.; Grinius, L.L. The SMR family: A novel family of multidrug efflux proteins involved with the efflux of lipophilic drugs. *Mol. Microbiol.* **1996**, *19*, 1167–1175. [CrossRef]

genes

MDPI

Article

Characterizing the Complete Mitochondrial Genomes of Three Bugs (Hemiptera: Heteroptera) Harming Bamboo

Wenli Zhu [1,2,3], Lin Yang [1,2,3], Jiankun Long [1,2,3], Zhimin Chang [1,2,3], Nian Gong [4], Yinlin Mu [1,2,3], Shasha Lv [1,2,3] and Xiangsheng Chen [1,2,3,*]

1 Institute of Entomology, Guizhou University, Guiyang 550025, China
2 The Provincial Special Key Laboratory for Development and Utilization of Insect Resources, Guizhou University, Guiyang 550025, China
3 The Provincial Key Laboratory for Agricultural Pest Management of Mountainous Regions, Guizhou University, Guiyang 550025, China
4 Engineering Research Center of Medicinal Resources and Health Care Products, Guiyang Healthcare Vocational University, Guiyang 550081, China
* Correspondence: xschen@gzu.edu.cn

Abstract: Herein, we report the mitochondrial genomic characteristics of three insect pests, *Notobitus meleagris*, *Macropes harringtonae*, and *Homoeocerus bipunctatus*, collected from bamboo plants in Guizhou Province, China. For the first time, the damaged conditions and life histories of *M. harringtonae* and *H. bipunctatus* are described in detail and digital photographs of all their life stages are provided. Simultaneously, the mitochondrial genome sequences of three bamboo pests were sequenced and analyzed. *Idiocerus laurifoliae* and *Nilaparvata lugens* were used as outgroups, and the phylogenetic trees were constructed. The mitochondrial genomes of the three bamboo pests contained 37 classical genes, including 13 protein-coding genes (PCGs), two ribosomal RNA genes (rRNAs), 22 transfer RNAs (tRNAs), and a control region, with a total length of 16,199 bp, 15,314 bp, and 16,706 bp, respectively. The A+T values of the three bamboo pests were similar, and trnS1 was a cloverleaf structure with missing arms. The phylogenetic analyses, using the Bayesian inference (BI) and Maximum likelihood (ML), supported that *N. meleagris* and *H. bipunctatus* belonged to the Coreoidea family, whereas *M. harringtonae* belonged to the Lygaeoidea family with high support values. This study involves the first complete sequencing of the mitochondrial genomes of two bamboo pests. By adding these newly sequenced mitochondrial genome data and detailed descriptions of life histories, the database of bamboo pests is improved. These data also provide information for the development of bamboo pest control methods by quick identification techniques and the use of detailed photographs.

Keywords: bamboo pests; Coreoidae; Lygaeoidae; mitogenome; hazard condition; life history

Citation: Zhu, W.; Yang, L.; Long, J.; Chang, Z.; Gong, N.; Mu, Y.; Lv, S.; Chen, X. Characterizing the Complete Mitochondrial Genomes of Three Bugs (Hemiptera: Heteroptera) Harming Bamboo. *Genes* **2023**, *14*, 342. https://doi.org/10.3390/genes14020342

Academic Editor: Zhiteng Chen

Received: 2 January 2023
Revised: 20 January 2023
Accepted: 25 January 2023
Published: 28 January 2023

1. Introduction

Bamboo is a common plant that belongs to the Gramineae family. Bamboo resources are an essential part of the terrestrial forest ecosystem with typical characteristics of wide distribution, rapid growth, high yield, strong regeneration ability, wide use, and high economic value. Bamboo has high economic, ecological, and social benefits, and is widely used in the construction, ornamental, and food industries. Additionally, the bamboo extracts exhibited excellent anti-free radicals, antioxidant, anti-aging, antibacterial, insecticidal, lipid regulation, cardiovascular and cerebrovascular protection, and pharmacological effects [1]. There are more than 120 genera and over 1500 species of bamboo worldwide, and approximately 44 genera and 762 species of bamboo in China [2]. With increased bamboo planting areas, bamboo pests become more common, and damage becomes more severe, which hinders the sustainable development of the bamboo industry. While investigating bamboo pests in Guizhou, the authors found three species that are serious bamboo pests: *N.*

meleagris, *M. harringtonae*, and *H. bipunctatus*, which belong to the Hemiptera (Heteroptera) order [3–5]. At the end of the 19th century and the beginning of the 20th century, Japanese researchers studied the behavior of *N. meleagris*. They noticed that *N. meleagris* have the habit of multi-male mating aggregation [6,7]. Many Chinese researchers have also reported the habits of *N. meleagris*. Adults and nymphs of *N. meleagris* are harmful because they suck sap through thorns. Bamboo forests were severely damaged by these pests, which killed 70% of the bamboo plants. The harmed bamboo withered, bamboo whips rotted, and dried without osmotic fluid. This led to the following year's reduction of bamboo shoots and the decline of the forest [8]. According to our study, the damage of *N. meleagris* in Guizhou is becoming increasingly severe, and there are over20 insects on each bamboo shoot simultaneously. No scholars have studied the biological characteristics of *M. harringtonae* and *H. bipunctatus* before. This study is the first to report these insects' effects on bamboo; they suck sugar and other nutrients from the bamboo rod's basal membrane and inner wall with their piercing mouthparts, and the damaged sites often become reddish-brown or black. The bottom of the node becomes corroded and withered. *M. harringtonae* can cause damage to every type of bamboo. Therefore, the damage and life histories of *M. harringtonae* and *H. bipunctatus* are described in this study.

Mitochondrial genome sequences are widely used in biogeographical, molecular, and systematic studies [9,10]. Mitochondrial genome research includes explaining species' origins and exploring insects' phylogeny, revealing the geographical distribution of species polymorphisms. This relationship provides several genome level characteristics, including changes in genomic diversity, control patterns for transcription and replication, and RNA secondary structures (such as cloverleaf structures) [11,12]. Because of the higher base replacement rate than nuclear genes, due to the lack of rearrangement during cell meiosis, these characteristics make mitochondrial DNA a focal genetic marker for evolutionary studies [13–16]. So far, only a few mitochondrial bamboo pests (such as *Notobitus montanus*, *Pirkimerus japonicus*, *Hippotiscus dorsalis*, and *Yemmalysus parallelus*) have been sequenced and are available on NCBI [17–19]. There were no complete mitochondrial sequences of *N. meleagris* and *H. bipunctatus* on NCBI. This study presents the complete mitochondrial genome of three bamboo pests (*N. meleagris*, *M. harringtonae*, and *H. bipunctatus*), which provides the basis for developing bamboo-pest gene bank data and supports prevention and management. This is the first study to investigate bamboo pests by using the mitochondrial genomes data and life histories with detailed molecular and morphological datasets. We also discuss their mitochondrial genome structures and analyze their tRNA's shamrock structure. This study aims to provide a reference for the identification, control, and phylogenetic analysis of bamboo pests.

2. Materials and Methods

2.1. Observation of the Damaged Condition and Occurrence Regularity

From May 2021 to August 2022, we observed the damaged conditions and occurrence regularity of three bamboo pests (*N. meleagris*, *M. harringtonae*, and *H. bipunctatus*) in Guizhou Province. The observations were madeon sunny days, at an interval of once every ten days. The damages were recorded and photographed using visual inspection and sweep net techniques [20].

2.2. Sample Isolation and DNA Extraction

N. meleagris was collected in July 2021 from the Baizi Bridge, Duyun, and Guizhou, whereas *M. harringtonae* and *H. bipunctatus* were collected in August 2021 from Huaxi Park, Guiyang, and Sajinriver, Fuquan, Guizhou, respectively. Identification was made based on external body morphology and genitalia with the help of theavailable literature [21–23]. After 48 h of starvation, fresh individuals were preserved in 95% ethanol at −40 °C at the Guizhou Provincial Key Laboratory for Agricultural Pest Management of the Mountainous Regions, Guizhou University. Total DNA was extracted from the entire body using the Genomic Extraction Kit [24–27].

2.3. Genome Assembly, Annotation, and Analysis

The DNA quality before sequencing was evaluated using agarose (1%) gel electrophoresis. Mitogenomes were sequenced using a next-generation sequencing platform with Illumina Hiseq 2500 at BerryGenomics (Beijing, China). BerryGenomics (Beijing, China) and FastQC (www.bioinformatics.babraham.ac.uk/projects/fastqc accessed on 17 March 2022) were used to evaluate the quality of the raw sequences. Then, the clean sequences were assembled using MitoZ v2.4 software [28] with default parameters and the mitogenomes of *Riptortus pedestris* (Alydidaeidae; NC_012462), *Aeschyntelus notatus* (Coreidae; NC_012446), and *Geocoris pallidipennis* (Geocoridae; NC_012424) were used as references. The three mitogenomes were initially annotated using the MITOS web server (http://mitos.bioinf.uni-leipzig.de/index.py, accessed on 1 January 2023) [29] with invertebrate genetic codes. Using the MITOS web server, we identified and predicted 22 tRNA genes' locations and secondary structures. The 13 protein-coding genes (PCGs) were predicted by determining their open reading frames using the invertebrate mitochondrial genetic codons [30–32]. The skews of AT and GC were calculated according to the following formulas: AT skew = $(A - T)/(A + T)$ and GC skew = $(G - C)/(G + C)$ [29,33,34]. The nucleotide composition and relative synonymous codon usage (RSCU) were obtained using PhyloSuite v1.2.2 [35], and RSCU figures were created using the package [36] of R 3.6.1 [37].

2.4. Phylogenetic Analysis

In addition to the three mitogenomes sequenced in this study, phylogenetic analyses were conducted based on an additional 22 complete mitogenomes of the Hemiptera species from NCBI. The Hemiptera species belonged to 6 superfamilies: Lygaeoidea (11 species), Coreoidea (8 species), Pentatomoidea (2 species), Reduvioidea (2 species), Fulgoroidea (1 species), and Membracoidea (1 species). The mitogenomes of *I. laurifoliae* (Cicadellidae) and *N. lugens* (Delphacidae) from Auchenorrhyncha were selected as the outgroup (Table A1). Accession numbers and detailed information on these mitogenomes are listed in Table A1. We used MEGA v6 [38] to align the nucleotide sequences of 13 PCGs with Muscle [39,40], and used SequenceMatrix v1.7 [41] to concatenate individual genes. Model testing and selection was completed by using the software PartitionFinder v2.1.1 [42] with the greedy algorithm [43]. Maximum likelihood (ML) analyses were employed using IQ-TREE v1.6.3 [44] with 10,000 replicates of ultrafast likelihood bootstrapping [45]. Bayesian inference (BI) analyses were employed using MrBayes 3.2 [46] under the matrix of two simultaneous operations of 1,000,000 generations, sampling every 1000 generations, with a burn-in of 25%. When the splitting frequency drops steadily to 0.01, the sample is considered to have converged. Finally, using FigTree v.1.4.3 [47], we viewed and beautified the resulting phylogenetic trees [48,49].

3. Results

3.1. Hazard Condition and Occurrence Regularity

According to our investigation of the three bamboo pests in Guizhou Province from 2021 to 2022, the preliminary observation showed that *N. meleagris* was significantly harmful to bamboo shoots, mainly harming them with clusters; *M. harringtonae* can harm whole bamboo tree, including bamboo poles and bamboo joints, and have a wide distribution range; *H. bipunctatus* often inhabits the growth of newly emerging leaves and feed on them. Furthermore, the insect pests of bamboo species often aggregate and harm the bamboo plants. The adults and nymphs are also harmful as they suck sap through the stylets (Figures 1–3).

Figure 1. Dorsal habitus of *M. harringtonae*. (**A**). First instar; (**B**). Second instar; (**C**). Third instar; (**D**). Fourth instar; (**E**). Fifth instar; (**F**). Adult. Scale bar = 1000 μm.

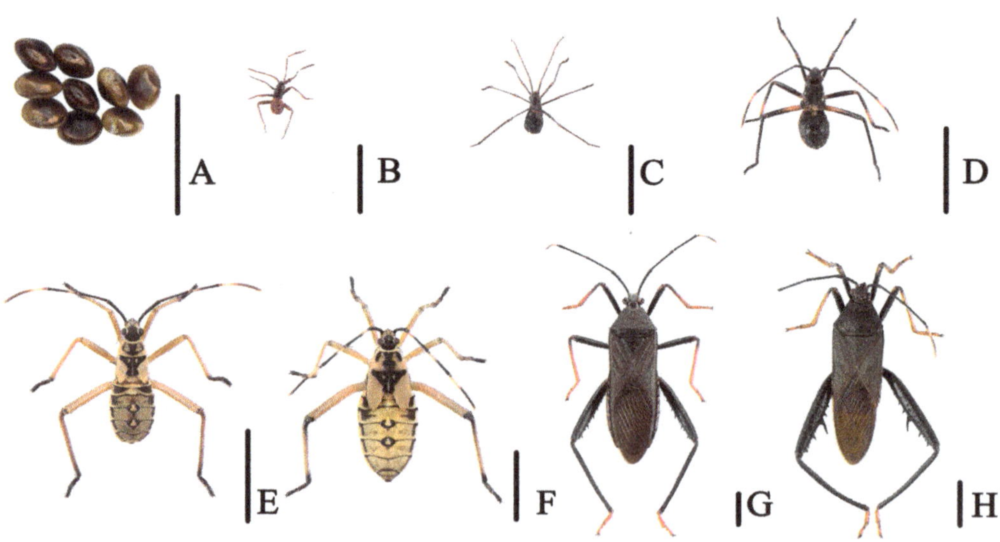

Figure 2. (**A**). Egg mass of *N. meleagris*; (**B–F**). Dorsal habitus of first to fifth nymphal instars; (**G**). Adult, female; (**H**). Adult, male. Scale bars = (**A**) (500 μm); (**B–F**) (5000 μm).

Figure 3. (**A**,**B**). Living dorsal habitus of adult *N. melegaris*; (**C**,**D**). Nymphal stages of *N. melegaris*; (**E**,**G**). *M. harringtonae*, adults; (**F**,**H**). *H. bipunctatus*, adults.

N. meleagris, five instars, mass egg production with at least 20 eggs per time, obvious generation overlap, >3 generations a year, and adults overwinter in dry trees; *M. harringtonae*, five instars, mass egg production with at least 20 eggs laid per time, >3 generations a year, overwintering as adults in bamboo nodes; *H. bipunctatus*, five instars, single egg production, >2 generations a year, and adults overwintering in weeds. (Table A2).

3.2. Mitogenomic Organization and Composition

The mitogenomes of the three bamboo pests, *N. meleagris* (GenBank No. OP442510; length: 16,199 bp), *M. harringtonae* (GenBank No. OP442511; length: 15,314 bp), and *H. bipunctatus* (GenBank No. OP442512; length: 16,706 bp) were double-stranded closed circular molecules (Figure 4). The newly sequenced mitogenomes of three bamboo pests presented 37 typical metazoan mitochondrial genes. These were similar to the mitogenomic sequences of other Hemipteran insects [50–53], containing 13 PCGs, 22 tRNA genes, two rRNA genes, and a control region (Table A3). Each sequence of the three bamboo pests included nine PCGs and 14 tRNAs encoded on the major (J-strand), and the minor (N-strand) consisted of four PCGs, eight tRNA, and two rRNAs. In addition, there were some differences between the overlapping regions and intergenic spacers of the three mitogenomes. There were seven overlapping regions and 12 intergenic spacers of *N. meleagris*, the largest overlapping region was 7 bp located between atp8 and atp6, and the largest intergenic spacer was 37 bp located between trnY and cox1. In addition, there were 11 overlapping regions and 10 intergenic spacers in *M. harringtonae*; the largest overlapping region was 7 bp between atp8 and atp6, and the largest intergenic spacer was 71 bp between trnH and nad4. There were nine overlapping regions and 13 intergenic spacers in *H. bipunctatus*; the largest overlapping region was 8 bp between trnW and trnC, and the largest intergenic spacer was 37 bp located between trnY and cox1. The nucleotide compositions of *N. meleagris*, *M. harringtonae*, and *H. bipunctatus* are shown in Table A4. The AT nucleotide content of the three mitogenomes was similar: In the range of 73–74.5%, the content occupoed a substantial proportion of the entire sequence. The AT skew of all three genomes is a positive number; on the contrary, the GC skew of all three genomes is a negative number.

3.3. PCGs and Codon Usage

The mitogenomes of the three bamboo pests belong to the Hemipteran order [47], which includes 13 PCGs. Their lengths in *N. meleagris*, *M. harringtonae*, and *H. bipunctatus* were 11,008 bp, 10,957 bp, and 11,010 bp, respectively. In the three sequences, the nine PCG genes (nad2, cox1, cox2, atp8, atp6, cox3, nad3, nad6, and cytb) were encoded on the major strand (J-strand), and four PCG genes (nad5, nad4, nad4L, and nad1) were encoded on the minor strand (N-strand). All 13 PCGs started with ATN. The stop codon of *N. meleagris* is the same as that of *H. bipunctatus*, atp8 and nad6 had TAA as the stop codon, and the other ten had incomplete T. The stop codon of *M. harringtonae* is special, except for the same features as the other two sequences, nad4L had TAA as the stop codon, and nad3 had TAG as the stop codon.

Except for the stop codon, the total number of codons was 3663 (*N. meleagris*), 3645 (*M. harringtonae*), and 3664 (*H. bipunctatus*). In descending order, the three most abundant amino acids, Leu2, Ile, and Phe, in *N. meleagris* are the same as *M. harringtonae*. In addition, Leu2, Ile, and Met were the most abundant amino acids in *H. bipunctatus* (Figure 5). According to Figure 6, the four most prevalent codons were Leu2 (UUA), Ile (AUU), Phe (UUU), and Met (AUA). The RSCU values of the PCGs indicated a pattern toward more A and T than G and C.

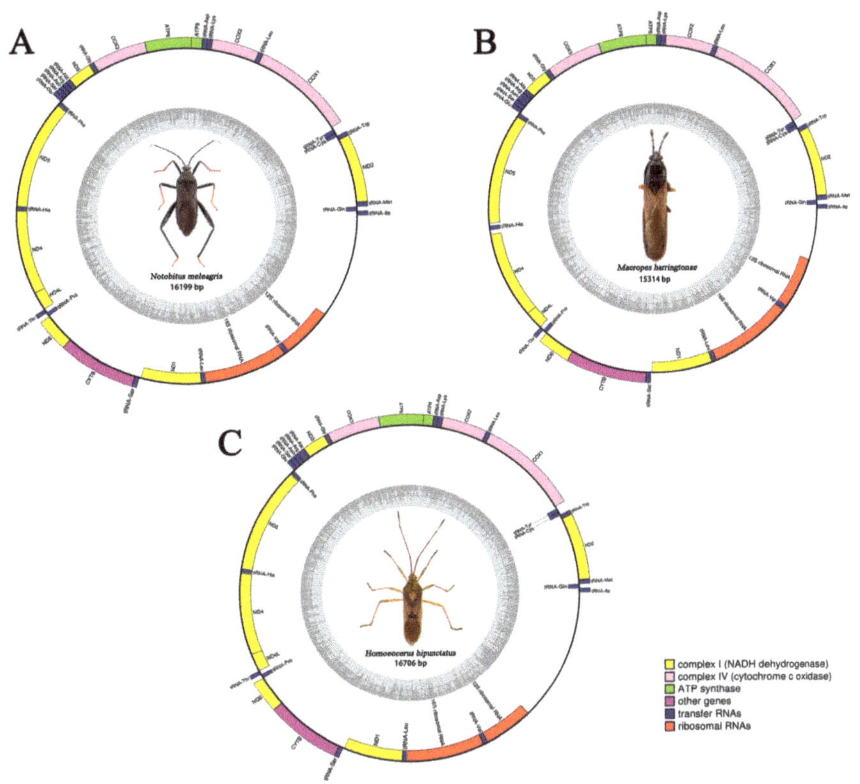

Figure 4. Circular maps of the mitogenomes of *N. meleagris* (**A**), *M. harringtonae* (**B**), and *H. bipunctatus* (**C**). The pink, green, yellow, and purple show PCGs, blue shows tRNAs, red shows the rRNAs, and blank shows the control region. Insect morphology and sequence length are indicated in the center of the circle diagram.

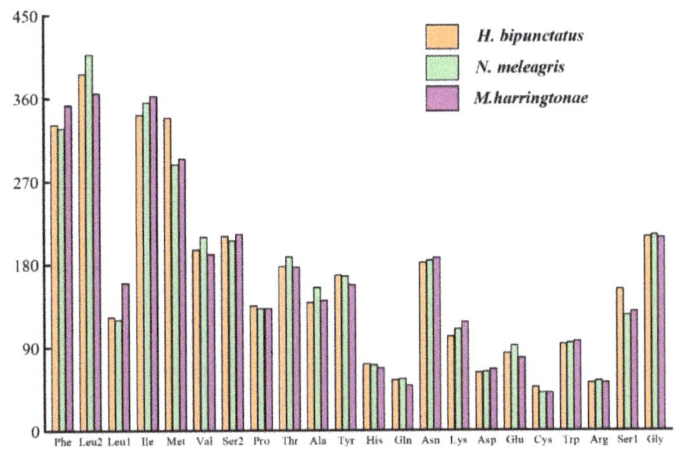

Figure 5. Codon distribution in three bamboo pest species: the color-filled orange blocks indicate *H. bipunctatus*, the filled green blocks represent *N. meleagris*, and the filled violet blocks show *M. harringtonae*. The total number of the codons are presented as numbers at the *Y*-axis and codon families are shown at the *X*-axis.

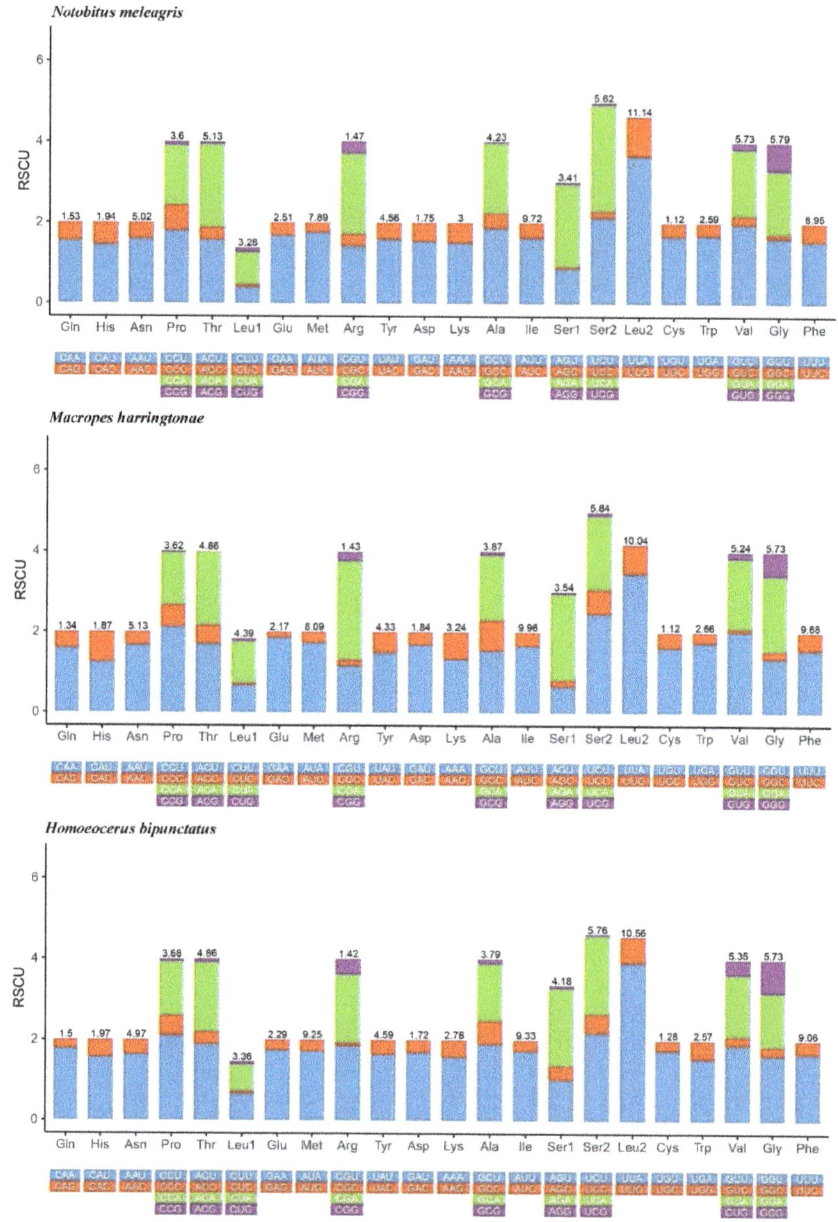

Figure 6. Relative synonymous codon usage (RSCU) within *N. meleagris*, *M. harringtonae* and *H. bipunctatus*. Codon families are shown on the *X*-axis and the frequency of RSCU on the *Y*-axis.

3.4. Transfer and Ribosomal RNA Genes

The rrnL (16S) and rrnS (12S) genes on the N-strand were located between the trnL1 and trnV and the control region in the mitogenome of three bamboo pests (Table A3). The total lengths of rrnL and rrnS of the three sequences were similar, in the range of 2036 bp to 2067 bp, and displayed a negative AT skew and a positive GC skew (Table A4).

The mitogenomes of *N. meleagris*, *M. harringtonae*, and *H. bipunctatus* included 22 transfer RNA genes, as in most invertebrates. The total lengths of tRNAs were 1449 bp,

1439 bp, and 1446 bp, and these tRNA genes ranged from 62–75 bp. In the three sequences, the 14 tRNA genes (trnI, trnM, trnW, trnL2, trnK, trnD, trnG, trnA, trnR, trnN, trnS1, trnE, trnT, and trnS2) were encoded on the major strand (J-strand), and the eight tRNA genes (trnQ, trnC, trnY, trnF, trnH, trnP, trnL1, trnV) were encoded on the minor strand (N-strand) (Table A3). We found that only trnS1 lacked the dihydrouridine (DHU) arm, and the remaining 21 tRNA genes can form a typical cloverleaf structure (Figures A1–A3). In addition to the typical base pairing (G-C and A-U), there was some wobble G-U pairs in these secondary structures, which could form stable chemical bonds between U and G.

3.5. Control Region

The control region, also called the A+T rich region, is the longest noncoding region with many genes involved in mitogenic replication and transcription. In the three bamboo pests, this region was located between the rrnS and trnL. The length of the control region was 1627 bp (*N. meleagris*), 772 bp (*M. harringtonae*), and 2138 bp (*H. bipunctatus*). The AT-rich region had the highest AT content with67.2% in *N. meleagris*, 79.3% in *M. harringtonae*, and 68.4% in *H. bipunctatus*, with positive AT skew (0.099–0.117) and negative CG skew (−0.329 to −0.188) (Table A4).

3.6. Phylogenetic Analyses

Phylogenetic relationships among 23 species of the heteropteran (including the three sequenced mitogenomes of the bamboo pests, two of them newly sequenced) and two outgroups (*I. laurifoliae* and *N. lugens*) were reconstructed based on 13 PCGs using ML and BI analyses under the partitioning scheme and models selected by PartitionFinder. The two resulting trees (Figure 7) had similar topologies, receiving strong support in most nodes. These phylogenetic relationships were consistent with previous studies [54,55]. The phylogenetic trees of *N. meleagris* and *H. bipunctatus* from sister group relationships, *M. harringtonae* with *M. robustus*, and *M. dentipes* also from sister group relationships, showed a high confidence value. The sister groups' relationship of Coreoidea and Lygaeoidea located in the middle of phylogenetic trees was also confirmed [56–58].

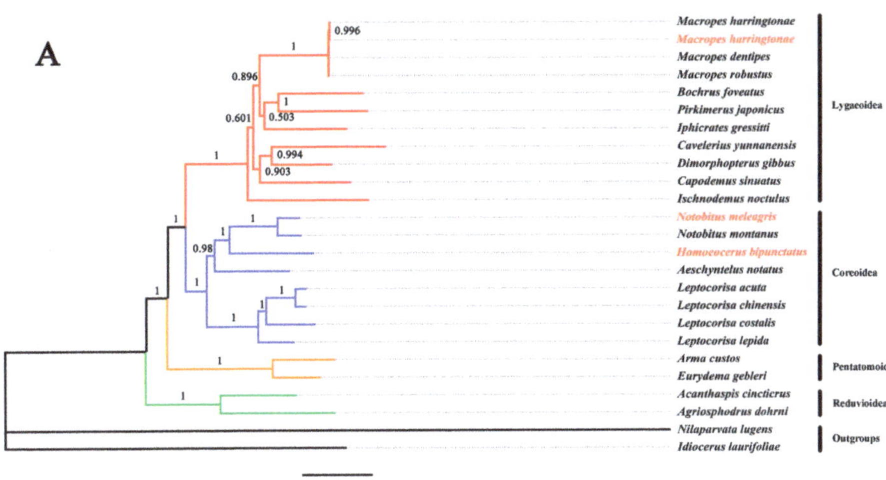

Figure 7. *Cont.*

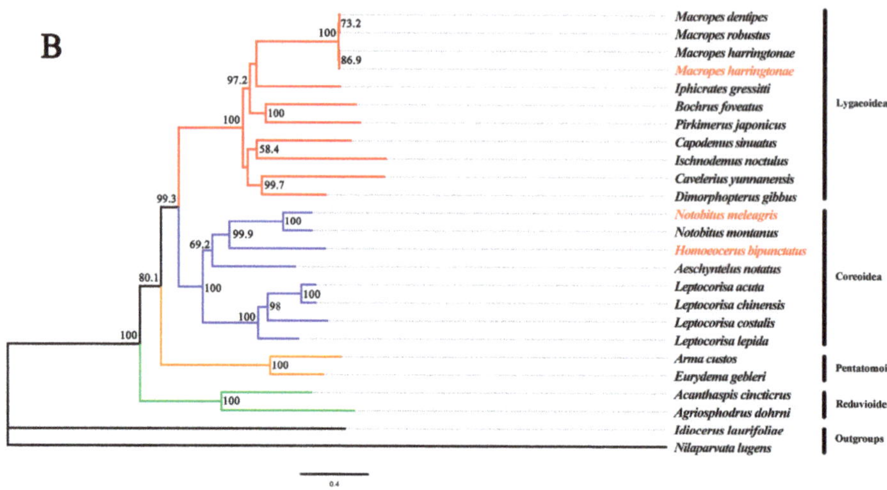

Figure 7. Phylogenetic relationships based on Bayesian inference (**A**) and maximum likelihood (**B**) analyses. Bayesian posterior probabilities are shown on each node. Bootstrap support values are shown on each node. The tree was rooted using *I. laurifoliae* and *N. lugens* as outgroups. The three bamboo pest species sequenced in this study are marked in red.

4. Discussion

N. meleagris are typical insects that attack and harm bamboo plants [59]. *M. harringtonae*, belonging to the family Blissidae, and many genera (*Macropes* and *Pirkimerus*), have been reported to have harmed bamboo, but this is the first time that *M. harringtonae* has been reported to have harmed bamboo seriously [60,61]. *Homoeocerus*, belonging to the Coreidae family, have previously been reported to harm only leguminous plants [62]; however, *H. bipunctatus* collected from the bamboo plants showed normal physiological activities, such as mating and oviposition, when the species were fed with bamboo. This study describes the extent of damages and life histories of *M. harringtonae* and *H. bipunctatus*. This paper updates and supplements the data on *N. meleagris*, *M. harringtonae*, and *H. bipunctatus* in Guizhou. Research in 2009 showed that only two generations of *N. meleagris* were present in Guizhou in one year [63]. However, according to our research observations, at least three generations of *N. meleagris* were present in Guizhou in one year, and the generations overlapped significantly. Previously, no study was performed on the biological characteristics of *M. harringtonae* and *H. bipunctatus*. Our research shows that at least three generations of *M. harringtonae* occur in a year in Guizhou, and at least two generations of *H. bipunctatus* occur in a year in Guizhou. This study fills a gap in the understanding of the biological characteristics of *M. harringtonae* and *H. bipunctatus*.

In this study, three complete mitogenomes (*N. meleagris*, *M. harringtonae*, and *H. bipunctatus*) were sequenced and analyzed for their genome size, base content, AT nucleotide bias, AT skew, GC skew, the codon usage of protein genes, and secondary structure of tRNA. Despite differences in the sequence length of three bamboo pest species, the mitochondrial genome order of *N. meleagris*, *M. harringtonae*, and *H. bipunctatus* was identical and conserved with the alignment to that of known ancestral taxa regarding the organization and composition of the genome [64–68]. The size, AT skew, and GC skew of genome and PCGs of *M. harringtonae* in our study has little difference with previous research. However, the positive and negative of AT skew and GC skew of rRNAs and tRNAs are opposite [17]. The analysis of synonymous codon usage showed that the occurrence of synonymous codons ending in A or T was much higher than those of other synonymous codons; that is, codons rich in AT were frequently used. Whether the abundant AT content in the control region affects transcription and replication of the mitogenome and indirectly affects the feeding behavior of the bamboo pests is unclear; further studies are required to verify the function of the conserved control region of the mitogenome of the bamboo pests. In the insect mitochondrial genome, the stem-loop structure with dihydrouracil deletion of trnS1 in the tRNA secondary structure is a typical feature [69–71]. The trnS1 secondary structures of the three mitochondrial genome sequences of bamboo pests are stem-loop structures with dihydrouracil deletion; the other 21 tRNA secondary structures were typical clover structures.

The phylogenetic tree was constructed using 13 protein genes, and the phylogenetic relationships of 25 species were analyzed. The results showed that *N. meleagris* and *H. bipunctatus* belonged to the Coreoidea, and *M. harringtonae* belonged to the Lygaeoidea. There was no dispute on the taxonomic status, consistent with the results of morphological identification [17,72,73]. The three mitogenomes sequenced enriched the database of Heteroptera and laid a foundation for better resolving the controversy of the taxonomic status of bugs. The study of the predator web of natural enemy insects is to determine the prey of natural enemy insects by measuring the DNA fragments of the intestinal contents of natural enemy insects and comparing them in the database [74]. However, before that, it is necessary to establish a database of insect pests and surrounding arthropod species. Therefore, this study will also provide essential information for subsequent research on analyzing predator nets of natural enemies of bamboo pests.

Author Contributions: Methodology, W.Z. and Y.M.; software, W.Z. and S.L.; validation, N.G. and Z.C.; investigation, W.Z. and L.Y.; writing—original draft preparation, W.Z.; writing—review and editing, X.C. and J.L. All authors have read and agreed to the published version of the manuscript.

Funding: This study was supported by the Science and Technology Support Program of Guizhou Province (No. 20201Y129) and the Program of Excellent Innovation Talents, Guizhou Province (No. 20154021).

Institutional Review Board Statement: No special permits were required to retrieve and process the samples because the study did not involve any live vertebrates, nor regulated invertebrates.

Informed Consent Statement: Not applicable.

Data Availability Statement: The *N. meleagrare*, *M. harringtonae*, and *H. bipunctatus* mitogenome sequence was submitted to NCBI (Acc. number).

Acknowledgments: We thank Feng-E Li (Institute of Entomology, Guizhou University) for her help on genome assembly and annotation. Muhammad Asghar Hassan, Guizhou University for his help on revision of manuscript. Cui-Q Gao (Nanjing Forestry University) for her help on identification.

Conflicts of Interest: The authors declare no conflict of interest.

Appendix A

Table A1. List of mitogenomes used for phylogenetic analysis in this study.

	Superfamily	Species	NCBI No.	Length
Ingroups	Lygaeoidea	*M. harringtonae*	NC065820	14,942
	Lygaeoidea	*M. harringtonae*	OP442511	15,314
	Lygaeoidea	*Macropes dentipes*	NC065821	14,923
	Lygaeoidea	*Macropes robustus*	NC065822	15,041
	Lygaeoidea	*Bochrus foveatus*	NC065814	14,738
	Lygaeoidea	*P. japonicus*	NC065823	15,440
	Lygaeoidea	*Iphicrates gressitti*	NC065818	15,288
	Lygaeoidea	*Cavelerius yunnanensis*	NC065816	15,330
	Lygaeoidea	*Dimorphopterus gibbus*	NC065817	14,988

Table A1. *Cont.*

	Superfamily	Species	NCBI No.	Length
	Lygaeoidea	*Capodemus sinuatus*	NC065815	15,199
	Lygaeoidea	*Ischnodemus noctulus*	NC065819	15,291
	Coreoidea	*N. meleagris*	OP442510	16,199
	Coreoidea	*Notobitus montanus*	NC065112	16,209
	Coreoidea	*H. bipunctatus*	OP442512	16,706
	Coreoidea	*A. notatus*	NC012446	14,532
	Coreoidea	*Leptocorisa acuta*	NC061738	15,373
	Coreoidea	*Leptocorisa chinensis*	NC061737	15,433
	Coreoidea	*Leptocorisa costalis*	NC061680	15,353
	Coreoidea	*Leptocorisa lepida*	NC061739	15,129
	Pentatomoidea	*Arma custos*	NC051562	15,629
	Pentatomoidea	*Eurydema gebleri*	NC027489	16,005
	Reduvioidea	*Acanthaspis cincticrus*	NC037735	15,686
	Reduvioidea	*Agriosphodrus dohrni*	NC015842	16,470
Outgroups	Fulgoroidea	*N. lugens*	NC021748	17,619
	Membracoidea	*I. laurifoliae*	NC039741	16,811

The sequence number marked in red is the research subject of this paper.

Table A2. List of annual life history.

M. harringtonae								
Generation	**Mouth**							
	Mar.	**Apr.**	**May**	**Jun.**	**Jul.**	**Aug.**	**Sep.**	**Oct.−Feb.**
Overwintering generation	(−)	−	−					
	(+)	+	+					
First generation		•	•	•				
		−	−	−				
				+	+	+		
Second generation				•	•	•		
				−	−	−		
				+	+	+		
Third generation				•	•	•		
					−	−	−	(−)
							+	(+)

H. bipunctatus								
Generation	**Mouth**							
	Apr.	**May**	**Jun.**	**Jul.**	**Aug.**	**Sep.**	**Oct.**	**Nov.−Mar.**
Overwintering generation	(+)	+	+	+				
First generation		•	•	•				
		−	−	−				

Table A2. *Cont.*

H. bipunctatus								
Generation	**Mouth**							
	Apr.	**May**	**Jun.**	**Jul.**	**Aug.**	**Sep.**	**Oct.**	**Nov.–Mar.**
Second generation				+ • –	+ • – +	+ • – +	+ • – +	(+)

"+" means adult, "•" means egg, "–" means nymph, "()" means overwintering.

Table A3. List of mitogenomes' configurations of the three bamboo pests.

N. meleagris						
Gene	**Direction**	**Location**	**Size (bp)**	**Start Codon**	**Stop Codon**	**INC**
trnI	J	1–65	65	-	-	0
trnQ	N	63–131	69	-	-	−3
trnM	J	131–199	69	-	-	−1
nad2	J	201–1200	1000	ATG	T	1
trnW	J	1201–1264	64	-	-	0
trnC	N	1257–1319	63	-	-	6
trnY	N	1320–1382	63	-	-	0
cox1	J	1420–2917	1498	ATT	T	37
trnL2	J	2918–2984	67	-	-	0
cox2	J	2985–3663	679	ATC	T	0
trnK	J	3663–3738	75	-	-	−1
trnD	J	3739–3803	65	-	-	0
atp8	J	3804–3965	162	ATA	TAA	0
atp6	J	3959–4630	672	ATG	TAA	−7
cox3	J	4630–5416	787	ATG	T	−1
trnG	J	5417–5480	64	-	-	0
nad3	J	5481–5832	352	ATA	T	0
trnA	J	5834–5897	64	-	-	1
trnR	J	5898–5961	64	-	-	0
trnN	J	5962–6027	66	-	-	0
trnS1	J	6027–6096	70	-	-	−1
trnE	J	6096–6160	65	-	-	−1
trnF	N	6161–6225	65	-	-	0
nad5	N	6228–7938	1711	ATG	T	2
trnH	N	7940–8002	63	-	-	1
nad4	N	8004–9318	1315	ATG	T	1
nad4L	N	9314–9602	289	ATT	T	−5
trnT	J	9605–9667	63	-	-	2
trnP	N	9668–9730	63	-	-	0
nad6	J	9736–10,218	483	ATC	TAA	5
cytb	J	10,218–11,352	1135	ATG	T	−1
trnS2	J	11,353–11,421	69	-	-	0
nad1	N	11,443–12,367	925	ATT	T	21
trnL1	N	12,368–12,432	65	-	-	0
rrnL	N	12,437–13,707	1271	-	-	4
trnV	N	13,709–13,776	68	-	-	1
rrnS	N	13,777–14,572	796	-	-	0
D-−loop	J	14,573–16,199	1627	-	-	0

Table A3. *Cont.*

M. harringtonae						
Gene	**Direction**	**Location**	**Size (bp)**	**Start codon**	**Stop codon**	**INC**
trnI	J	1–62	62	-	-	0
trnQ	N	60–128	69	-	-	−3
trnM	J	128–195	68	-	-	−1
nad2	J	196–1186	991	ATT	T	0
trnW	J	1187–1249	63	-	-	0
trnC	N	1249–1312	64	-	-	−1
trnY	N	1313–1377	65	-	-	0
cox1	J	1380–2913	1543	TTG	T	2
trnL2	J	2914–2978	65	-	-	0
cox2	J	2979–3657	679	ATA	T	0
trnK	J	3658–3729	72	-	-	0
trnD	J	3730–3792	63	-	-	0
atp8	J	3793–3948	156	ATT	TAA	0
atp6	J	3942–4607	666	ATG	TAA	−7
cox3	J	4607–5390	784	ATG	T	−1
trnG	J	5391–5456	66	-	-	0
nad3	J	5457–5810	354	ATT	TAG	0
trnA	J	5810–5871	62	-	-	−1
trnR	J	5872–5935	64	-	-	0
trnN	J	5939–6005	66	-	-	3
trnS1	J	6005–6073	69	-	-	−1
trnE	J	6073–6136	64	-	-	−1
trnF	N	6137–6201	65	-	-	0
nad5	N	6199–7870	1672	ATT	T	−3
trnH	N	7910–7972	63	-	-	39
nad4	N	8044–9358	1315	ATG	T	71
nad4L	N	9352–9630	279	ATT	TAA	−7
trnT	J	9633–9694	62	-	-	2
trnP	N	9695–9757	63	-	-	0
nad6	J	9759–10,232	474	ATA	TAA	1
cytb	J	10,232–11,363	1132	ATG	T	−1
trnS2	J	11,364–11,433	70	-	-	0
nad1	N	11,450–12,370	921	ATA	T	16
trnL1	N	12,371–12,435	65	-	-	0
rrnL	N	12,436–13,684	1249	-	-	4
trnV	N	13,686–13,753	68	-	-	1
rrnS	N	13,756–14,542	787	-	-	2
D-−loop	J	14,543–15,314	772	-	-	0
H. bipunctatus						
Gene	**Direction**	**Location**	**Size (bp)**	**Start codon**	**Stop codon**	**INC**
trnI	J	1–65	65	-	-	0
trnQ	N	63–131	69	-	-	−3
trnM	J	131–198	68	-	-	−1
nad2	J	199–1201	1003	ATG	T	0
trnW	J	1202–1269	68	-	-	0
trnC	N	1262–1324	63	-	-	−8
trnY	N	1325–1386	62	-	-	0
cox1	J	1424–2921	1498	ATT	T	37
trnL2	J	2922–2986	65	-	-	0
cox2	J	2987–3665	679	ATC	T	0

Table A3. *Cont.*

H. bipunctatus						
Gene	**Direction**	**Location**	**Size (bp)**	**Start codon**	**Stop codon**	**INC**
trnK	J	3666–3739	74	-	-	0
trnD	J	3741–3803	63	-	-	1
atp8	J	3804–3962	159	ATC	TAA	0
atp6	J	3956–4627	672	ATG	TAA	−7
cox3	J	4627–5413	787	ATG	T	−1
trnG	J	5414–5474	61	-	-	0
nad3	J	5475–5826	352	ATT	T	0
trnA	J	5828–5889	62	-	-	1
trnR	J	5895–5958	64	-	-	5
trnN	J	5959–6025	67	-	-	0
trnS1	J	6025–6093	69	-	-	−1
trnE	J	6093–6155	63	-	-	−1
trnF	N	6165–6231	67	-	-	9
nad5	N	6238–7948	1711	ATG	T	6
trnH	N	7950–8015	66	-	-	1
nad4	N	8019–9333	1315	ATG	T	3
nad4L	N	9329–9614	286	ATT	T	−5
trnT	J	9617–9679	63	-	-	2
trnP	N	9680–9742	63	-	-	0
nad6	J	9746–10,237	492	ATG	TAA	5
cytb	J	10,237–11,368	1132	ATG	T	−1
trnS2	J	11,369–11,437	69	-	-	0
nad1	N	11,459–12,382	924	ATC	T	21
trnL1	N	12,383–12,449	67	-	-	0
rrnL	N	12,456–13,718	1263	-	-	6
trnV	N	13,719–13,786	68	-	-	1
rrnS	N	13,787–14,568	782	-	-	0
D-−loop	J	14,569–16,706	2138	-	-	0

Note: "J" means majority strand, and "N" means minority strand. Positive sign indicates the interval in base pairs between genes, and the negative sign indicates overlapping base pairs between genes.

Table A4. Base composition of three mitochondrial whole genomes.

N. meleagris									
	Size (bp)	**T**	**C**	**A**	**G**	**A + T%**	**G + C%**	**AT Skew**	**GC Skew**
Genome	16,199	30.90	16.40	42.10	10.60	73.00	27.00	0.153	−0.215
PCGs	11,008	40.80	13.30	32.10	13.90	72.90	27.10	−0.119	0.022
rRNA	2067	45.30	8.30	30.70	15.80	76.00	24.00	−0.192	0.276
tRNAs	1449	37.40	10.10	38.80	13.70	76.20	23.80	0.018	0.151
Control region	1627	29.70	21.80	37.60	11.00	67.20	32.80	0.117	−0.329
M. harringtonae									
	Size (bp)	**T**	**C**	**A**	**G**	**A + T%**	**G + C%**	**AT skew**	**GC skew**
Genome	15,314	31.80	15.30	42.60	10.30	74.50	25.50	0.145	−0.195
PCGs	10,957	41.20	13.70	31.80	13.20	73.00	27.00	−0.129	−0.019
rRNA	2036	46.80	8.10	31.60	13.60	78.40	21.60	−0.194	0.253
tRNAs	1439	39.50	9.30	37.70	13.50	77.20	22.80	−0.023	0.184
Control region	772	35.50	12.30	43.80	8.40	79.30	20.70	0.105	−0.188
H. bipunctatus									
	Size (bp)	**T**	**C**	**A**	**G**	**A+T%**	**G+C%**	**AT skew**	**GC skew**
Genome	16,706	32.40	15.00	41.10	11.60	73.40	26.60	0.118	−0.128
PCGs	11,010	41.80	12.90	31.70	13.60	73.50	26.50	−0.137	0.026
rRNA	2045	46.40	8.60	31.40	13.60	77.80	22.20	−0.193	0.225
tRNAs	1446	37.30	10.50	36.80	15.40	74.10	25.90	−0.007	0.189
Control region	2138	30.80	19.70	37.60	11.90	68.40	31.60	0.099	−0.247

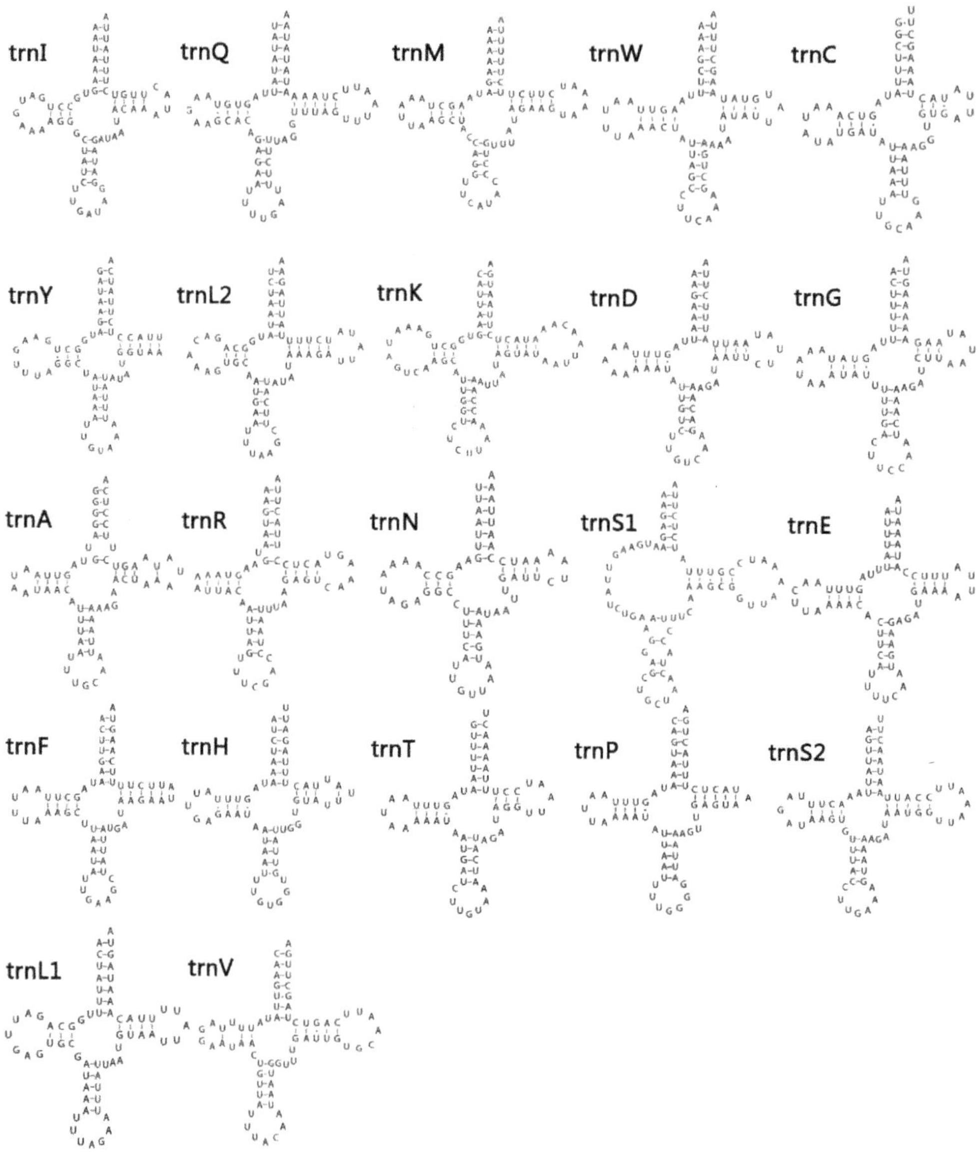

Figure A1. Secondary structures of all tRNAs of *N. meleagris*.

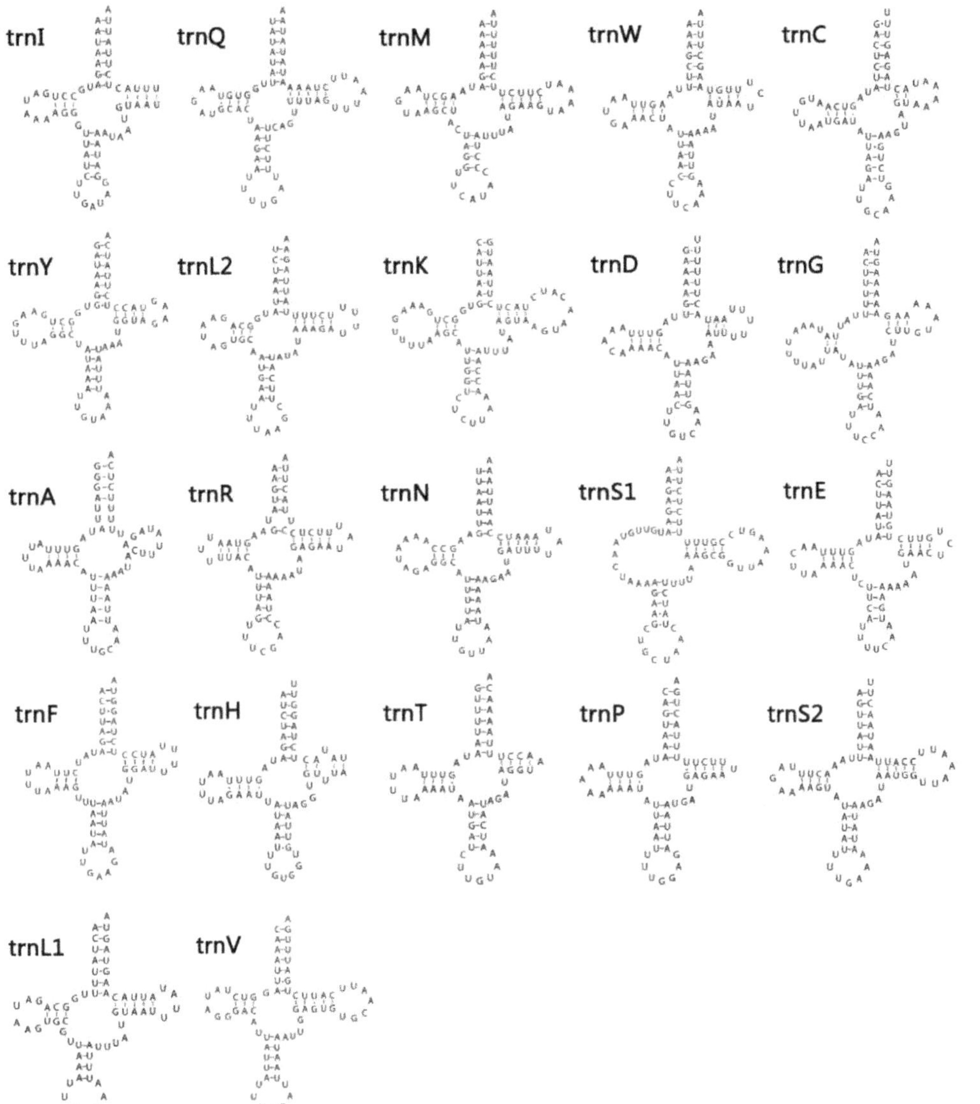

Figure A2. Secondary structures of all tRNAs of *M. harringtonae*.

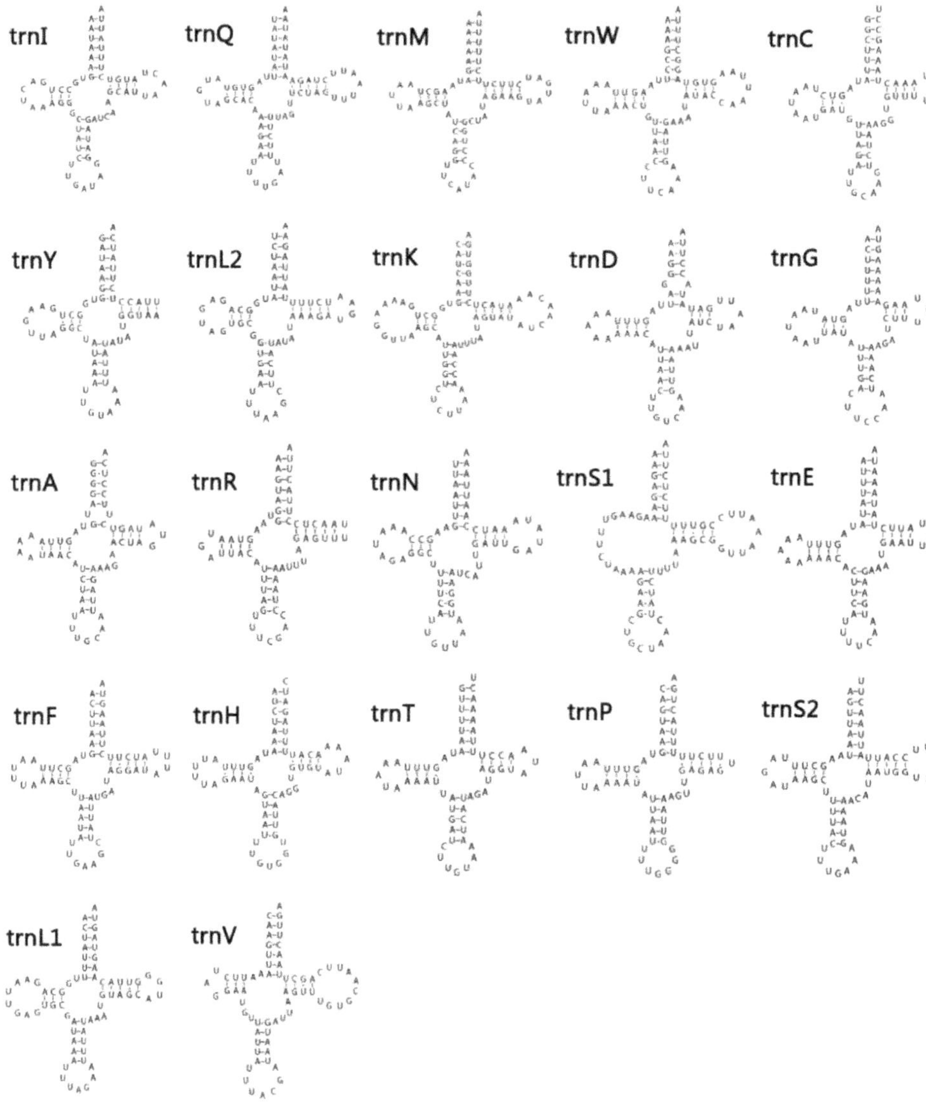

Figure A3. Secondary structures of all tRNAs of *H. bipunctatus*.

References

1. Yue, Y.D.; Cao, H.Q.; Tang, F. Advance in bamboo chemical ingredients and its utilizations. *J. Anhui Agric. Univ.* **2007**, *34*, 328–333.
2. Shi, J.Y.; Zhou, D.Q.; Ma, L.S.; Yao, J.; Zhang, D. Diversity of bamboo species in China. *World Bamboo Rattan.* **2020**, *18*, 55–65.
3. Xiao, C.Y. *Handbook for Identification of Hemiptera-Heteroptera in China Volume 1*; Science Press: Beijing, China, 1977; pp. 198–268.
4. Gao, C.Q.; Bu, W.J. A review of the Macropes Motschulsky (Hemiptera: Lygaeoidea: Blissidae) from China, with descriptions of three new species. *Zootaxa* **2010**, *2366*, 55–68. [CrossRef]
5. Xiao, C.Y.; Ren, S.Z.; Zheng, L.Y.; Jing, X.L.; Zou, H.G.; Liu, S.L. *Handbook for Identification of Hemiptera-Heteroptera in China Volume 2*; Science Press: Beijing, China, 1981; pp. 43–53.
6. Miyatake, T. Territorial mating aggregation in the bamboo bug, *Notobitus Meleagris*, Fabricius (Heteroptera: Coreidae). *J. Ethol.* **1995**, *13*, 185–189. [CrossRef]
7. Miyatake, T. Multi-male mating aggregation in *Notobitus meleagris* (Hemiptera: Coreidae). *Ann. Entomol. Soc. Am.* **2002**, *95*, 340–344. [CrossRef]
8. Chen, Z.Y. Studies on the biology of *Notobitus meleagris*. *J. Appl. Entomol.* **1989**, *26*, 226–228.

9. Ding, Y.S.; Chang, Z.M.; Yang, L.; Chen, X.S. Morphological and molecular identification of *Kallitaxila sinica* (Walker, 1851). A New Pest of Chinese Tallow Tree. *For. Res.* **2018**, *31*, 69–75.
10. Gong, N.; Yang, L.; Chen, X.S. Structural features and phylogenetic implications of four new mitogenomes of Caliscelidae (Hemiptera: Fulgoromorpha). *Int. J. Mol. Sci.* **2021**, *22*, 1348. [CrossRef]
11. Zhou, Z.C.; Liu, Y.Y.; Chen, X.S. Structural features and phylogenetic implications of three new mitochondrial genomes of blister geetles (Coleoptera: Meloidae). *J. Insect Sci.* **2021**, *21*, 19. [CrossRef] [PubMed]
12. Liu, D.; Guo, H.; Zhu, J.; Qu, K.; Chen, Y.; Guo, Y.T.; Ding, P.; Yang, H.P.; Xu, T.; Jing, Q.; et al. Complex physical structure of complete mitochondrial genome of *Quercus acutissima* (Fagaceae): A significant energy plant. *Genes* **2022**, *13*, 1321. [CrossRef]
13. Curole, J.P.; Kocher, T.D. Mitogenomics: Digging deeper with complete mitochondrial genomes. *Trends Ecol. Evol.* **1999**, *14*, 394–398. [CrossRef] [PubMed]
14. Saccone, C.; De Giorgi, C.; Gissi, C.; Pesole, G.; Reyes, A. Evolutionary genomics in Metazoa: The mitochondrial DNA as a model system. *Gene* **1999**, *238*, 195–209. [CrossRef]
15. Liu, Y.Y.; Zhou, Z.C.; Chen, X.S. Characterization of the complete mitochondrial genome of *Epicauta impressicornis* (Coleoptera: Meloidae) and its phylogenetic implications for the infraorder Cucujiformia. *J. Insect Sci.* **2020**, *20*, 16. [CrossRef] [PubMed]
16. Jiang, P.; Li, H.; Song, F.; Cai, Y.; Wang, J.; Liu, J.; Cai, W. Duplication and remolding of tRNA genes in the mitochondrial genome of *Reduvius tenebrosus* (Hemiptera: Reduviidae). *Int. J. Mol. Sci.* **2016**, *17*, 951. [CrossRef] [PubMed]
17. Wang, S.; Zhu, R.; Xue, H.; Li, Y.; Bu, W. Mitogenomics of Chinch Bugs from China and Implications for Its Coevolutionary Relationship with Grasses. *Insects* **2022**, *13*, 643. [CrossRef] [PubMed]
18. Xu, S.; Wu, Y.; Liu, Y.; Zhao, P.; Chen, Z.; Song, F.; Li, H.; Cai, W. Comparative Mitogenomics and Phylogenetic Analyses of Pentatomoidea (Hemiptera: Heteroptera). *Genes* **2021**, *12*, 1306. [CrossRef] [PubMed]
19. Hua, J.; Li, M.; Dong, P.; Cui, Y.; Xie, Q.; Bu, W. Comparative and phylogenomic studies on the mitochondrial genomes of Pentatomomorpha (Insecta: Hemiptera: Heteroptera). *BMC Genom.* **2008**, *9*, 1–15. [CrossRef] [PubMed]
20. Lv, S.S.; Chen, X.S. Morphology and DNA barcode of *Nisia fuliginosa* Yang & Hu, 1985, A New Pest of Schoenoplectus tabernaemontani. *Sichuan J. Zool.* **2021**, *40*, 130–140.
21. Xiao, C.Y. A brief introduction to the species of Cloresmini in China (Hemiptera Coreidae). *Acta Entomol. Sin.* **1963**, *12*, 124–128.
22. Brailovsky, H.; Barrera, E. A revision of the Costa Rican species of *Stenoeurilla Brailovsky* Barrera (Hemiptera: Heteroptera: Coreidae: Stenoscelideini), with the description of two new species, new distributional records, synonymical note, and key to the known species. *Zootaxa* **2019**, *4550*, 545–556. [CrossRef]
23. Tian, X.; Su, X.; Li, C.; Zhou, Y.; Li, S.; Guo, J. Draft genome of the blister beetle, *Epicauta chinensis*. *Int. J. Biol. Macromol.* **2021**, *193*, 1694–1706. [CrossRef] [PubMed]
24. Li, H.M.; Deng, R.Q.; Wang, J.W.; Chen, Z.Y.; Jia, F.L.; Wang, X.Z. A preliminary phylogeny of the Pentatomomorpha (Hemiptera: Heteroptera) based on nuclear 18S rDNA and mitochondrial DNA sequences. *Mol. Phylogenet. Evol.* **2005**, *37*, 313–326. [CrossRef]
25. Li, H.; Durbin, R. Fast and accurate short read alignment with Burrows-Wheeler transform. *Bioinformatics* **2009**, *25*, 1754–1760. [CrossRef] [PubMed]
26. Liang, A.P.; Gao, J.; Zhao, X. Characterization of the complete mitochondrial genome of the treehopper *Darthula hardwickii* (Hemiptera: Aetalionidae). *Mitochondrial DNA A* **2016**, *27*, 3291. [CrossRef]
27. Mu, Y.L.; Zhang, C.H.; Zhang, Y.J.; Yang, L.; Chen, X.S. Characterizing the complete mitochondrial genome of Arma custos and *Picromerus lewisi* (Hemiptera: Pentatomidae: Asopinae) and conducting phylogenetic analysis. *J. Insect Sci.* **2022**, *22*, 6–15. [CrossRef] [PubMed]
28. Meng, G.; Li, Y.; Yang, C.; Liu, S. MitoZ: A toolkit for animal mitochondrial genome assembly, annotation and visualization. *Nucleic Acids Res.* **2019**, *47*, e63. [CrossRef] [PubMed]
29. Bernt, M.; Donath, A.; Jühling, F.; Externbrink, F.; Florentz, C.; Fritzsch, G.; Pütz, J.; Middendorf, M.; Stadler, P.F. MITOS: Improved de novo metazoan mitochondrial genome annotation. *Mol. Phylogenet. Evol.* **2013**, *69*, 313–319. [CrossRef] [PubMed]
30. Li, H.; Leavengood, J.M.; Chapman, E.G.; Burkhardt, D.; Song, F.; Jiang, P. Mitochondrial phylogenomics of Hemiptera reveals adaptive innovations driving the diversification of true bugs. *Proc. Biol. Sci.* **2017**, *284*, 20171223. [CrossRef] [PubMed]
31. Yu, T.H.; Zhang, Y.L. Two complete mitochondrial genomes of Mileewinae (Hemiptera: Cicadellidae) and a phylogenetic analysis. *Insects* **2021**, *12*, 668. [CrossRef]
32. Gong, N.; Yang, L.; Chen, X.S. Comparative analysis of twelve mitogenomes of Caliscelidae (Hemiptera: Fulgoromorpha) and their phylogenetic implications. *PeerJ* **2021**, *9*, e12465. [CrossRef] [PubMed]
33. Yang, C.P.; Shan, B.B.; Liu, Y.; Wang, L.M.; Wu, Q.E.; Luo, Z.L.; Sun, D.R. Complete mitochondrial genome of two Ectoparasitic Capsalids (Platyhelminthes: Monogenea: Monopisthocotylea): Gene content, composition, and rearrangement. *Genes* **2022**, *13*, 1376. [CrossRef] [PubMed]
34. Ruiz-Mena, A.; Mora, P.; Montiel, E.E.; Palomeque, T.; Lorite, P. Complete nucleotide sequence of the mitogenome of *Tapinoma ibericum* (Hymenoptera: Formicidae: Dolichoderinae), gene organization and phylogenetics implications for the Dolichoderinae subfamily. *Genes* **2022**, *13*, 1325. [CrossRef]
35. Zhang, D.; Gao, F.; Jakovlić, I.; Zou, H.; Zhang, J.; Li, W.X.; Wang, G.T. PhyloSuite: An integrated and scalable desktop platform for streamlined molecular sequence data management and evolutionary phylogenetics studies. *Mol. Ecol. Resour.* **2020**, *20*, 348–355. [CrossRef]

36. Bivand, R.; Altman, M.; Anselin, L.; Assunção, R.; Berke, O.; Bernat, A. Spdep: Spatial Dependence: Weighting Schemes, Statistics and Models. R Package Version 1.1–2. 2019. Available online: https://cran.r-project.org/web/packages/spdep/ (accessed on 11 August 2022).
37. R Core Team. *R: A Language and Environment for Statistical Computing*; R Foundation for Statistical Computing: Vienna, Austria, 2019.
38. Tamura, K.; Stecher, G.; Peterson, D.; Filipski, A.; Kumar, S. MEGA6: Molecular evolutionary genetics analysis version 6.0. *Mol. Biol. Evol.* **2013**, *30*, 2725–2729. [CrossRef] [PubMed]
39. Edgar, R.C. Muscle: Multiple sequence alignment with high accuracy and high throughput. *Nucleic Acids Res.* **2004**, *32*, 1792–1797. [CrossRef] [PubMed]
40. Hao, J.; Liang, Y.Y.; Su, Y.J.; Wang, T. The complete mitochondrial genome of *Ophioglossum vulgatum* L. is with highly repetitive sequences: Intergenomic fragment transfer and phylogenetic analysis. *Genes* **2022**, *13*, 1287. [CrossRef]
41. Vaidya, G.; Lohman, D.J.; Meier, R. Sequence matrix: Concatenation software for the fast assembly of multi-gene datasets with character set and codon information. *Cladistics* **2011**, *27*, 171–180. [CrossRef]
42. Lanfear, R.; Frandsen, P.B.; Wright, A.M.; Senfeld, T.; Calcott, B. PartitionFinder 2: New methods for selecting partitioned models of evolution for molecular and morphological phylogenetic analyses. *Mol. Biol. Evol.* **2017**, *34*, 772–773. [CrossRef]
43. Lanfear, R.; Calcott, B.; Ho, S.Y.; Guindon, S. PartitionFinder: Combined selection of partitioning schemes and substitution models for phylogenetic analyses. *Mol. Biol. Evol.* **2012**, *29*, 1695–1701. [CrossRef]
44. Nguyen, L.T.; Schmidt, H.A.; von, H.A.; Minh, B.Q. IQ-TREE: A fast and effective stochastic algorithm for estimating maximum-likelihood phylogenies. *Mol. Biol. Evol.* **2014**, *32*, 268–274. [CrossRef] [PubMed]
45. Minh, B.Q.; Nquyen, M.A.; von, H.A. Ultrafast approximation for phylogenetic bootstrap. *Mol. Biol. Evol.* **2013**, *30*, 1188–1195. [CrossRef]
46. Ronquist, F.; Teslenko, M.; van der, M.P.; Ayres, D.L.; Darling, A.; Höhna, S.; Larget, B.; Liu, L.; Suchard, M.A.; Huelsenbeck, J.P. MrBayes 3.2: Efficient Bayesian phylogenetic inference and model choice across a large model space. *Syst. Biol.* **2012**, *61*, 539–542. [CrossRef]
47. Mousavi, S.A.; Österman, J.; Wahlberg, N.; Nesme, X.; Lavire, C.; Vial, L.; Paulin, L.; de Lajudie, P.; Lindström, K. Phylogeny of the Rhizobium-Allorhizobium-Agrobacterium clade supports the delineation of Neorhizobium gen. nov. *Syst. Appl. Microbiol.* **2014**, *37*, 208–215. [CrossRef]
48. Therese, A.C.; Christopher, H.D. Molecular phylogeny of the grassland leafhopper tribe Hecalini (Hemiptera: Cicadellidae: Deltocephalinae). *Ann. Entomol. Soc. Am.* **2017**, *111*, 68–72.
49. Zheng, Y.L.; Bourgoin, T.; Yang, L.; Chen, X.S.; Luo, X.Q.; Luo, G.J. Complete mitochondrial genome of the planthopper *Orthopagus splendens* (Germar, 1830) (Hemiptera: Fulgoromorpha: Dictyopharidae). *Mitochondrial DNA B* **2021**, *6*, 2667–2668. [CrossRef] [PubMed]
50. Zhang, H.; Yu, X.; Jiang, W.; Gao, H.; Tan, W.; Wang, W. The complete mitochondrial genome sequence of *Cletus punctiger* (Heteroptera: Coreidae). *Mitochondrial DNA B* **2019**, *4*, 3421–3422. [CrossRef]
51. Cao, Y.; Wu, H.T.; Li, M.; Chen, W.T.; Yuan, M.L. The complete mitochondrial genome of *Nysius fuscovittatus* (Hemiptera: Lygaeidae). *Mitochondrial DNA B* **2020**, *5*, 3483–3484. [CrossRef] [PubMed]
52. Sureshan, S.C.; Tanavade, R.V.; Ghosh, S.; Ghosh, S.; Sella, R.N.; Mohideen, H.S. Complete mitochondrial genome sequencing of *Oxycarenus laetus* (Hemiptera: Lygaeidae) from two geographically distinct regions of India. *Sci. Rep.* **2021**, *11*, 23738. [CrossRef]
53. Forthman, M.; Miller, C.W.; Kimball, R.T. Phylogenomic analysis with improved taxon sampling corroborates an Alydidae+Hydarinae+Pseudophloeinae clade (Heteroptera: Coreoidea: Alydidae, Coreidae). *Org. Divers. Evol.* **2022**, *22*, 669–679. [CrossRef]
54. Song, N.; Liang, A.P.; Bu, C.P. A molecular phylogeny of Hemiptera inferred from mitochondrial genome sequences. *PLoS ONE* **2012**, *7*, e48778. [CrossRef]
55. Yao, Y.Z.; Ren, D. Phylogeny, origin and evolution of Pentatomomorpha. Abstract Volume. In Proceedings of the 10th National Congress of Palaeontological Society of China (PSC)-The 25th Annual Conference of PSC, Nanjing, China, 13–17 October 2009.
56. Zhu, G.L.; Wang, Q.R.; Zheng, Z.M. Phylogenetic relationship among ten species of *Riptortus linearis* in Coreinae (Hemiptera: Coreidae) based on EST Isozyme. *J. Anhui Agric. Sci.* **2011**, *39*, 11462–11463, 11466.
57. Zhang, Q.L.; Feng, R.Q.; Li, M.; Guo, Z.L.; Zhang, L.J.; Luo, F.Z.; Cao, Y.; Yuan, M.L. The complete mitogenome of *Pyrrhocoris tibialis* (Hemiptera: Pyrrhocoridae) and phylogenetic implications. *Genes* **2019**, *10*, 820. [CrossRef] [PubMed]
58. Dong, X.; Wang, K.B.; Tang, Z.C.; Zhang, Y.; Yi, W.; Xue, H. Phylogeny of Coreoidea based on mitochondrial genomes show the paraphyly of Coreidae and Alydidae. *Arch. Insect Biochem. Physiol.* **2022**, *110*, e21878. [CrossRef] [PubMed]
59. Zhang, S.M. *Fauna Editorial Committee Academia Sinica. Economic Insect Fauna of China Fasc. 31 Hemiptera (1)*; Science Press: Beijing, China, 1985; pp. 117–159.
60. Xu, G.Y.; Yang, A.N.; Yang, S.D.; Zhang, Y.H.; Chen, C.S. Studies on biology characteristic observation and control experiment in *Pirkimerus japonicus* Hidaka. *J. Anhui Agric. Sci.* **2007**, *35*, 4884–4958.
61. Gao, Z.W. Studies on the morphology, life cycle and control methods of *Pirkimerus japonicus*. *J. Zhejiang For. Sci. Technol.* **1980**, *000*, 5–7.
62. Haldhar, S.M. Report of *Homoeocerus variabilis* (Hemiptera: Coreidae) on khejri (Prosopis cineraria) in Rajasthan, India: Incidence and Morphometric Analysis. *Fla. Entomol. Int. J. Am.* **2012**, *95*, 848–853. [CrossRef]

63. Long, Z.Q. Bionomics and occurrence of *Notobius Meleagris* in Guizhou. *Chin. Bull. Entomol.* **2009**, *46*, 133–135.
64. Wang, Y.; Huang, X.L.; Qiao, G.X. Comparative analysis of mitochondrial genomes of five aphid species (Hemiptera: Aphididae) and phylogenetic implications. *PLoS ONE* **2013**, *8*, e77511. [CrossRef]
65. Li, T.; Yang, J.; Li, Y.W.; Cui, Y.; Xie, Q.; Bu, W.J.; Hillis, D.M. A mitochondrial genome of Rhyparochromidae (Hemiptera: Heteroptera) and a comparative analysis of related mitochondrial genomes. *Sci. Rep.* **2016**, *6*, 35175. [CrossRef]
66. Wang, J. Comparative Mitogenomics and Genetic Diversity of Mirid Bugs (Hemiptera: Miridae). Master's Thesis, Lanzhou University, Lanzhou, China, 2017.
67. Zhao, L.; Wei, J.F.; Zhao, W.Q.; Chen, C.; Gao, X.Y.; Zhao, Q. The complete mitochondrial genome of *Pentatoma rufipes* (Hemiptera, Pentatomidae) and its phylogenetic implications. *ZooKeys* **2021**, *1042*, 51–72. [CrossRef]
68. Jiang, Y.; Li, H.X.; Yu, X.F.; Yang, M.F. Comparative analysis of mitochondrial genomes among twelve sibling species of the genus *Atkinsoniella* Distant, 1908 (Hemiptera: Cicadellidae: Cicadellinae) and phylogenetic analysis. *Insects* **2022**, *13*, 254. [CrossRef]
69. Song, F.; Li, H.; Shao, R.F.; Shi, A.M.; Bai, X.S.; Zheng, X.R.; Heiss, E.; Cai, W.Z. Rearrangement of mitochondrial tRNA genes in flat bugs (Hemiptera: Aradidae). *Sci. Rep.* **2016**, *6*, 25725. [CrossRef]
70. Liu, H.L.; Chen, Q.D.; Chen, S.; Pu, D.Q.; Chen, Z.T.; Liu, Y.Y.; Liu, X. The highly rearranged mitochondrial genomes of three economically important scale insects and the mitochondrial phylogeny of Coccoidea (Hemiptera: Sternorrhyncha). *PeerJ* **2020**, *8*, e9932. [CrossRef]
71. Ye, F.; Li, H.; Xie, Q. Mitochondrial genomes from two specialized subfamilies of Reduviidae (Insecta: Hemiptera) reveal novel gene rearrangements of true bugs. *Genes* **2021**, *12*, 1134. [CrossRef] [PubMed]
72. Li, H.; Liu, H.Y.; Cao, L.G.; Shi, A.M.; Yang, H.L.; Cai, W.Z. The complete mitochondrial genome of the damsel bug *Alloeorhynchus bakeri* (Hemiptera: Nabidae). *Int. J. Biol. Sci.* **2012**, *8*, 93–107. [CrossRef]
73. Li, H.U.; Liu, H.; Song, F.; Shi, A.; Zhou, X.; Cai, W. Comparative mitogenomic analysis of damsel bugs representing three tribes in the family Nabidae (Insecta: Hemiptera). *PLoS ONE* **2012**, *7*, 1–9. [CrossRef]
74. Schmidt, J.M.; Barney, S.K.; Williams, M.A.; Bessin, R.T.; Coolong, T.W.; Harwood, J.D. Predator–prey trophic relationships in response to organic management practices. *Mol. Ecol.* **2014**, *23*, 3777–3789. [CrossRef]

genes

MDPI

Article

Characterization of the Complete Mitochondrial Genome of a Flea Beetle *Luperomorpha xanthodera* (Coleoptera: Chrysomelidae: Galerucinae) and Phylogenetic Analysis

Jingjing Li [1,2], Bin Yan [1,2], Hongli He [1,2], Xiaoli Xu [1,2], Yongying Ruan [3] and Maofa Yang [1,2,*]

1 Institute of Entomology, Guizhou University, Guiyang 550025, China
2 Guizhou Provincial Key Laboratory for Agricultural Pest Management of the Mountainous Region, Guiyang 550025, China
3 Plant Protection Research Center, Shenzhen Polytechnic, Shenzhen 518055, China
* Correspondence: gdgdly@126.com; Tel.: +86-13984073566

Abstract: In this study, the mitochondrial genome of *Luperomorpha xanthodera* was assembled and annotated, which is a circular DNA molecule including 13 protein-coding genes (PCGs), 22 transfer RNA (tRNA) genes, 2 ribosomal RNA genes (12S rRNA and 16S rRNA), and 1388 bp non-coding regions (A + T rich region), measuring 16,021 bp in length. The nucleotide composition of the mitochondrial genome is 41.3% adenine (A), 38.7% thymine (T), 8.4% guanine (G), and 11.6% cytosine (C). Most of the protein-coding genes presented a typical ATN start codon (ATA, ATT, ATC, ATG), except for ND1, which showed the start codon TTG. Three-quarters of the protein-coding genes showed the complete stop codon TAR (TAA, TAG), except the genes COI, COII, ND4, and ND5, which showed incomplete stop codons (T- or TA-). All the tRNA genes have the typical clover-leaf structure, except $tRNA^{Ser1}$ (AGN), which has a missing dihydrouridine arm (DHU). The phylogenetic results determined by both maximum likelihood and Bayesian inference methods consistently supported the monophyly of the subfamily Galerucinae and revealed that the subtribe Luperina and genus *Monolepta* are polyphyletic groups. Meanwhile, the classification status of the genus *Luperomorpha* is controversial.

Keywords: leaf beetles; mitochondrion; monophyly; next-generation sequencing; phylogeny

Citation: Li, J.; Yan, B.; He, H.; Xu, X.; Ruan, Y.; Yang, M. Characterization of the Complete Mitochondrial Genome of a Flea Beetle *Luperomorpha xanthodera* (Coleoptera: Chrysomelidae: Galerucinae) and Phylogenetic Analysis. *Genes* **2023**, *14*, 414. https://doi.org/10.3390/genes14020414

Academic Editor: Zhiteng Chen

Received: 16 December 2022
Revised: 27 January 2023
Accepted: 2 February 2023
Published: 4 February 2023

1. Introduction

The insect mitochondrial DNA (mtDNA) is a double-stranded circular DNA molecule, with the maternal inheritance features of a relatively small molecular mass, a relatively conservative gene arrangement, and a rapid rate of gene evolution, etc. [1]. The mitochondrial genes have been widely used in species identification, the estimation of evolutionary relationships, and the recognition of population structure and phylogeography [2]. The mitogenome contains 13 protein-coding genes (PCGs), 2 ribosomal RNA genes (rRNAs), 22 transfer RNA genes (tRNAs), and 1 A + T-rich region [3]. Nowadays, mitochondrial DNAs (mtDNAs) have been commonly used as molecular markers for reconstructing phylogenetic relationships, revealing population genetic structures, estimating divergence times, identifying relatedness between recently diverged species, etc. [4,5].

The invasive flea beetle, *Luperomorpha xanthodera* (Coleoptera: Chrysomelidae: Galerucinae), was originally described by Fairmaire [6] in Jiangxi Province of China [7]. It has previously been recorded in 16 provinces in China, distributing throughout Jilin, Gansu, Shanxi, Shandong, Jiangsu, Zhejiang, Hubei, Jiangxi, Hunan, Fujian, Taiwan, Guangdong, Guangxi, Sichuan, Guizhou, and Yunnan. It has also been found in many other countries, such as Korea, Japan, Italy, France, the Netherlands, and Poland [8–12]. This flea beetle is a crucial pest of economic crop, such as roses (Rosa) and *Zanthoxylum* spp. Their larvae are commonly found concealed in the roots of plants, while the adults feed on grasses, shrubs, and trees flowers. *Luperomorpha xanthodera* has been reported to feed on 21 different plants, including white mustard (*Sinapis alba* L., Brassicaceae), hyssop (*Hyssopus officinalis* L.), and *Monarda* spp. (Lamiaceae), etc. [8]. To our knowledge, in the natural population of *Luperomorpha xanthodera*, the females constituted the vast majority [13]. This means that this species is highly susceptible to outbreaks.

In the past, the species of the genus *Luperomorpha* Weise were placed among the Alticinae (currently known as Alticini) due to the presence of the metafemoral spring. However, some scholars [14,15] studied the morphological characters of Galerucinae and Alticinae (currently known as Galerucini and Alticini), and they discovered that the metafemoral spring could not be the sole trait that divides the two subfamilies. In a later paper, it was noted that although *Luperomorpha* has a metafemoral spring, it only has one basal patch at the base of the ventral side of the sheath wing and should be transferred to Galerucinae. However, the classification and taxonomic position of the genus *Luperomorpha* has continued to prove problematic. Nowadays, only one species of *Luperomorpha* has been reported for a complete mitogenome.

For the present study, we sequenced and annotated *Luperomorpha xanthodera*'s complete mitogenome and studied its characteristics. This study aimed to acquire the data on mitochondrial genome, additional analyses of the composition of the *Luperomorpha xanthodera* mitochondrial genome, the values of the relative synonymous codon usage (RSCU), the evolutionary rate of *Luperomorpha* and the phylogenetic relationship, etc. We believe such results will be helpful in studies of the population genetics and phylogenetic relationships of this species, as well as studies on other flea beetles in the future. We adopted Nie's classification system [16] to conduct the subsequent analyses.

2. Materials and Methods

2.1. Sample Collection and DNA Extraction

The invasive flea beetle, *Luperomorpha xanthodera*, was collected on the *Zanthoxylum planispinum* from Zhenfeng county, Guizhou, China (25°24′ N, 105°35′ E), on 23 June 2021. The total DNA was extracted from the entire body without the abdomen using the DNeasy Blood & Tissue Kit (Cat. No. 69504) (Qiagen, Hilden, Germany), as per the manufacturer's protocol. The voucher specimen's genome DNA and male genitalia were deposited at the Institute of Entomology, Guizhou University, Guiyang, China (GUGC). The identification of adult specimens was based on morphological characteristics [8,10].

2.2. Genome Sequencing, Assembly and Annotation

Illumina TruSeq libraries with an average insert size of 300 bp were prepared and sequenced on the Illumina NovaSeq 6000 platform (Beijing Berry Bioinformatics Technology Co., Ltd., Beijing, China), obtaining 150 bp paired-end reads. The mitogenome was preliminarily annotated by Mitoz 2.4-α [17], with the invertebrate mitochondrial genetic codes. The MITOS web server (http://mitos.bioinf.uni-leipzig.de/index.py, accessed on 14 December 2021) [18] and the tRNAscan-SE search server (http://lowelab.ucsc.edu/tRNAscan-SE/, accessed on 14 December 2021) [19,20] were used to reconfirm and predict the locations and secondary structures of the tRNA genes, and then the tRNA secondary structures were visualized using Adobe Illustrator. The circular mitogenome maps were visualized using Geneious Prime [21].

2.3. Mitogenome Sequence Analyses

MEGA 7.0 [22] was used to obtain the nucleotide composition statistics, relative synonymous codon usage (RSCU) of all the protein-coding genes (PCGs), and the base composition. The following formula was used to calculate the strand asymmetry: AT skew = [A − T]/[A + T] and GC skew = [G − C]/[G + C] [23]. In the A + T rich region, the tandem repeat elements were verified using the Tandem Repeats Finder program (http://tandem.bu.edu/trf/trf.html, accessed on 23 December 2021) [24]. The non-synonymous substitutions (Ka) and synonymous substitutions (Ks) of 13 PCGs of 3 species (*A. zanthoxylumi*, *A. planipennis*, and *A. ornatus*) were calculated using DnaSP v 5 [25].

2.4. Phylogenetic Analysis

For the phylogenetic analysis, 24 additional mitogenome sequences were downloaded from GenBank, and we selected *Chrysomela vigintipunctata* and *Chrysolina aeruginosa* as the outgroups (Table 1). The L-INS-I strategy in MAFFT software was used for the sequences of amino acids of 13 PCGs, trimmed using trimAl v1.4.1 [26] with the heuristic method 'automated1' to eliminate the gap-only and ambiguous-only locations, and was concatenated using FASconCAT g v1.04 [27].

Table 1. Mitochondrial genomes were used for the phylogenetic analyses in this research.

Subfamily	Tribe	Species	Length (bp)	G + C%	A + T%	AT-Skew	GC-Skew	GenBank No.
Galerucinae	Alticini	*Agasicles hygrophila*	15,917	24.9	75.2	0.05	−0.19	NC_028332
		Altica cirsicola	15,864	22.2	77.8	0.03	−0.21	NC_042876
		Altica fragariae	16,220	22.0	78	0.03	−0.22	NC_042875
		Altica viridicyanea	16,706	22.3	77.8	0.04	−0.23	NC_048472
		Argopistes tsekooni	16,552	20.5	79.5	0.04	−0.19	NC_045929
		Chaetocnema pelagica	16,331	23.1	76.9	0.06	−0.20	NC_041170
		Macrohaltica subplicata	15,840	22.0	78	0.02	−0.22	NC_041169
		Phyllotreta striolata	15,689	24.2	75.8	0.05	−0.26	NC_045901
		Psylliodes chlorophana	14,561	23.7	76.3	0.05	−0.22	NC_053362
		Luperomorpha hainana	16,081	22.2	77.8	0.05	−0.25	MF960124
		Luperomorpha xanthodera	16,021	20	80	0.03	−0.16	ON631248
	Incertae sedis	*Paleosepharia posticata*	15,729	20.2	79.8	0.02	−0.17	NC_033532
		Podagricomela nigricollis	16,756	21.3	78.8	0.05	−0.18	NC_041423
	Galerucini	*Diorhabda carinata*	16,232	22.1	77.9	0.03	−0.19	NC_042945
		Diorhabda carinulata	16,298	21	78.9	0.03	−0.18	NC_042946
		Galeruca daurica	16,615	21.8	78.1	0.04	−0.20	NC_027114
		Ophraella communa	16,553	22.6	77.4	0.06	−0.18	NC_039710
	Luperini	*Aulacophora indica*	16,246	21.4	78.6	0.03	−0.20	NC_047467
		Aulacophora lewisii	15,691	21.1	78.9	0.03	−0.18	NC_039712
		Diabrotica barberi	16,366	20.8	79.2	0.04	−0.18	NC_022935
		Hoplasoma unicolor	14,568	22.2	77.8	0.05	−0.23	NC_041168
		Monolepta hieroglyphica	15,963	19.7	80.3	0.02	−0.18	NC_057489
		Monolepta occifluvis	15,998	20.7	79.2	0.03	−0.20	NC_045838
		Monolepta quadriguttata	16,130	19.7	80.3	0.02	−0.19	NC_039711
	Oidini	*Oides livida*	16,127	20.0	79.9	0.05	−0.14	MF960098
Chrysomelinae	Chrysomelini	*Chrysomela vigintipunctata*	17,474	21.7	78.2	0.04	−0.21	NC_050933
		Chrysolina aeruginosa	16,335	24.5	75.5	0.07	−0.25	NC_052915

Using the amino acid sequence with 13 PCGs in the phylogenetic analysis, the maximum likelihood (ML) and Bayesian inference (BI) were used to create phylogenetic trees. The ML analyses were carried out using IQ-TREE V1.6.3 [28] and 1000 ultrafast bootstrap [29] and 1000 SHaLRT replicates [30]. The best model was determined through the AIC and BIC scores. The PMSF model [31] was used for the amino acid sequence matrix with 13 PCGs by specifying the profile mixture model with the option "-mtInv + C60 + FO + R". Under the CAT + GTR model [32], the amino acid sequence matrix with 13 PCGs was further analyzed with Bayesian inference using PhyloBayes MPI v1.8b [33] and independently run for 10,000 generations for two separate chains. We calculated the largest discrepancy (maxdiff) across all the bipartitions using the bpcomp program. When the maxdiff was <0.3, runs were considered to achieve a good convergence. The tracecomp program was also used to evaluate the discrepancy between two independent runs. The minimum effective size >50 was recognized to be a good run. In this study, values greater than 98 percent (SH-aLRT, UFBoot2) and 0.99 (posterior probability) are considered to be of a "high" support; values of 80 percent to 98 percent for SH-aLRT, 95 percent to 98 percent for UFBoot2, and 0.95 percent to 0.99 percent for the posterior probability are considered to be of a "moderate" support; and values of 95 percent for UFBoot2 and 0.95 percent for the posterior probability are considered to be of a "low" support. The resulting trees were visualized and edited using FigTree v.1.4.2.

3. Results and Discussion

3.1. Mitogenome Organization and Nucleotide Composition

In this study, the complete mitogenome of *Luperomorpha xanthodera* was successfully obtained and uploaded to GenBank under the accession number ON631248 (Table 1; Figure 1). It has a total length of 16,021 bp and includes 37 genes, including 13 protein-coding genes (PCGs), 2 ribosomal RNA (rRNA) genes, and 22 transfer RNA (tRNA) genes, as well as a noncoding control region (A + T-rich region). The gene order of the *Luperomorpha xanthodera* mitogenome was consistent with the ancestral gene order of *Drosophila yakuba*, this is thought to be the ground pattern for insect mitogenomes. As with most galerucine species, 23 genes were on the majority strand (F-strand), and the other 14 genes (tRNA^{Gln}, tRNA^{Cys}, tRNA^{Tyr}, tRNA^{Phe}, ND5, tRNA^{His}, ND4, ND4L, tRNA^{Pro}, ND1, tRNA^{Leu}, 16S rRNA, tRNA^{Val}, 12S rRNA) were on the minority strand (R-strand) (Table 2).

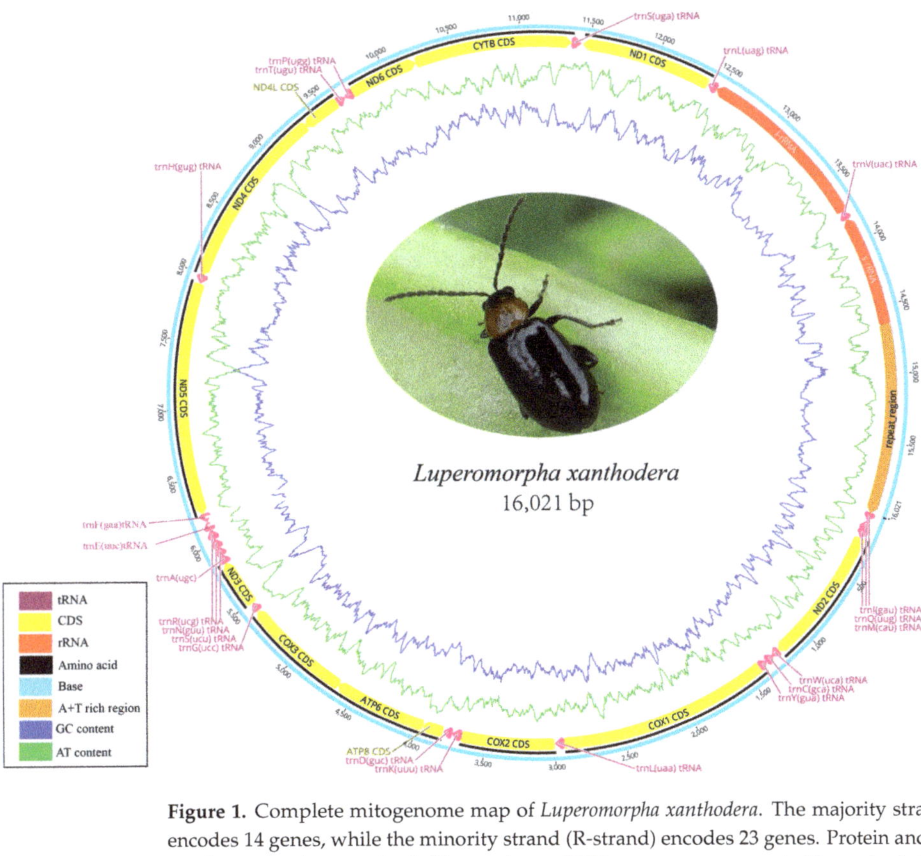

Figure 1. Complete mitogenome map of *Luperomorpha xanthodera*. The majority strand (F-strand) encodes 14 genes, while the minority strand (R-strand) encodes 23 genes. Protein and rRNA genes are denoted using standard abbreviations. tRNA genes are represented by a single letter for each amino acid, with two leucine tRNAs and two serine tRNAs separated by numbers.

Table 2. Organization of the *Luperomorpha xanthodera* mitochondrial genome. IGN: intergenic nucleotides. Negative values refer to overlapping nucleotides. dots (•) indicate no value, hyphens (-) represent the size of overlapping regions.

Gene	Direction	Location (bp)		Size (bp)	Codon		Anti-Codon	IGN
		from	to		Start	Stop		
tRNA^{Ile}	F	1	62	62	•	•	GAT	
tRNA^{Gln}	R	63	131	69	•	•	TTG	0
tRNA^{Met}	F	131	199	69	•	•	CAT	-1
ND2	F	200	1210	1011	ATT	TAA	•	0
tRNA^{Trp}	F	1211	1274	64	•	•	TCA	0
tRNA^{Cys}	R	1267	1328	62	•	•	GCA	-8
tRNA^{Tyr}	R	1329	1391	63	•	•	GTA	0
COI	F	1384	2926	1543	ATT	T	•	-8
tRNA^{Leu}	F	2927	2991	65	•	•	TAA	0
COII	F	2992	3676	685	ATG	T	•	0
tRNA^{Lys}	F	3677	3746	70	•	•	TTT	0
tRNA^{Asp}	F	3747	3809	63	•	•	GTC	0
ATP8	F	3810	3965	156	ATC	TAA	•	0
ATP6	F	3959	4633	675	ATG	TAA	•	-7
COIII	F	4633	5421	789	ATG	TAA	•	-1
tRNA^{Gly}	F	5424	5488	65	•	•	TCC	2
ND3	F	5489	5842	354	ATT	TAG	•	0
tRNA^{Ala}	F	5841	5904	64	•	•	TGC	-2
tRNA^{Arg}	F	5905	5968	64	•	•	TCG	0
tRNA^{Asn}	F	5966	6033	68	•	•	GTT	-3
tRNA^{Ser}	F	6034	6100	67	•	•	TCT	0
tRNA^{Glu}	F	6102	6165	64	•	•	TTC	1
tRNA^{Phe}	R	6165	6227	63	•	•	GAA	-1
ND5	R	6228	7932	1705	ATT	T	•	0
tRNA^{His}	R	7933	7995	63	•	•	GTG	0
ND4	R	8005	9325	1321	ATG	T	•	9
ND4L	R	9319	9600	282	ATG	TAA	•	-7
tRNA^{Thr}	F	9603	9665	63	•	•	TGT	2
tRNA^{Pro}	R	9666	9729	64	•	•	TGG	0
ND6	F	9732	10,235	504	ATT	TAA	•	2
CYTB	F	10,235	11,374	1140	ATG	TAG	•	-1
tRNA^{Ser}	F	11,373	11,439	67	•	•	TGA	-2
ND1	R	11,457	12,407	951	TTG	TAG	•	17
tRNA^{Leu}	R	12,409	12,473	65	•	•	TAG	1
16S rRNA	R	12,474	13,753	1280	•	•	•	0
tRNA^{Val}	R	13,754	13,821	68	•	•	TAC	0
12S rRNA	R	13,822	14,633	812	•	•	•	0
A + T rich region	F	14,634	16,021	1388	•	•	•	0

The *Luperomorpha xanthodera* mitogenome contained 41 bp overlapping regions in 11 pairs of neighboring genes ranging in length from 1 to 8 bp. The longest overlapping region of 8 bp was located between tRNA^{Trp} and tRNA^{Cys} and tRNA^{Tyr} and COI, respectively. A total of 34 bp intergenic nucleotides (IGN) ranging in size from 1 to 17 bp were distributed across seven locations. The longest intergenic spacer was found between the tRNA^{Ser} and ND1 (Table 2). These overlapping and intergenic regions are very frequent in flea beetle mitochondrial genomes [34–36]. The start codons and termination codons of all the PCGs genes coincided with the typical coleopteran [37] (truncated termination codons).

The overall nucleotide composition of *Luperomorpha xanthodera* is 41.3% A, 38.7% T, 11.6% C, and 8.4% G and includes a high A + T content (80%) and positive AT skew (0.032) and negative GC skew values (−0.160), this is comparable to other flea beetles. For instance, all the 27 flea beetles we analyzed had positive AT skews and negative GC skews, indicating that the base compositions of the flea beetle mitogenomes were, overall, biased towards A and C (Table 1). Furthermore, the A + T content of the third codon position was the highest (92.2%), while the A + T content of the second codon position was the lowest (69.0%). The AT skews of all the codon positions are positive, indicating a slightly higher occurrence of a likeness to T nucleotides. Meanwhile, the GC skews of the first codon position are positive, but the other two codon positions are negative.

3.2. Protein-Coding Genes and Codon Usage

The PCGs include seven NADH dehydrogenases (ND1-ND6 and ND4L), three cytochrome c oxidases (COI-COIII), two ATPases (ATP6 and ATP8), and one cytochrome b (CYTB). The total length of the 13 PCGs of *Luperomorpha xanthodera* was 11,116 bp, with an A + T content of 78.2%. Additionally, nine of its genes (ND2, COI, COII, ATP8, ATP6, COIII, ND3, ND6, CYTB) are located on the major strand (F-strand), and the others (ND5, ND4, ND4L, ND1) are located on the minor strand (R-strand). The ND5 gene was the longest at 1705 bp, and the smallest was the ATP8 gene at 156 bp. The base compositions of the 13 PCG genes were as follows: 40.2% A, 38.0% T, 9.4% G, and 12.5% C (Table 3). Most PCGs of *Luperomorpha xanthodera* were initiated with the typical ATN codon (except ND1 with TTG). Most PCGs (ND2, ATP8, ATP6, COIII, ND3, ND4L, ND6, CYTB, and ND1) had the complete stop codon of TAA or TAG, whereas four PCGs (COI, COII, ND4, and ND5) ended with the incomplete stop codon (T-). Some scholars believe that incomplete termination codons will be repaired by polyadenylation after transcription [38,39]. Based on 3698 codons, the 13 PCGs' relative synonymous codon usage (RSCU) values were determined. UUU (Phe), UUA (Leu2), AUU (Ile), and AUA (Met) were the four most common codons. The preferred codons all ended in A or U, contributing to the overall A + T bias of the mitogenomes. (Table 4; Figure 2).

Table 3. Nucleotide composition of the *Luperomorpha xanthodera* mitogenomes.

Genes	Size (bp)	Nucleotides Composition						ATskew	GCskew
		A (%)	T (%)	G (%)	C (%)	A + T (%)	G + C (%)		
Complete mitogenome	16,021	41.3	38.7	8.4	11.6	80	20	0.032	−0.160
PCGs	11,125	33.9	44.3	11.1	10.7	78.2	21.8	−0.133	0.018
1st codon position	3711	35.0	38.2	16.7	10.1	73.2	26.8	−0.043	0.247
2nd codon position	3707	21.1	47.9	13.4	17.5	69.0	29	−0.388	−0.135
3rd codon position	3707	45.5	46.7	3.2	4.6	92.2	7.8	−0.014	−0.181
tRNA	1432	41.9	38.8	8.7	10.6	80.7	19.3	0.038	−0.098
rRNA	2092	44.6	39.6	5.5	10.3	84.2	15.8	0.059	−0.304
16S rRNA	1280	45.1	38.7	5.7	10.5	83.8	16.2	0.076	−0.296
12S rRNA	812	43.8	41.0	5.2	10.0	84.8	15.2	0.033	−0.316
A + T rich region	1388	44.1	43.2	5.0	7.7	87.3	12.7	0.010	−0.213

Table 4. Relative synonymous codon usage (RSCU) of *Luperomorpha xanthodera* mitochondrial PCGs. The asterisk (*) represent Termination codon.

Amino Acid	Codon	Count	RSCU	Amino Acid	Codon	Count	RSCU
Phe	UUU	325	1.86	Tyr	UAU	162	1.79
	UUC	24	0.14		UAC	19	0.21
Leu2	UUA	524	5.22	His	CAU	59	1.66
	UUG	21	0.21		CAC	12	0.34
Leu1	CUU	22	0.22	Gln	CAA	57	1.84
	CUC	3	0.03		CAG	5	0.16
	CUA	27	0.27	Asn	AAU	190	1.83
	CUG	5	0.05		AAC	18	0.17

Table 4. *Cont.*

Amino Acid	Codon	Count	RSCU	Amino Acid	Codon	Count	RSCU
Ile	AUU	391	1.87	Lys	AAA	102	1.87
	AUC	27	0.13		AAG	7	0.13
Met	AUA	241	1.85	Asp	GAU	62	1.8
	AUG	20	0.15		GAC	7	0.13
Val	GUU	66	1.8	Glu	GAA	69	1.86
	GUC	5	0.14		GAG	5	0.14
	GUA	69	1.88	Cys	UGU	30	1.88
	GUG	7	0.19		UGC	2	0.13
Ser2	UCU	107	2.58	Trp	UGA	88	1.91
	UCC	7	0.17		UGG	4	0.09
	UCA	88	2.12	Arg	CGU	19	1.36
	UCG	3	0.07		CGC	1	0.07
Pro	CCU	66	2.06		CGA	34	2.43
	CCC	8	0.25		CGG	2	0.14
	CCA	53	1.66	Ser1	AGU	27	0.56
	CCG	1	0.03		AGC	3	0.07
Thr	ACU	90	2.03		AGA	85	2.05
	ACC	11	0.25		AGG	12	0.29
	ACA	76	1.72	Gly	GGU	50	1.05
	ACG	0	0		GGC	6	0.13
Ala	GCU	63	1.81		GGA	110	2.3
	GCC	16	0.46		GGG	25	0.52
	GCA	59	1.7	*	UAA	0	0
	GCG	1	0.03		UAG	0	0

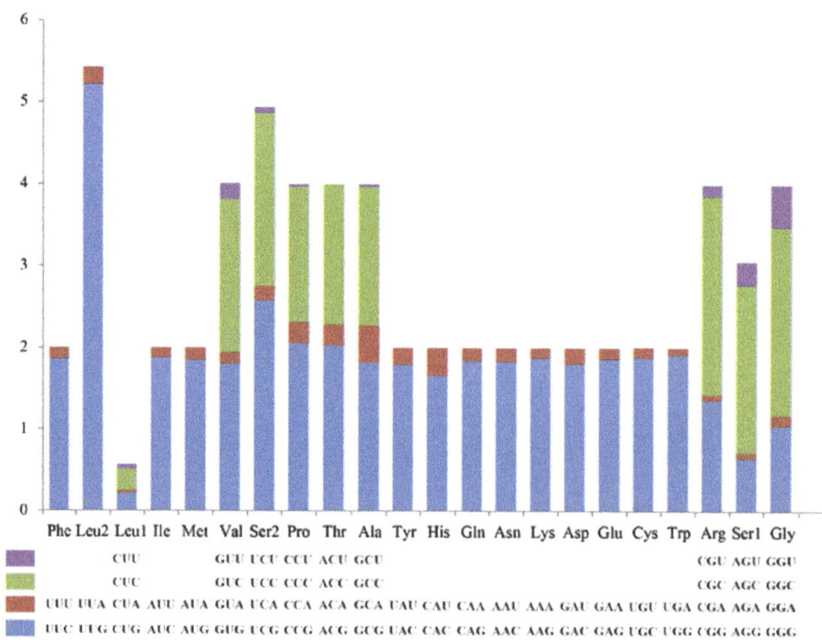

Figure 2. Relative synonymous codon usage (RSCU) of mitogenome of *Luperomorpha xanthodera*. Codon families are depicted in boxes below the x-axis; the box colors represent the stacked columns; and the values of RSCU are shown on the y-axis.

We calculated the nonsynonymous substitution rate (Ka), synonymous substitution rate (Ks), and the ratio of Ka/Ks for each PCG in three species (Figure 3). In all PCGs, the highest Ks value was exhibited by ATP6, whereas the Ka value of ATP8 was distinctly higher than others (Figure 3). The average Ka/Ks ranged from 0.181 (ATP6) to 1.037 (ND4). It is noteworthy that with the exception of ND4, all other PCGs of Ka/Ks values are lower than 1, demonstrating that the mutations were exchanged out by synonymous substitutions (Figure 3). The ND4 gene reached 1.037; this result provides an indication that positive selection had a dominating impact on the evolution of ND4 gene. The smallest Ka/Ks -values were 0.181 for COX1 and ATP6, which was considered to be a strong purifying selection.

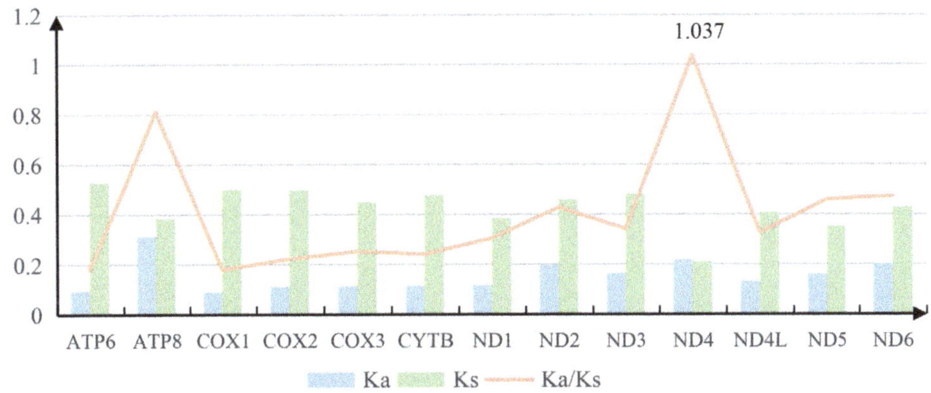

Figure 3. Evolutionary rates of 13 PCGs in between *Luperomorpha xanthodera* and *Luperomorpha hainana*. A blue boxes, green boxes, and orange line indicate the non-synonymous substitutions rate (Ka), the synonymous substitutions rate (Ks), and the ratio of the rate of non-synonymous substitutions to the rate of synonymous substitutions (Ka/Ks), respectively.

3.3. *Transfer and Ribosomal RNA Genes*

The length of the 22 tRNA genes of *Luperomorpha xanthodera* range from 62 bp ($tRNA^{Ile}$) to 70 bp ($tRNA^{Lys}$), comprising 8.94% of the complete mitogenome (Table 2), of which all the tRNA genes can be folded into the typical cloverleaf structure, except for trnS1(AGN), whose dihydrouridine (DHU) arm is converted to a simple loop (Figure 4)—this characteristic is frequent in flea beetle mitogenomes [40]. The anticodon of $tRNA^{Lys}$ mutations is TTT, and when we identified the insects of Chrysomelidae, using this would be one of the quickest methods [16].

The *Luperomorpha xanthodera* of ribosomal RNA (rRNA) is composed of 16S rRNA (situated between $tRNA^{Leu}$ and $tRNA^{Val}$) and 12S rRNA (situated between $tRNA^{Val}$ and the A + T rich region). The size of the two genes was 1280 bp (16S rRNA) and 812 bp (12S rRNA), respectively. The A + T content reached 84.2% in *Luperomorpha xanthodera*. Furthermore, the two rRNAs contained a positive AT skew and negative GC skew (Table 3).

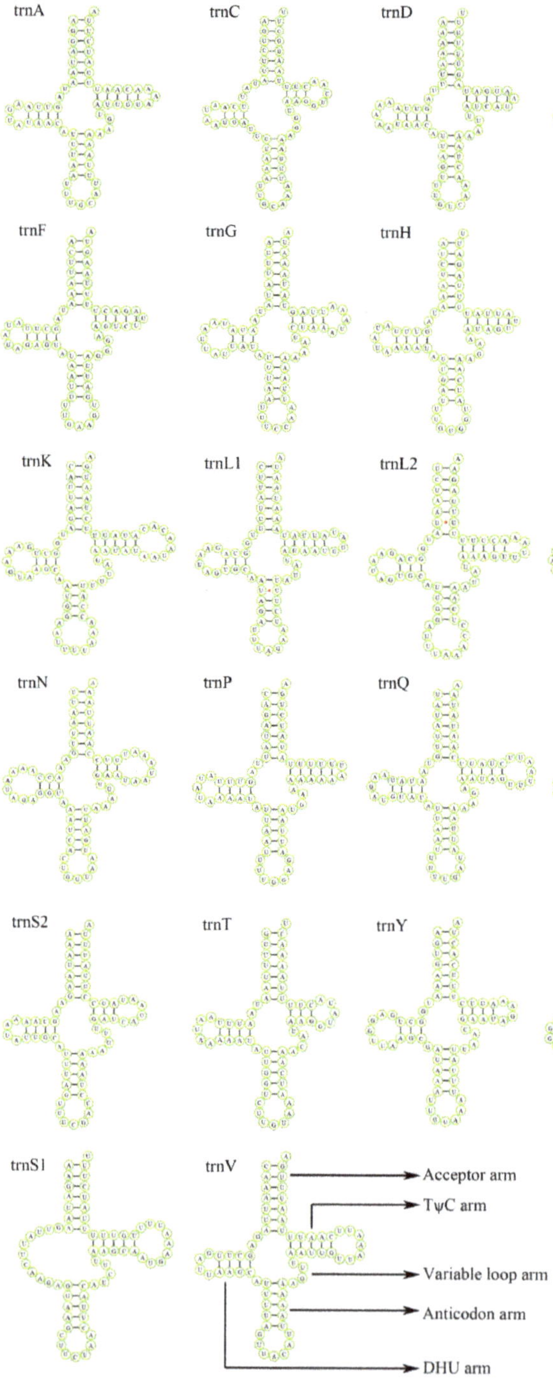

Figure 4. Predicted secondary structures of 22 tRNAs in *Luperomorpha xanthodera*. Lines (-) indicate Watson–Crick base pairings, whereas dots (•) indicate unmatched base pairings. The uppercase letter indicate the identity of each tRNA genes. Structural elements in tRNA arms and loops are illustrated as for trnV.

3.4. A + T Rich Region

The A + T-rich region is an important noncoding region in insect mitogenome, which regulates the mitochondrial DNA (mtDNA) transcription and replication [41–43]. The A + T rich region of *Luperomorpha xanthodera* is located between 12S rRNA and $tRNA^{Ile}$ and is 1388 bp in length (Table 2). Meanwhile, the A + T content was 87.3%, with positive AT skews (0.010) and negative GC skews (−0.213) (Table 3). Of note, we found four tandem repeat units in the control region, ranging from 13 to 35 bp. Furthermore, a poly-A region and three poly-T regions were found in the control region (Figure 5).

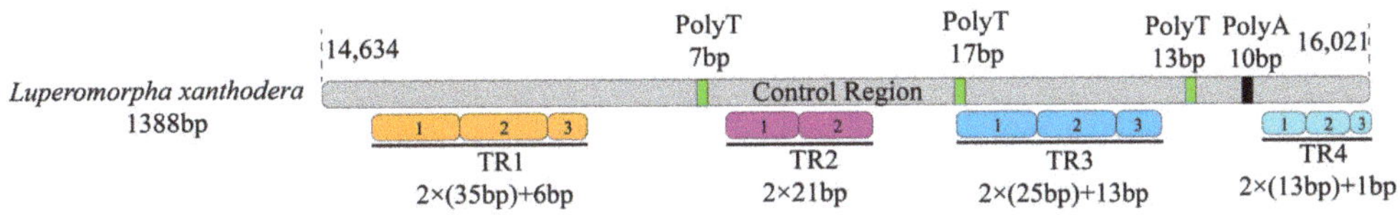

Figure 5. A + T-rich region of *Luperomorpha xanthodera* mitogenome. Gray blocks represent the A + T rich region. The numbers at the beginning and end of the gray blocks show the position of the control region in the *Luperomorpha xanthodera* mitogenome. R stands for repeat unit, and the number denotes the number of repeats. The black and green blocks represent the structures of poly-A and poly-T, respectively.

3.5. Phylogenetic Relationships

In the past two decades, molecular systematic approaches are widely used for disclosing unsettled classification and phylogenetic relationships in Insecta [44,45]. The phylogenetic trees of 37 mitochondrial genes (including one newly generated sequence, 24 other Chrysomelidae sequences from GenBank and two outgroup sequences) were generated from a single dataset (amino acids sequence with the 13 PCGs) using maximum likelihood (ML) and Bayesian inference; the BI tree and ML tree are shown in this paper (Figures 6 and 7). The two phylogenetic trees produced consistent results: (1) the Galerucinae is represented as a monophyletic group; (2) the Alticini were clustered into a monophyletic clade and formed sister groups with other subtribes of the Galerucini; (3) the Luperina was the polyphyletic group, which was consistent with the results of Gillespie et al. [46]; and (4) *Monolepta occifluvis* does not cluster with the other two species of the genus *Monolepta* but is instead sister to *Paleosepharia posticata* with high support values (0.99/98.7/100). These findings are consistent with earlier mitogenome-based research [47], which indicates that *Monolepta* is a polyphyletic genus. There are distinctions between the two trees as well. In the BI tree, the *Hoplasoma unicolor* is recovered as the sister group of the *Oises livida*, and their combined clade is grouped with *Luperomorpha* Weise. In the ML tree, *Luperomorpha* and Luperina showed a closer relationship.

According to our phylogenetic tree, the two *Luperomorpha* species are nested in Galerucini. The cladistic taxonomic position of several genera (including *Luperomorpha* Weise) has been rotating between Galerucini and Alticini in evolutionary studies of Chrysomelidae [48–50], and these genera are known as the 'problematic' genera [15]. As research on these two tribes has progressed, their taxonomic basis has shifted from a single character analysis (metafemoral spring) to a combination of characters (metafemoral spring, hind wing venation, female spermatheca, male aedeagus, etc.), resulting in *Luperomorpha* Weise being removed from Alticinae (currently Alticini) and added to Galerucinae (currently Galerucini) [51]. Some phylogenetic studies also support the *Luperomorpha* Weise being attributed to Galerucini [16,52].

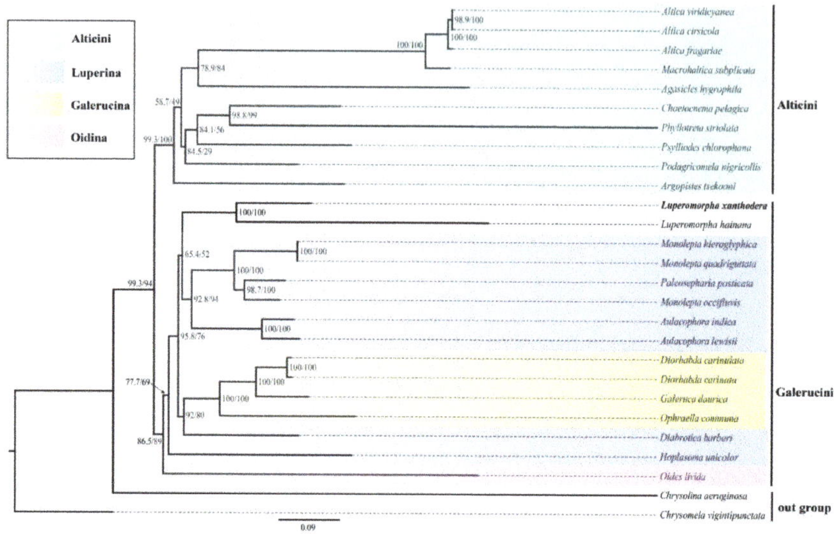

Figure 6. ML analysis was conducted using IQ-TREE with the dataset of amino acids of 13 PCGs. Note: Name in black bold show the phylogenetic position of *Luperomorpha xanthodera* that we sequenced in this research. The numbers on the nodes are the SH-aLRT support (%) and ultrafast bootstrap support (%).

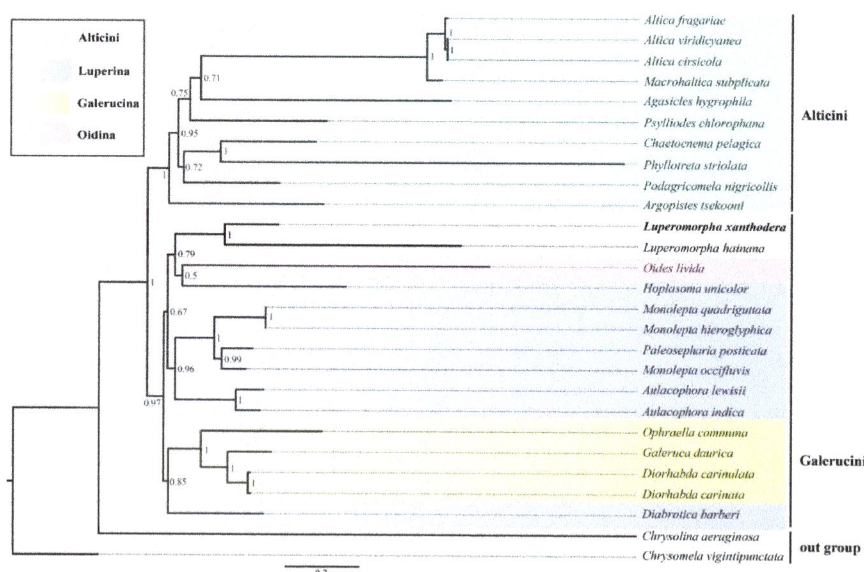

Figure 7. BI analysis was conducted using PhyloBayes with the dataset of amino acids of 13 PCGs. Note: Name in black bold show the phylogenetic position of *Luperomorpha xanthodera* that we sequenced in this research. The numbers on the nodes are the posterior probability.

4. Conclusions

At present, we sequenced and analyzed *Luperomopha xanthodera*'s complete mitogenome. The complete mitochondrial genome of *Luperomopha xanthodera* has a final size of 16,021 bp, with 80% AT content. The circular mitogenome encoded a typical set of 37 genes (13 PCGs, two rRNAs, 22 tRNAs, and one control region). The majority of PCGs use ATN (except

ND1 with TTG) as their start codon and TAA/TAG/T- as the stop codons. For the analysis of the selective pressure, we discovered that most of the 13 PCGs of *Luperomorpha* were less than one, particularly COX1 and ATP6 genes, which had the lowest value, indicating a high conservation. This revealed that PCGs were subjected to purifying selection in the genus, while the ND4 gene has a high value, demonstrating that it may have been mutated during the evolution process. Phylogenetic trees based on the mitogenomes of 27 species contributed to the scientific classification of *Luperomopha xanthodera*. Our study identified a close relationship between *Luperomopha* Weise and Luperini (currently Luperina). Overall, our results provide a hint for the phylogenetic position of *Luperomorpha*, and they have provided basic genetic information for understanding the phylogeny and evolution of leaf beetles.

Author Contributions: Conceptualization: J.L., B.Y. and M.Y.; methodology: J.L. and B.Y.; software: J.L., B.Y., H.H. and X.X.; data curation: J.L. and B.Y.; writing—original draft preparation: J.L. and B.Y.; writing—review and editing: J.L., Y.R., B.Y. and M.Y. All authors have read and agreed to the published version of the manuscript.

Funding: This study was supported by the Guizhou Province Science and Technology Plan Project (Qian Ke He Zhicheng [2021] Yiban221) and Guizhou Province Science and Technology Innovation Talent Team Project (Qian Ke He Pingtai Rencai-CXTD [2021]004).

Institutional Review Board Statement: Not applicable.

Informed Consent Statement: Informed consent was obtained from all subjects involved in the study.

Data Availability Statement: The *Luperomorpha xanthodera* sequenced mitogenome was submitted to the GenBank database (ON631248), the multiple sequence alignment used for phylogentic analysis are avalable on FigShare at this weblink: https://doi.org/10.6084/m9.figshare.21957440.v1, accessed on 23 December 2021.

Conflicts of Interest: The authors declare no conflict of interest.

References

1. Zhongying, Q.; Huihui, C.; Hao, Y.; Yuan, H.; Huimeng, L.; Xia, L.; Xingchun, G. Comparative mitochondrial genomes of four species of *Sinopodisma* and phylogenetic implications (Orthoptera, Melanoplinae). *Zookeys* **2020**, *2020*, 23–42. [CrossRef] [PubMed]
2. Liu, Q.N.; Zhu, B.J.; Dai, L.S.; Liu, C.L. The complete mitogenome of *Bombyx mori* strain Dazao (Lepidoptera: Bombycidae) and comparison with other lepidopteran insects. *Genomics* **2013**, *101*, 64–73. [CrossRef] [PubMed]
3. Yi, M.R.; Hsu, K.C.; Gu, S.; He, X.B.; Luo, Z.S.; Lin, H.D.; Yan, Y.R. Complete mitogenomes of four *Trichiurus* species: A taxonomic review of the *T.lepturus* species complexe. *ZooKeys* **2022**, *1084*, 1–26. [CrossRef] [PubMed]
4. Wang, B.; Nishikawa, K.; Matsui, M.; Nguyen, T.Q.; Xie, F.; Li, C.; Khatiwada, J.R.; Zhang, B.; Gong, D.; Mo, Y.; et al. Phylogenetic surveys on the newt genus *Tylototriton* sensu lato (Salamandridae, Caudata) reveal cryptic diversity and novel diversification promoted by historical climatic shifts. *PeerJ* **2018**, *2018*, e4384. [CrossRef]
5. Rancilhac, L.; Irisarri, I.; Angelini, C.; Arntzen, J.W.; Babik, W.; Bossuyt, F.; Kunzel, S.; Luddecke, T.; Pasmans, F.; Sanchez, E.; et al. Phylotranscriptomic evidence for pervasive ancient hybridization among Old World salamanders. *Mol. Phylogenet Evol* **2021**, *155*, 106967. [CrossRef]
6. Fairmaire, L. Coléoptères de l'intérieur de la Chine. *Annls. Soc. Ent. Belg* **1887**, *31*, 87–136.
7. Bieňkowski, A.O.; Orlova-Bienkowskaja, M.J. Quick spread of the invasive rose flea beetle *Luperomorpha Xanthodera* (Fairmaire, 1888) in Europe and its first record from Russia: (Coleoptera, Chrysomelidae, Galerucinae, Alticini). *Spixiana* **2018**, *41*, 99–104.
8. Del Bene, G.; Conti, B. Notes on the biology and ethology of *Luperomorpha Xanthodera*, a flea beetle recently introduced into Europe. *Bull. Insectology* **2009**, *62*, 61–68.
9. Beenen, R.; Roques, A. Leaf and seed beetles (Coleoptera, Chrysomelidae). *BioRisk* **2010**, *4*, 267–292. [CrossRef]
10. Kozlowski, M.W.; Legutowska, H. The invasive flea beetle *Luperomorpha Xanthodera* (Coleoptera: Chrysomelidae: Alticinae), potentially noxious to ornamental plants-first record in Poland. *J. Plant Prot. Res.* **2014**, *54*, 106–107. [CrossRef]
11. Yang, X.; Siqin, G.; Nie, R.; Ruan, Y.; Li, W. *Chinese Leaf Beetles*; Science Press: Beijing, China, 2015; 504p.
12. Fagot, J.; Libert, P. Entretiens sur les Chrysomelidae de Belgique et des régions limitrophes 6. *Luperomorpha Xanthodera* (Fairmaire, 1888), espèce nouvelle pour la faune belge (Chrysomelidae, Alticinae). *Entomol. Faun. Entomol.* **2016**, *69*, 81–82.
13. Sady, E.A.; Kiełkiewicz, M.; Kozłowski, M.W. The rose flea beetle (*Luperomorpha xanthodera*, Coleoptera: Chrysomelidae), an alien species in central Poland – from an episodic occurrence in an established population. *J. Plant Prot. Res.* **2020**, *60*, 86–97. [CrossRef]

14. Furth, D.G. Relationships of Palearctic and Nearctic genera of Alticinae (Coleoptera: Chrysomelidae). *Entomography* **1985**, *3*, 375–392.
15. Furth, D.G.; Suzuki, K. Character correlation studies of problematic genera of Alticinae in relation to Galerucinae (Coleoptera: Chrysomelidae). *Proc. Third Int. Symp. Chrysomelidae Beijing* **1992**, *1994*, 116–135.
16. Nie, R.E.; Breeschoten, T.; Timmermans, M.J.T.N.; Nadein, K.; Xue, H.J.; Bai, M.; Huang, Y.; Yang, X.K.; Vogler, A.P. The phylogeny of Galerucinae (Coleoptera: Chrysomelidae) and the performance of mitochondrial genomes in phylogenetic inference compared to nuclear rRNA genes. *Cladistics* **2018**, *34*, 113–130. [CrossRef]
17. Meng, G.; Li, Y.; Yang, C.; Liu, S. MitoZ: A toolkit for animal mitochondrial genome assembly, annotation and visualization. *Nucleic Acids Res.* **2019**, *47*, 1–7. [CrossRef]
18. Bernt, M.; Donath, A.; Jühling, F.; Externbrink, F.; Florentz, C.; Fritzsch, G.; Pütz, J.; Middendorf, M.; Stadler, P.F. MITOS: Improved de novo metazoan mitochondrial genome annotation. *Mol. Phylogenet. Evol.* **2013**, *69*, 313–319. [CrossRef]
19. Lowe, T.M.; Eddy, S.R. tRNAscan-SE: A program for improved detection of transfer RNA genes in genomic sequence. *Nucleic Acids Res.* **1997**, *25*, 955–964. [CrossRef]
20. Lowe, T.M.; Chan, P.P. tRNAscan-SE On-line: Integrating search and context for analysis of transfer RNA genes. *Nucleic Acids Res.* **2016**, *44*, W54–W57. [CrossRef]
21. Kearse, M.; Moir, R.; Wilson, A.; Stones-Havas, S.; Cheung, M.; Sturrock, S.; Buxton, S.; Cooper, A.; Markowitz, S.; Duran, C.; et al. Geneious Basic: An integrated and extendable desktop software platform for the organization and analysis of sequence data. *Bioinformatics* **2012**, *28*, 1647–1649. [CrossRef] [PubMed]
22. Kumar, S.; Stecher, G.; Tamura, K. MEGA7: Molecular evolutionary genetics analysis version 7.0 for bigger datasets. *Mol. Biol. Evol.* **2016**, *33*, 1870–1874. [CrossRef]
23. Perna, N.T.; Kocher, T.D. Patterns of nucleotide composition at fourfold degenerate sites of animal mitochondrial genomes. *J. Mol. Evol.* **1995**, *41*, 353–358. [CrossRef] [PubMed]
24. Benson, G. Tandem repeats finder: A program to analyze DNA sequences. *Nucleic Acids Res.* **1999**, *27*, 573–580. [CrossRef]
25. Librado, P.; Rozas, J. DnaSP v5: A Software for Comprehensive Analysis of DNA Polymorphism Data. *Bioinformatics* **2009**, *25*, 1451–1452. [CrossRef] [PubMed]
26. Capella-Gutiérrez, S.; Silla-Martínez, J.M.; Gabaldón, T. TrimAl: A tool for automated alignment trimming in large-scale phylogenetic analyses. *Bioinformatics* **2009**, *25*, 1972–1973. [CrossRef]
27. Kück, P.; Longo, G.C. FASconCAT-G: Extensive functions for multiple sequence alignment preparations concerning phylogenetic studies. *Front. Zool.* **2014**, *11*, 1–8. [CrossRef]
28. Nguyen, L.T.; Schmidt, H.A.; Von Haeseler, A.; Minh, B.Q. IQ-TREE: A fast and effective stochastic algorithm for estimating maximum-likelihood phylogenies. *Mol. Biol. Evol.* **2015**, *32*, 268–274. [CrossRef]
29. Hoang, D.T.; Chernomor, O.; Von Haeseler, A.; Minh, B.Q.; Vinh, L.S. UFBoot2: Improving the ultrafast bootstrap approximation. *Mol. Biol. Evol.* **2018**, *35*, 518–522. [CrossRef]
30. Guindon, S.; Dufayard, J.F.; Lefort, V.; Anisimova, M.; Hordijk, W.; Gascuel, O. New algorithms and methods to estimate maximum-likelihood phylogenies: Assessing the performance of PhyML 3.0. *Syst. Biol.* **2010**, *59*, 307–321. [CrossRef]
31. Wang, H.C.; Minh, B.Q.; Susko, E.; Roger, A.J. Modeling site heterogeneity with posterior mean site frequency profiles accelerates accurate phylogenomic estimation. *Syst. Biol.* **2018**, *67*, 216–235. [CrossRef] [PubMed]
32. Lartillot, N.; Philippe, H. A Bayesian mixture model for across-site heterogeneities in the amino-acid replacement process. *Mol. Biol. Evol.* **2004**, *21*, 1095–1109. [CrossRef] [PubMed]
33. Lartillot, N.; Rodrigue, N.; Stubbs, D.; Richer, J. PhyloBayes MPI: Phylogenetic reconstruction with infinite mixtures of profiles in a parallel environment. *Syst. Biol.* **2013**, *62*, 611–615. [CrossRef] [PubMed]
34. Stahlke, A.R.; Ozsoy, A.Z.; Bean, D.W.; Hohenlohe, P.A. Mitochondrial genome sequences of *Diorhabda Carinata* and *Diorhabda Carinulata*, two beetle species introduced to North America for biological control. *Microbiol. Resour. Announc.* **2019**, *8*, 7–8. [CrossRef] [PubMed]
35. Li, J.H.; Shi, Y.X.; Song, L.; Zhang, Y.; Zhang, H.F. The complete mitochondrial genome of an important agricultural pest *Monolepta Hieroglyphica* (Coleoptera: Chrysomelidae: Galerucinae). *Mitochond. DNA Part B* **2020**, *5*, 1820–1821. [CrossRef]
36. Nie, R.E.; Wei, J.; Zhang, S.K.; Vogler, A.P.; Wu, L.; Konstantinov, A.S.; Li, W.Z.; Yang, X.K.; Xue, H.J. Diversification of mitogenomes in three sympatric Altica flea beetles (Insecta, Chrysomelidae). *Zool. Scr.* **2019**, *48*, 657–666. [CrossRef]
37. Sheffield, N.C.; Song, H.; Cameron, S.L.; Whiting, M.F. A comparative analysis of mitochondrial genomes in coleoptera (Arthropoda: Insecta) and genome descriptions of six new beetles. *Mol. Biol. Evol.* **2008**, *25*, 2499–2509. [CrossRef]
38. Ojala, D.; Montoya, J.; Attardi, G. tRNA punctuation model of RNA processing in human mitochondria. *Nature* **1981**, *290*, 470–474. [CrossRef]
39. Boore, J.L. Complete mitochondrial genome sequence of Urechis caupo, a representative of the phylum Echiura. *BMC Genom.* **2004**, *5*, 67. [CrossRef]
40. Wang, Q.; Tang, G. The mitochondrial genomes of two walnut pests, *Gastrolina Depressa Depressa* and *G. Depressa Thoracica* (Coleoptera: Chrysomelidae), and phylogenetic analyses. *PeerJ* **2018**, *6*, e4919. [CrossRef] [PubMed]
41. Cameron, S.L. Insect mitochondrial genomics: Implications for evolution and phylogeny. *Annu. Rev. Entomol.* **2014**, *59*, 95–117. [CrossRef]
42. Boore, J.L. Animal mitochondrial genomes. *Nucleic Acids Res.* **1999**, *27*, 1767–1780. [CrossRef] [PubMed]

43. Zhang, D.X.; Hewitt, G.M. Insect mitochondrial control region: A review of its structure, evolution and usefulness in evolutionary studies. *Biochem. Syst. Ecol.* **1997**, *25*, 99–120. [CrossRef]
44. Lee, S.; Leschen, R.A.B.; Lee, S. Evolution of feeding habits of sap beetles (Coleoptera: Nitidulidae) and placement of Calonecrinae. *Syst. Entomol.* **2020**, *45*, 911–923. [CrossRef]
45. Gimmel, M.L.; Bocakova, M.; Gunter, N.L.; Leschen, R. Comprehensive phylogeny of the Cleroidea (Coleoptera: Cucujiformia). *Syst. Entomol.* **2019**, *44*, 527–558. [CrossRef]
46. Gillespie, J.J.; Tallamy, D.W.; Riley, E.G.; Cognato, A.I. Molecular phylogeny of rootworms and related Galerucine beetles (Coleoptera: Chrysomelidae). *Zool. Scr.* **2008**, *37*, 195–222. [CrossRef]
47. Lee, C.F. The genus *Paleosepharia Laboissière*, 1936 in Taiwan: Review and nomenclatural changes (Coleoptera, Chrysomelidae, Galerucinae). *Zookeys* **2018**, *19*, 19–41. [CrossRef]
48. Ge, D.; Gómez-Zurita, J.; Chesters, D.; Yang, X.; Vogler, A.P. Suprageneric systematics of flea Beetles (Chrysomelidae: Alticinae) inferred from multilocus sequence data. *Mol. Phylogenet. Evol.* **2012**, *62*, 793–805. [CrossRef]
49. Gillespie, J.J.; Kjer, K.M.; Duckett, C.N.; Tallamy, D.W. Convergent evolution of cucurbitacin feeding in spatially isolated rootworm taxa (Coleoptera: Chrysomelidae; Galerucinae, Luperini). *Mol. Phylogenet. Evol.* **2003**, *29*, 161–175. [CrossRef]
50. Kim, S.J.; Kjer, K.M.; Duckett, C.N. Comparison between molecular and morphological-based phylogenies of Galerucine/Alticine leaf beetles (Coleoptera: Chrysomelidae). *Insect Syst. Evol.* **2003**, *34*, 53–64. [CrossRef]
51. Ge, D.; Chesters, D.; Zhang, L.; Yang, X.; Alfried, P. Anti-predator defence drives parallel morphological evolution in flea beetles. *Proc. R. Soc. B Biol. Sci.* **2011**, *278*, 2133–2141. [CrossRef] [PubMed]
52. Chaboo, C.S.; Clark, S. Beetles (Coleoptera) of Peru: A survey of the families. Chrysomelidae: Galerucinae (not including Alticini). *J. Kansas Entomol. Soc.* **2015**, *88*, 361–367. [CrossRef]

genes

MDPI

Article

Complete Mitochondrial Genome of *Piophila casei* (Diptera: Piophilidae): Genome Description and Phylogenetic Implications

Shenghui Bi [1,2], Yanfei Song [2], Linggao Liu [1], Jing Wan [1], Ying Zhou [1], Qiujin Zhu [1,*] and Jianfeng Liu [2,*]

1 School of Liquor and Food Engineering, Guizhou University, Guiyang 550025, China; lgliu_0127@163.com (L.L.)
2 Scientific Observing and Experimental Station of Crop Pest in Guiyang, Guizhou Provincial Key Laboratory for Agricultural Pest Management of the Mountainous Region, Institute of Entomology, Guizhou University, Ministry of Agriculture, Guiyang 550025, China; gs.songyf20@gzu.edu.cn
* Correspondence: ls.qjzhu@gzu.edu.cn (Q.Z.); jfliu3@gzu.edu.cn (J.L.)

Abstract: *Piophila casei* is a flesh-feeding Diptera insect that adversely affects foodstuffs, such as dry-cured ham and cheese, and decaying human and animal carcasses. However, the unknown mitochondrial genome of *P. casei* can provide information on its genetic structure and phylogenetic position, which is of great significance to the research on its prevention and control. Therefore, we sequenced, annotated, and analyzed the previously unknown complete mitochondrial genome of *P. casei*. The complete mt genome of *P. casei* is a typical circular DNA, 15,785 bp in length, with a high A + T content of 76.6%. It contains 13 protein-coding genes (PCG), 2 ribosomal RNA (rRNA) genes, 22 transfer RNA (tRNA) genes, and 1 control region. Phylogenetic analysis of 25 Diptera species was conducted using Bayesian and maximum likelihood methods, and their divergence times were inferred. The comparison of the mt genomes from two morphologically similar insects *P. casei* and *Piophila megastigmata* indicates a divergence time of 7.28 MYA between these species. The study provides a reference for understanding the forensic medicine, taxonomy, and genetics of *P. casei*.

Keywords: Piophilidae; mitochondrial DNA; phylogenetic analysis; comparative analyses

Citation: Bi, S.; Song, Y.; Liu, L.; Wan, J.; Zhou, Y.; Zhu, Q.; Liu, J. Complete Mitochondrial Genome of *Piophila casei* (Diptera: Piophilidae): Genome Description and Phylogenetic Implications. *Genes* **2023**, *14*, 883. https://doi.org/10.3390/genes14040883

Academic Editor: Michael J. Palladino

Received: 12 March 2023
Revised: 4 April 2023
Accepted: 6 April 2023
Published: 8 April 2023

1. Introduction

P. casei (Linnaeus, 1758) (Diptera: Piophilidae) belongs to the holometabolous insects, and its growth cycle includes four stages: egg, larva, pupa, and adult. The life cycle of *P. casei* lasts 12–30 days, and it can usually breed 7–9 generations per year [1]. *P. casei* exhibits a highly developed olfactory system, displaying a specific attraction to fishy and putrid odors [2,3], and has been identified as the most harmful pest to dry-cured ham. Its voracious appetite enables it to quickly consume ham, leading to the formation of cavities, darkening of the flesh, and the production of a putrid smell, ultimately resulting in significant damage to ham quality and considerable economic losses for ham manufacturers [4]. Similarly, it has been found in cheese and fish, as well as in human corpses in advanced stages of decomposition. It is widely distributed in the world and has been documented not only in Guizhou, Yunnan, and Zhejiang provinces in China but also in many parts of Europe and North America [5]. In Italy, *P. casei* is used to make cheese products with a unique flavor [6]. However, ingestion of the larvae by humans can lead to intestinal myiasis, so safety remains a concern [7,8].

P. casei is generally not favored in the food industry, but it is highly valued in the fields of forensic medicine and forensic entomology. In forensic medicine, the Piophilidae family is extremely important, especially in the advanced stages of decomposition, as it can provide valuable information when determining the minimum post-mortem interval [5,9].

Piophilidae is a small family of flies that comprises several species, including *P. casei* and *Piophila megastigmata* (McAlpine 1978) [10,11]. In the early years, the high similarity in appearance between *P. casei* and *P. megastigmata*, coupled with the lack of research on *P. megastigmata*, made it challenging to differentiate between the two species, with *P. megastigmata* often being misidentified as *P. casei*. Until recent times, *P. megastigmata* was found to be more prevalent on cadavers than *P. casei* [12]. However, the genetic information available on *P. casei* is still limited, and forensic entomologists rely mostly on morphological observations rather than genetic comparisons to distinguish and compare the appearance of *P. casei* and *P. megastigmata* [13,14].

Mitochondria (mt) genes are powerful molecular markers in phylogenetic and population genetics studies. The complete mt genomes that contain more variants provide more information than partial mt gene sequences to help scientists resolve phylogenetic relationships and infer species evolution [15–17]. The mt genome of insects and mites generally contains 13 protein genes, 22 tRNAs, 2 rRNAs, and a control region [18–20]. In this study, the complete mt genome was assembled, and the basic characteristics of *P. casei* mt genes were studied, focusing on the comparison of the PCGs between *P. casei* and *P. megastigmata*. The known mt genome data of 25 Diptera insects from GenBank was selected for the construction of phylogenetic relations and evolution time, laying the foundation for a more in-depth study at the molecular level and providing theoretical bases and ideas for the accurate identification of forensic insects *P. casei* and *P. megastigmata*.

2. Materials and Methods

2.1. Sample Collection

Fifty adult specimens of *P. casei* were collected on 2 November 2021 from the dry-cured hams in Panxian (25°75.87′ N; 104°52.57′ E), Guizhou Province, China. The specimens were preserved in 100% alcohol for species identification and DNA extraction in the laboratory. Illustrations of *P. casei* were obtained from samples of eggs, larvae, pupae, and adults that were collected. Color images were captured using the Keyence VHX-6000 model (Keyence, Osaka, Japan) color-image analysis instrument.

2.2. DNA Extraction, Library Construction, and Sequencing

Genomic DNA was extracted from *P. casei* male adults using Qiagen DNeasy Blood and Tissue Kit (Qiagen, Hilden, Germany) according to the manufacturer's instructions. DNA degradation and contamination were monitored on 1% agarose gels. DNA purity was measured using a NanoDrop 2000 spectrophotometer (Thermo Fisher Scientific, Waltham, MA, USA), and DNA was quantified using Qubit DNA Assay Kit Fluorometer (Life Technologies, Carlsbad, CA, USA).

The Truseq Nano DNA HT Sample Preparation Kit (Illumina, San Diego, CA, USA) was used to generate sequencing libraries following the manufacturer's instructions, and index codes were added to assign the sequences to each sample. The DNA sample (1.5 μg DNA per sample) was fragmented by sonication to a size of 350 b. Then the DNA fragments were end-polished, A-tailed, and ligated to the full-length adapter from the sequencing kit with further PCR amplification. Last, the PCR products were purified (AMPure XP system) and analyzed for size distribution by an Agilent2100 Bioanalyzer. These constructed libraries were sequenced with the Illumina NovaSeq 6000 platform (BIOZERON, Shanghai, China), and 150 bp paired-end reads were generated with an insert size of around 350 bp [21].

2.3. Genome Assembly and Gene Annotation

In this study, a script implemented in MitoZ was utilized to filter the extracted mitochondrial gene sequences, which effectively removed reads containing numerous 'N's, low-quality reads, or PCR duplicates (i.e., pairs of identical reads) [22]. Then the mt genome was assembled in MitoZ using De novo assembly. De novo assembly is an algorithm well suited to mt mitochondrial genome assembly, which, with the assistance of other algorithms and evaluation metrics in MitoZ, enables better assembly of genomic sequences [23]. The principle is that the average sequencing depth of mitochondrial genome reads is much higher than that of the nuclear genome, and different Kmer parameters are set to achieve the best possible assembly. The Kmer parameter was finally adjusted to 33 to achieve the best assembly results. Protein-coding genes were annotated using scripts in MitoZ. tBLAST in BLAST v2.2.19 [24] was used to find candidate PCG sequences by matching sequences to the protein database sequences (MitoZ has a built-in database of insect PCG annotations); Genewise v2.2.0 [25], which produces faster and more accurate results, was used to identify each PCG for annotation. tRNAs were annotated using MiTFi [26] of the covariance model (CM) for annotation, and rRNAs were annotated using infernal-1.1.1 [27] and rRNA CMs based on an extensive manually curated alignment [28]. The complete mitogenome sequence of *P. casei* has been submitted to GenBank (ON204020).

2.4. Genomic and Phylogenetic Analysis

The circular map of the *P. casei* mt genome was painted using proksee (https://proksee.ca/ (accessed on 10 September 2022)). The basic information about the genome was calculated using CodonW v1.4.2. The relative synonymous codon usage (RSCU) values were analyzed using ggplot2 v3.3.6 and plotted using aplot v0.1.6 packages in the R v4.0.4 [29].

To understand the taxonomic position of *P. casei* in Diptera, the mt genomes of 25 insects (including 2 outgroup mt genomes) were downloaded from GenBank and used to construct phylogenetic trees using both the maximum likelihood (ML) and Bayesian inference (BI) methods. The nucleotide diversity (Pi), the nonsynonymous substitution rate (Ka), and the synonymous substitution rate (Ks) were calculated using Launch DnaSP6 v6.12.3 [30], and the results were plotted using GraphPad Prism v8.0.2. Base sequence substitution saturation was analyzed and plotted using DAMBE v7.3.11 [31].

The 22 tRNA genes, 13 PCGs, and 2 rRNA genes of each of the 25 insects were extracted from NCBI and imported into PhyloSuite v1.2.2 [32]. The sequences were trimmed using Gblocks v0.91 b [33] to remove redundant codons and then concatenated using concatenate sequence to form a single concatenated sequence. All sequences were then aligned using MAFFT v7.313 and optimized using MACSE v2.0.1 [34]. The secondary structure of 22 tRNAs was predicted with ARWEN and compared manually. The MFE (minimum free energy) structures of two ribosomal RNA genes (rrnL and rrnS) were obtained by RNAfold [35] prediction. For the tandem sequences, the optimal GTR + F + I + G4 and GTR + F + G4 molecular phylogenetic model was found using ModelFinder [36] and imported into MrBayes v3.2.6 [37] and IQ-TREE v1.6.8 [38] for BI and ML phylogenetic tree construction with 104 and 2 × 106 bootstraps, respectively. iTol (https://itol.embl.de/itol.cgi (accessed on 10 September 2022)) was used to visualize the phylogenetic trees, and then the developmental trees constructed by the two methods were compared.

2.5. Divergence Time Estimate

The divergence time was estimated from the BI phylogenetic trees with timetree5 (http://www.timetree.org/ (accessed on 10 September 2022)) and BEAST v1.10.4 [39] according to the related literature and using age 70 MYA as the divergence time calibration [40]. The estimation of divergence models for subfamilies was based on complete mt genomes according to the strict clock log-normal model in BEAST. A calibrated Yule model was used and spliced using the GTR + F + I + G4 and GTR + F + G4 models. After confirming the convergence of the chains with Trancer v1.7.2 [41], the first 10% of generations were burn-in as ageing every 1000 generations sampled at 10^7 conditions. We summarized the subsample trees in a maximum clade credibility tree with mean heights using Tree Annotator v1.10.4. The mean heights and 95% highest probability density (95% HPD) were displayed in Figtree v1.4.3. The labeling and photo-compositing of the images in this article were made using Adobe Photoshop 2022 and Adobe Illustrator 2022.

3. Results and Discussion

3.1. General Features of *P. casei* mt Genome

The complete mitochondrial genome of *P. casei* is 15,785 bp long (GenBank No. ON204020) and contains 13 PCGs (Protein-coding genes) (11,206 bp), 22 tRNA genes (1467 bp), 2 rRNA genes (2107 bp), and 1 control region (744 bp). The circular map of the *P. casei* mt genome is shown in Figure 1a and is generally consistent with the mt genomes of eight other Diptera, such as *P. megastigmata* (15,410 bp) (Figure 1b), *Bactrocera dorsalis* (15,915 bp) (Figure 1c), *Liriomyza bryoniae* (16,183 bp) (Figure 1e), and *Anopheles oryzalimnetes* (15,422bp) (Figure 1f). The difference in mt genome length between these five species is mainly due to the differences in the size of the control region (from a minimum of 388bp to a maximum of 1354 bp), as previously observed between Diptera species [42]. In Diptera, the mt genomes of some species, such as *Zeugodacus caudatus* (15,311 bp) (Figure 1d), *Homoneura interstincta* (16,351 bp) (Figure 1g), and *Tropidia scita* (15,739 bp) (Figure 1f) are arranged differently than most species.

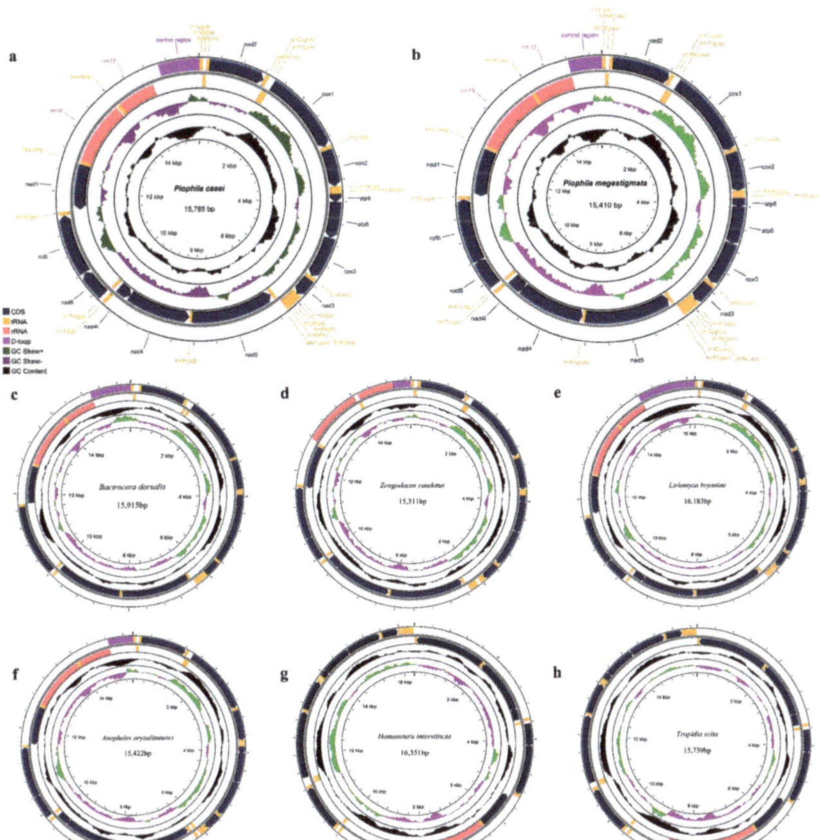

Figure 1. The mt genome of eight species. (**a**) *P. casei*, (**b**) *P. megastigmata*, (**c**) *B. dorsalis*, (**d**) *Z. caudatus*, (**e**) *L. bryoniae*, (**f**) *A. oryzalimnetes*, (**g**) *H. interstincta*, and (**h**) *T. scita*. The chain is marked with an arrow indicating the direction of gene transcription. Gene lengths correspond to nucleotide lengths in the diagram. The outermost layer is the J chain, and the second layer is the N chain. The third black peak indicates GC content, and the purple and green peaks indicate the ±skew of GC.

The mitochondrial genomes of *P. casei* and *P. megastigmata* exhibit a high degree of conservation, as evidenced by the presence of 37 genes in each genome, as well as the absence of any inversion events or swapping of gene coding directions.

In terms of ATCG content and AT-skew, CG-skew, the mt genome of *P. megastigmata* is only slightly different than that of *P. casei* (Table 1). The difference between the two is more pronounced in the AT% and AT-skew of the tRNA gene, indicating differences in amino acid content between the two species.

Table 1. The base composition of the complete mt genome of *P. casei* and *P. megastigmata*.

(*P. casei*) Region	**Size (bp)**	**A%**	**T%**	**AT-skew**	**C%**	**G%**	**CG-skew**
Mitogenome	15,785	39.5	37.1	0.03	13.8	9.6	−0.18
Protein-coding genes	11,206	31.2	43.6	−0.17	12.3	12.9	0.02
tRNA	1467	37.7	38.9	−0.02	10.2	13.2	0.13
rRNA	2107	38.2	41.0	−0.04	7.4	13.4	0.29
(*P. megastigmata*) Region	**Size (bp)**	**A%**	**T%**	**AT-skew**	**C%**	**G%**	**CG-skew**
Mitogenome	15,410	39.2	37.2	0.03	14.1	9.5	−0.2
Protein-coding genes	11,207	31.3	43.6	−0.16	12.4	12.7	0.01
tRNA	1463	38.3	38.8	−0.01	10.0	12.9	0.13
rRNA	2111	37.9	41.8	−0.05	7.1	13.2	0.3

A%, T%, C%, and G% represent the proportion of the respective nucleotides in the complete mitochondrial genome.

The complete *P. casei* mt genome consisted of 39.5% A, 37.1% T, 13.8% C, and 9.6% G. The total A + T content was 76.6%, which is relatively high among Tephritidae species. The mt genome had a positive AT skew (0.03) and a negative CG skew (−0.18); according to the previous reports, a bias against the use of Gs in all strands is characteristic of the metazoan mitochondrial genome [43]. Four PCGs (*nad5, nad4, nad4l, nad1*), eight tRNA genes (*trnQ, trnC, trnY, trnF, trnH, trnP, trnL, trnV*), and two rRNA genes (*rrn12, rrn16*) were found on the minority strand (N strand), while the remaining genes were on the majority strand (J strand) (Table 2). Ten genes overlap on the mt genome of *P. casei*, with a total of thirty-four overlapping bases. There are 14 gene spacers, with the largest spacer (16 bp) occurring between rrn12 and the control region.

In order to compare the AT% variations among different species within the same family, nine species in Tephritoidea were selected [41]. Overall, the AT content of each of their genes was found to be similar, with significant differences only in the proportion of codons used in the first, second, or third codons. The first codon position in both *P. casei* and *P. megastigmata* was found to have higher AT% compared with several other species in the Tephritoidea(Figure 2).

Table 2. Gene order and basic characteristics of the *P. casei* mt genome.

Gene	**Type**	**Strand**	**Position (Start–End)**	**Length (bp)**	**Intergenic Spacer**
trnl(gat)	tRNA	J	1–65	65	–
trnQ(ttq)	tRNA	N	63–131	69	−3
trnM(cat)	tRNA	J	131–199	69	−1
nad2	CDS	J	200–1228	1029	0
trnW(tca)	tRNA	J	1227–1294	68	−2
trnC(gca)	tRNA	N	1287–1351	65	−8
trnY(gta)	tRNA	N	1355–1422	68	3
cox1	CDS	J	1424–2954	1531	1
trnL(taa)	tRNA	J	2955–3020	66	0

Table 2. *Cont.*

Gene	Type	Strand	Position (Start–End)	Length (bp)	Intergenic Spacer
cix2	CDS	J	3027–3714	688	6
trnK(ctt)	tRNA	J	3715–3785	71	0
trnD(gtc)	tRNA	J	3788–3854	67	2
atp8	CDS	J	3855–4016	162	0
atp6	CDS	J	4010–4687	678	−7
cox3	CDS	J	4691–5479	789	3
trnG(tcc)	tRNA	J	5488–5552	65	8
nad3	CDS	J	5553–5906	63	0
trnA(tgc)	tRNA	J	5908–5971	66	1
trnR(tcg)	tRNA	J	5971–6033	68	−1
trnN(gtt)	tRNA	J	6034–6099	66	0
trnS(qct)	tRNA	J	6100–6167	68	0
trnE(ttc)	tRNA	J	6168–6233	66	0
trnF(gaa)	tRNA	N	6252–6317	66	18
nad5	CDS	N	6318–8052	1753	0
trnH(gtg)	tRNA	N	8053–8118	66	0
nad4	CDS	N	8119–9457	1339	0
nad4l	CDS	N	9451–9747	297	−7
trnT(tgt)	tRNA	J	9750–9814	65	2
trnP(tgg)	tRNA	N	9815–9880	66	0
nad6	CDS	J	9883–10,407	525	2
cytb	CDS	J	10,411–11,547	1137	3
trnS(tga)	tRNA	J	11,546–11,612	67	−2
nad1	CDS	N	11,628–12,576	949	5
trnL(tag)	tRNA	N	12,578–12,642	65	1
rrn16	rRNA	N	12,643–13,939	1297	0
trnV(tac)	tRNA	N	13,968–14,039	72	−2
rrn12	rRNA	N	14,039–14,825	787	−1
control region	D-loop	J	15,042–15,785	744	16

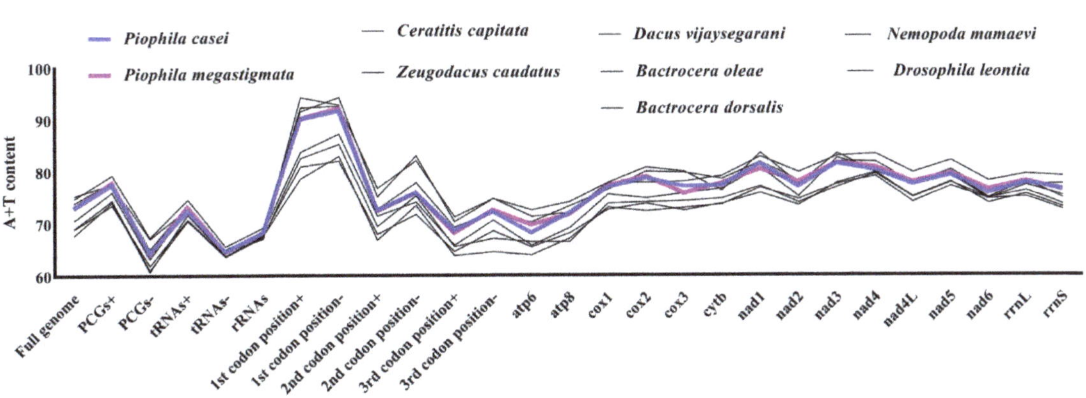

Figure 2. AT content of genes in various parts of the mt genome among the nine species in Tephritoidea.

3.2. Protein-CODING Genes and Codons

The PCGs accounted for 71% (11,206 bp) of *P. casei* mt genes. The start codon ATN was found in 11 of the 13 PCGs genes, including ATG in *cox2*, *atp6*, *cox3*, *nad4*, *nad4l*, and *cytb*; ATT in *nad2*, *atp8*, *nad3*, and *nad6*; and ATA in *nad1*. However, *cox1* uses CGA and *nad5* uses GTG as their start codon. The stop codons of 13 PCGs are mostly TAA (*nad2*, *cox1*, *atp8*, *atp6*, *cox3*, *nad3*, *nad4l*, *nad6*) [44] and include TAG (*cytb*) and T (*cox2*, *nad5*, *nad4*, *nad1*). The PCGs of the two species were measured to have only 863 different loci.

The nucleotide diversity (Pi) was compared for the PCGs of 26 species (Figure 3a), and the strength of polymorphisms was used to characterize the PCGs of 26 species. Overall, the Pi values ranged from 0.153 to 0.269. The *nad2* (Pi = 0.269) gene had the greatest variability among these genes, followed by *nad6* (Pi = 0.259), *atp8* (Pi = 0.255), *nad4l* (Pi = 0.202), and, finally, *cox1* (Pi = 0.153) with the least variability. It shows that *cox1* has high genetic stability among dipterans, while *nad2* has weak stability and is suitable for the study of the evolution of variation [45]. The Pi values of *P. casei* and *P. megastigmata* were also compared individually in each species (Figure 3b): *cox1* (Pi = 0.018) demonstrated the least variability, *nad3* (Pi = 0.120) showed the greatest variability, and *atp8* (Pi = 0.025) had a low degree of variability, unlike the results of the multispecies comparison.

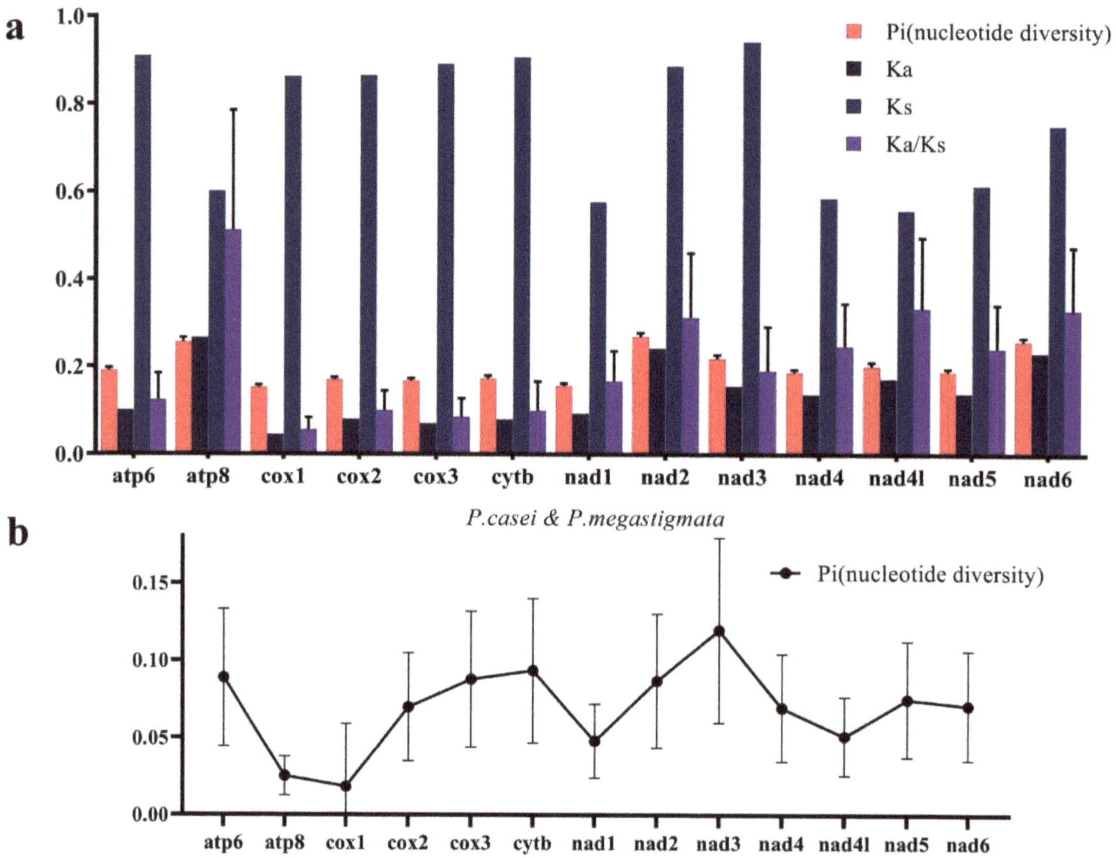

Figure 3. (**a**) Nucleotide diversity (Pi) values of *P. casei* and *P. megastigmata* PCGs; (**b**) Pi values and Ka/Ks values of PCGs in the mt genomes of 26 species.

The frequency of non-synonymous mutations (Ka), the frequency of synonymous mutations (Ks), and the ratio of the two (Ka/Ks) were calculated for 13 PCGs of 26 species and used to understand the rate at which synonymous and non-synonymous mutations occur (Figure 3d). The Ka/Ks values determine whether there is selection pressure on the PCGs and also the degree of conservation of the coding gene. The closer the Ka/Ks is to 1, the lower the selection pressure is on that gene. The Ka/Ks values range from 0.0561 to 0.5119, and the order of Ka/Ks values for PCGs from the smallest to the largest is as follows: *cox1, cox3, cytb, cox2, atp6, nad1, nad3, nad5, nad4, nad2, nad6, nad4l, atp8*. Among them, the genes of the cytochrome oxidase (cox) family *cox1, cox2, cox3,* and *cytb* have smaller Ka/Ks values, indicating that they face a stronger purifying selection, a slower evolutionary rate, especially *cox1*, which, because of the lowest Ka/Ks value, has been subjected to the highest purity selection. This may be why it has been used for the study of evolutionary biology as the molecular marker. Combined with Pi and Ka/Ks values, the evolutionary rate and variability of *cox1* are very low. Based on the computational information, the molecular weight (Da), isoelectric points, instability index, aliphatic index, GRAVY, and other indices of the proteins were predicted (Table 2).

The Pi and Ka/Ks values of *atp8* are high compared with other genes. In contrast, another atp synthase *atp6* shows a much slower evolutionary rate. The genes *atp6* and *atp8* are arranged in close proximity and overlap by 7 bp, indicating that *ap6* and *atp8* did not co-evolve as a gene cluster during the evolutionary process. The Pi of *atp8* within the dipteran range showed a large difference in Pi values between the two species, and the difference between the two results is a better indication that the two species share similarities in an evolutionary direction.

The 13 PCGs were further analyzed by calculating the effective codon number (ENc) (Figure 4b), with values closer to 20 indicating a higher expression and a greater codon preference. The GC of the silent third codon posit (GC3s) value was calculated to reflect the probability of synonymous codon use. Plotting the ENc-plot according to the formula Enc = 2 + GC3 + 29/(GC32 + (1 − GC3)2) with the ENc value as the vertical axis and GC3 as the horizontal axis revealed that most of the codons were distributed around the standard curve (Figure 4b). The results suggest that the formation of codon bias in the *P. casei* mt genome may be related to mutations alone and is less affected by other factors (e.g., natural selection) [46]. In comparison, *P. megastigmata* is more subject to external factors that produce codon bias (Figure 4f).

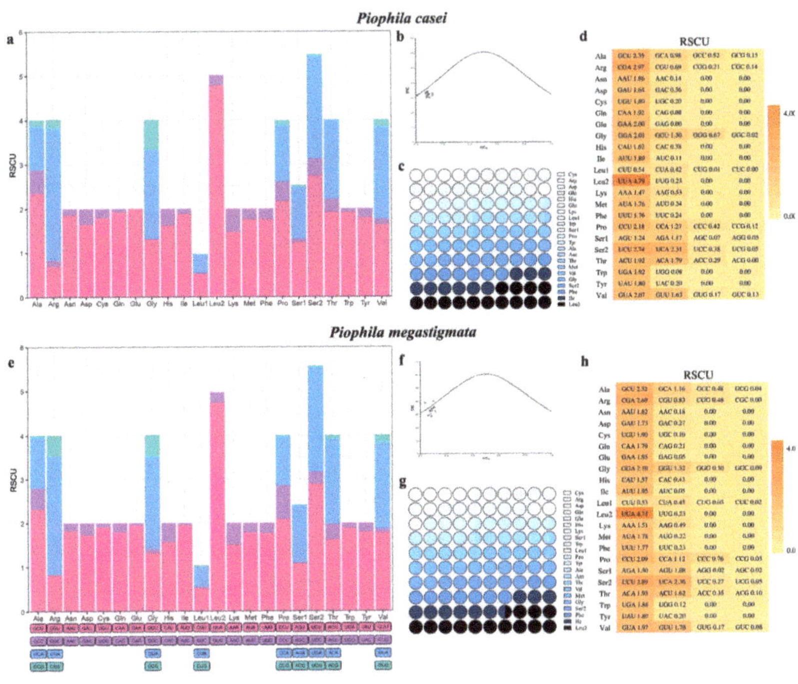

Figure 4. RSCU information of PCGs (**a**) Relative synonymous codon usage (RSCU) in the mtgenomes of *P. casei*; (**b**) ENc-plot of *P. casei*; (**c**) Amino acid usage graph for *P. casei*; (**d**) Heat map of RSCU results for *P. casei* calculated by CodonW software; (**e**) Relative synonymous codon usage (RSCU) in the mtgenomes of *P. megastigmata*; (**f**) ENc-plot of *P. megastigmata*; (**g**) Amino acid usage graph for *P. megastigmata*; (**h**) Heat map of RSCU results for *P. megastigmata* calculated by CodonW software.

The number of synonymous codons (L_sym) and the total number of amino acids (L_aa) were analyzed [47]. Based on the computational information, the molecular weight (Da), isoelectric points, instability index, aliphatic index, GRAVY, and other indices of the proteins were predicted. The results show that *cox1*, *cox2*, *cox3*, and *cytb* are structurally stable and can be used for species classification and mutation-related studies (Table 3), which is consistent with the Pi values as well as with the Ka/Ks values analysis [48].

Table 3. Basic information on the protein molecules of the PCGs gene in the complete mt genome of *P. casei*.

Gene	ENc	GC3s	L_sym	L_aa	Molecular Weight (Da)	Isoelectric Points	Instability Index	Aliphatic Index	GRAVY
atp6	33.95	0.083	218	220	25,145.94	7.09	41.84	123.91	0.884
atp8	32.74	0.08	50	50	6118.28	9.4	44.61	93.77	0.477
cytb	32.52	0.083	363	365	43,057.3	8.33	29	126.64	0.781
cox1	33.09	0.101	496	496	56,373.37	5.97	26.74	110.69	0.732
cox2	34.23	0.077	221	222	26,113.28	4.92	31.38	113.64	0.346
cox3	33.01	0.114	246	250	30,025.75	6.2	32.14	98.21	0.516
nad1	31.27	0.082	305	308	36,067.56	8.45	32.83	127.06	1.085
nad2	33.72	0.127	332	333	39,520.5	9.1	39.9	121.78	0.924
nad3	31.66	0.105	114	114	13,510.31	5.66	45.26	134.19	0.998
nad4	30.26	0.061	426	436	51,075.7	8.3	38.68	127.65	1.068
nad4l	30.22	0.000	96	97	11,502.86	5.99	24.97	121.22	1.093
nad5	31.64	0.089	560	564	65,550.86	6.5	33.48	122.94	0.972
nad6	31.64	0.076	172	173	20,277.72	8.64	26.59	132.82	1.032

In *P. casei*, the five codons with the highest relative synonymous codon usage rate (RSCU) [49] are UUA (4.79), CGA (2.97), UCU (2.74), GCU (2.35), and UCA (2.31) (Figure 4a,d). The RSCU values of these codons are all greater than one, indicating that they are more frequently used in encoding amino acids. The RSCU of *P. casei* and *P. megastigmata* show a similar bias, with only minor differences in the use of a few codons, such as CGC, CUC, and ACG. As with other Diptera, the mt genome of Piophilidae is more biased toward the use of amino acids encoded by codons with A or T in the third position [42]. The amino acid use frequencies of PCGs in the mt genomes of *P. casei* and *P. megastigmata* were counted and showed that the number of 20 amino acids varied, but both had the highest frequency of leucine (Leu) usage, followed by isoleucine (Ile), serine (Ser), and glycine (Gly) (Figure 4c,g). Overall, there is still a predominance of nonpolar amino acids (hydrophobic) and very little amount of basic amino acids (Argue, Lysol, His), with no bias toward acidic (Asp, Glu). By comparing the amino acid content with the codon, it was found that the codon content with the third codon being A or G had an effect on the amino acid content. The small difference in the frequency of amino acid use between the two species suggests that there is also a bias toward the use of amino acids.

3.3. Transfer and Ribosomal RNA Genes

The complete mt genome of *P. casei* included 22 tRNA genes and 2 rRNA genes. The 22 tRNAs had a length between 65 bp and 72 bp. A total of 21 typical trilobal secondary structures were predicted. The tRNA gene trnS was the only one missing the D-arm (Figure 5). It is very common in the fly family, and the absence of this arm does not affect the function of the tRNA genes such as *Chrysomya chloropyga* [43]. The predicted secondary structure maps of rrnL and rrnS by energy-minimization (MFE) are in general agreement with the secondary structures proposed for other insects and provide a reference for subsequent studies (Figure 6) [19,50,51]. The features of tRNAs were conserved between the *P. casei* and *P. megastigmata* genomes. The 16s-rRNA gene is 1320 bp and is located between trnL and trnV. The 12s-rRNA gene is 787 bp and is located between trnV and the control region. The complete mt genome of *P. casei* has a 744 bp control region, which is responsible for regulating DNA replication and transcription. The A + T content of this region is 92.61% between rrnS and trnL.

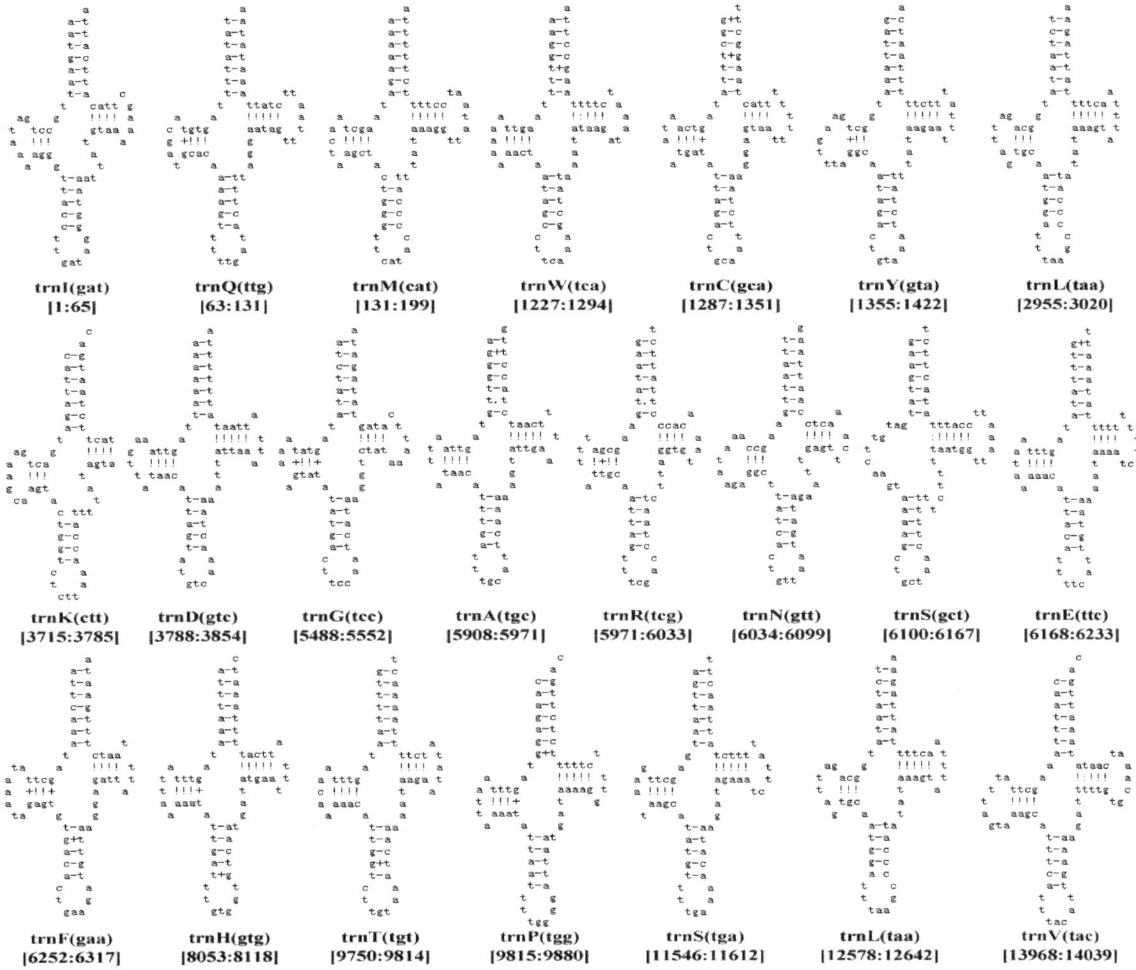

Figure 5. The 22 tRNA secondary structures of *P. casei*. The numbers below indicate their sequence position in the mt genome.

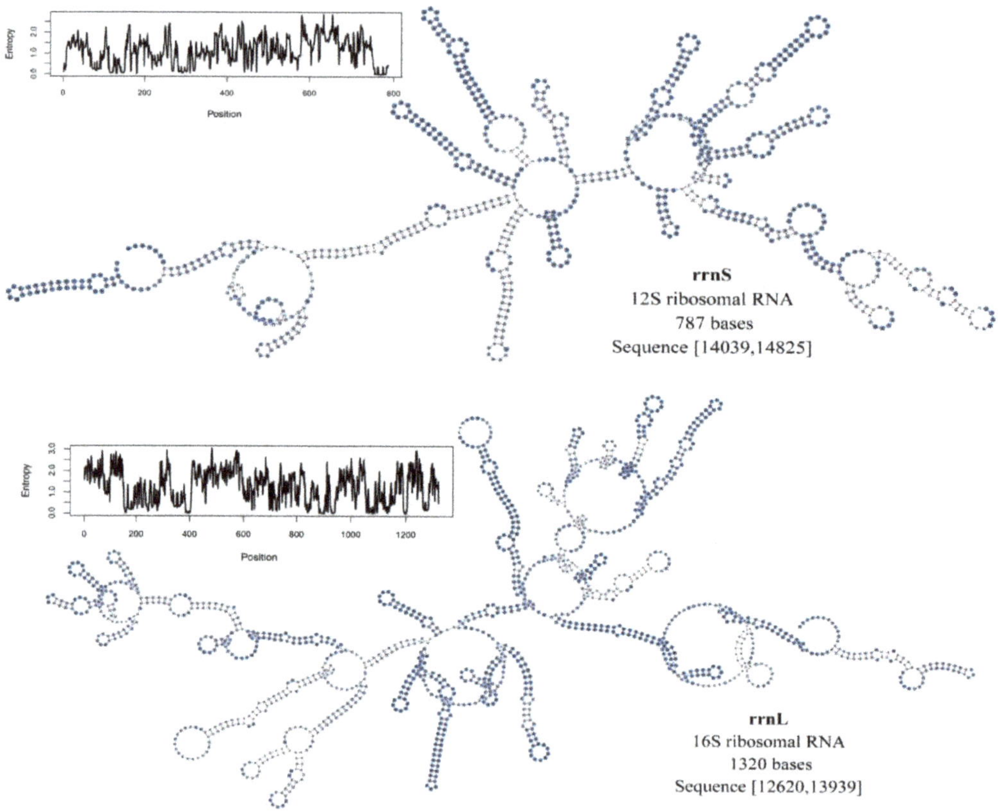

Figure 6. Secondary structure of rrnS and rrnL mt genome of *P. casei*.

3.4. Phylogenetic Analysis

We used the complete mt genome sequence of *P. casei* for phylogenetic analysis, starting with a tree-building assessment of all species' gene sequences (PCGs and RNAs). The saturation plots in Figure 7 show the relationships between general time reversible (GTR) distance and transition/transversion rates. No significant substitution saturation was detected for any codon positions of the 13 PCGs and RNAs in either the symmetrical or asymmetrical topology tests (Iss (0.2701, 0.3261) < Iss.c (0.8452, 0.8046), $p < 0.001$) [52]. The Iss values were all lower than the Iss.c values, indicating that the results of the tree construction using the complete mt genome of 26 species were meaningful for phylogenetic tree construction. Therefore, we subsequently used the complete mt genomes of 26 species for phylogenetic tree construction and analysis.

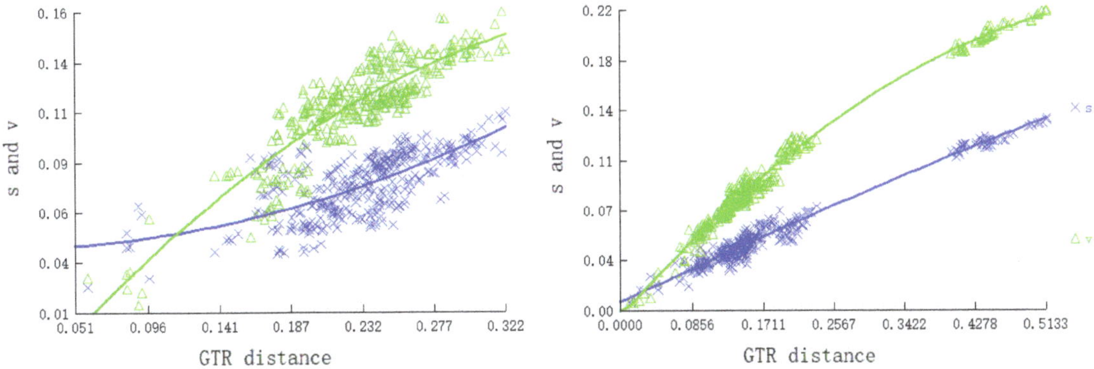

Figure 7. Substitution saturation plots of mt genomic PCGs and RNAs for 26 species. Plots in blue and green indicate transition and transversion, respectively.

In this study, phylogenetic trees were constructed using 12 PCG genes, 2 rRNA genes, and 22 tRNA genes from 26 species using different best-fit substitution models, the ML algorithm (Figure 8), and the BI algorithm (Figure 9). Two common outgroups (Culicoidea, *A. oryzalimnetes*, NC_030715, and Chironomoidea, *Simulium variegatum*, NC_033348) were used. Both ML and BI phylogenetic analyses produced a similar topology with slightly different node support values. The node support was generally higher for the BI tree than for the ML tree, which is common in studies of taxa [53–55]. High support was obtained throughout the phylogenetic tree, with Tephritoidea and Sciomyzoidea appearing distinctly separated from other families, and the abdomen of species within Tephritoidea is consistent with Hoi-Sen's study [56].

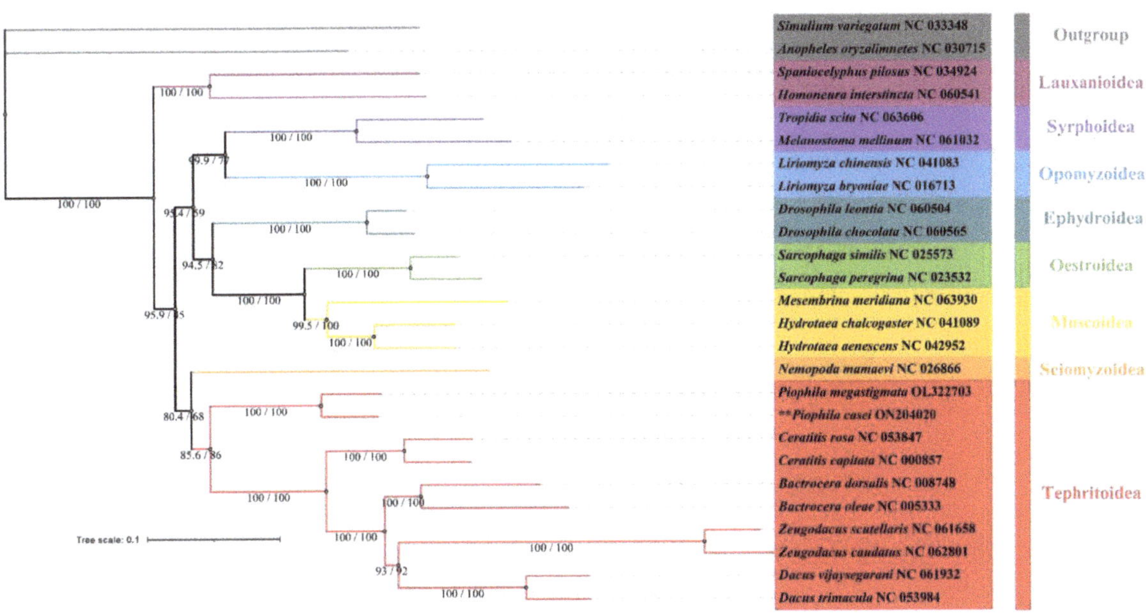

Figure 8. A genome-wide phylogenetic tree of the mitochondria of 26 dipteran insect species constructed based on ML analysis. Two values of developmental dendrites (SH-aLRT support/bootstrap support). (The specie marked with ** is the main subjects of study in this research).

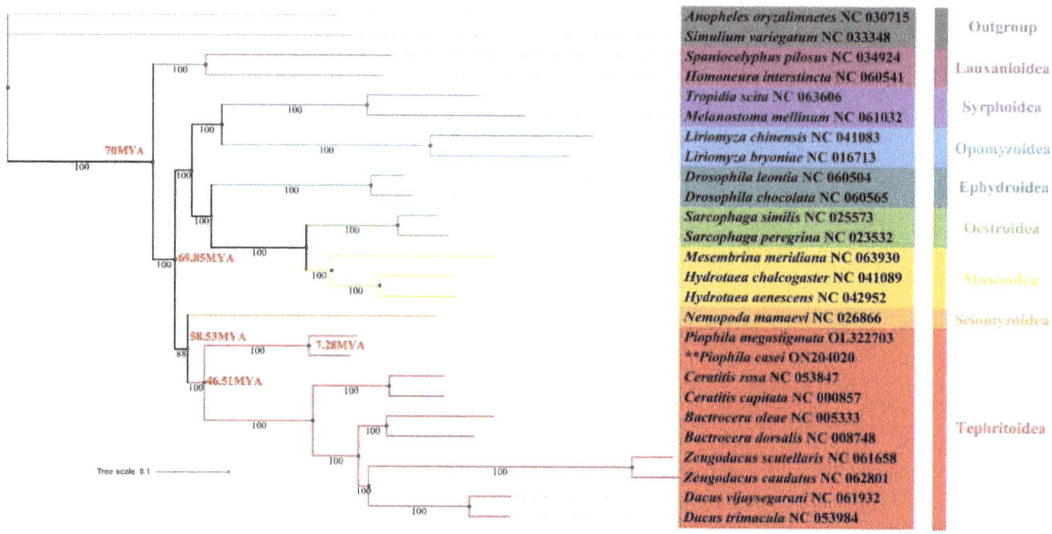

Figure 9. A genome-wide phylogenetic tree of the mitochondria of 26 dipteran insect species constructed based on BI analysis. The red values in the developmental tree indicate the time of divergence between species MYA (million years ago). (The specie marked with ** is the main subjects of study in this research).

We did not use the older calibration points (238.5–295.4 MYA) and instead obtained divergence time data that is more realistic [40]. Combined with the timing of divergence, the evolutionary history of Oestroidea species and Piophilidae species differ significantly, although they share similar feeding habits, with their ancestors diverging as early as 69.85 MYA (95% HPD: 68.69–70.89 MYA) around the late Cretaceous periods. The Piophilidae of the Tephritoidea diverged from other plant-feeding species at 46.51 MYA (95% HPD: 37.16–55.81 MYA) to produce two different diets, despite sharing similar morphological characters.

The results of the phylogenetic tree show that *P. casei* is very closely related to *P. megastigmata*; the mt genetic distance within the same family of 26 Piophilidae species is 0.062. Judging by the evolutionary branch lengths, the *P. megastigmata* has longer branch lengths, reflecting its greater degree of genetic variation and a higher degree of evolution. The divergence between *P. casei* and *P. megastigmata* was inferred at 7.28 MYA (95% HPD: 3.64–18.20 MYA) around the late Cenozoic Tertiary during the Miocene [42]. These results may provide a valuable basis for the study of the evolution of the *P. casei* gene family, population transmission, and biodefense potential [57–59].

4. Conclusions

Previous studies have attempted to describe the phylogenetic relationships of *P. casei* and to identify it taxonomically based on morphological characters or nucleotide fragments. In this study, we report the first complete mt genome of *P. casei* and document the phylogenetic relationships of *P. casei* with the mt genomes of other 26 fly species. This research provides the evidence for the accurate identification of two morphologically closely related forensic insects *P. casei* and *P. megastigmata*, and a theoretical basis for a more in-depth study of their genes. Unfortunately, the sample size of the species used in the article is limited, and further investigations are needed to compare the mitochondrial genome among families of Diptera as a whole, as well as to explore the species divergence time. A deeper exploration of the *P. casei* genome also merits further investigation.

Supplementary Materials: The following supporting information can be downloaded at: https://www.mdpi.com/article/10.3390/genes14040883/s1, Table S1: The optimal partition schemes and the best-fit replacement models for the Bayesian Inference (BI) and the maximum likelihood (ML); Figure S1: Genetic distances of PCGs between 26 Diptera species; Figure S2: Genetic distances of mt genomes between 26 Diptera species; Figure S3: AT-Skew and CG-Skew values for the mt genomes of 26 Diptera species; Figure S4: Box-and-whisker plots for nucleotide composition of each gene. (a) GC-skew; (b) AT-skew; Table S2. Divergence time estimates calculate information such as the ESS of the resulting value; Table S3. Summary of the representative species and their mitogenome information in this study. Figure S5. The life cycle states of *P. casei*.

Author Contributions: Conceptualization, J.L.; software, L.L.; validation, Y.S.; formal analysis, Y.Z. and J.W.; writing-original draft preparation, methodology, writing-review and editing, S.B.; funding acquisition, Q.Z. All authors have read and agreed to the published version of the manuscript.

Funding: This research was supported by the National Natural Science Foundation of China (No. 32060554), Guizhou Science and Technology Innovation Talent Team of Ecological Characteristic Meat Products (QKHPTRC [2020] 5004), the Natural Science Special Project of Guizhou University (Special post, [2020]-02), and the Guizhou High-Level Innovative Talent Training Project (Qianke Cooperation Platform Talent number [2016] 5662).

Data Availability Statement: The data presented in this study are available in the article and in the Supplementary Materials.

Acknowledgments: We sincerely thank the editors and reviewers for their valuable suggestions and comments on this study. The authors are very grateful to the Yang Laonai Dry-Cured Ham Factory in Panxian, Guizhou Province, Bingxia Feng, Zhuokun Liu from the Institute of Entomology, Guizhou University, and Beibei Li, Xiao Xiao, Yanhong Chen, Xinxin Zhang, Xi Yue, and Li Chen at the School of Liquor and Food Engineering for their support of this project. We thank Jing-Jiang Zhou of Guizhou University for critical revisions and proofreading.

Conflicts of Interest: The authors declare no conflict of interest. The funders had no role in the design of the study; in the collection, analyses, or interpretation of data; in the writing of the manuscript; or in the decision to publish the results.

References

1. Russo, A.; Cocuzza, G.E.; Vasta, M.C.; Simola, M.; Virone, G. Life Fertility Tables of *Piophila casei* L. (Diptera: Piophilidae) Reared at Five Different Temperatures. *Environ. Entomol.* **2006**, *35*, 194–200. [CrossRef]
2. Simmons, P. *The Cheese Skipper as a Pest in Cured Meats*; US Department of Agriculture: Washington, DC, USA, 1927.
3. Arnaldos, M.; Sánchez, F.; Álvarez, P.; García, M. A forensic entomology case from the Southeastern Iberian Peninsula. *Anil Aggrawal's Internet J. Forensic Med. Toxicol.* **2004**, *5*, 22–25.
4. Zhao, Y.; Abbar, S.; Amoah, B.; Phillips, T.; Schilling, M. Controlling pests in dry-cured ham: A review. *Meat Sci.* **2016**, *111*, 183–191. [CrossRef] [PubMed]
5. Martín-Vega, D. Skipping clues: Forensic importance of the family Piophilidae (Diptera). *Forensic Sci. Int.* **2011**, *212*, 1–5. [CrossRef] [PubMed]
6. Martín-Vega, D.; Gómez-Gómez, A.; Baz, A.; Díaz-Aranda, L. New piophilid in town: The first Palaearctic record of Piophila megastigmata and its coexistence with *Piophila casei* in central Spain. *Med. Vet. Entomol.* **2011**, *25*, 64–69. [CrossRef] [PubMed]
7. Garrood, J.R. Note on a case of intestinal myiasis. *Parasitology* **1910**, *3*, 315–318. [CrossRef]
8. Peckenschneider, L.E.; Pokorny, C.; Hellwig, C.A. Intestinal infestation with maggots of the "cheese fly" (*Piophila casei*). *J. Am. Med. Assoc.* **1952**, *149*, 262–263. [CrossRef]
9. Kirinoki, M.; Hitosugi, M.; Kato-Hayashi, N.; Iwasa, M.; Chigusa, Y. Discovery of *Liopiophila varipes* and *Protopiophila contecta* (Diptera: Piophilidae) from human cadavers. *Forensic Sci. Int.* **2015**, *248*, e8–e12. [CrossRef]
10. Ozerov, A. On the classification of the Family Piophilidae (Diptera). *Zool. Zhurnal* **2004**, *83*, 1353–1360.
11. Zajac, B.K.; Martin-Vega, D.; Feddern, N.; Fremdt, H.; e Castro, C.P.; Szpila, K.; Reckel, F.; Schuett, S.; Verhoff, M.A.; Amendt, J. Molecular identification and phylogenetic analysis of the forensically important family Piophilidae (Diptera) from different European locations. *Forensic Sci. Int.* **2016**, *259*, 77–84. [CrossRef]
12. e Castro, C.P.; Cunha, E.; Serrano, A.; García, M.D. *Piophila megastigmata* (Diptera: Piophilidae): First records on human corpses. *Forensic Sci. Int.* **2012**, *214*, 23–26. [CrossRef]
13. Paños, A.; Arnaldos, M.; García, M.; Ubero-Pascal, N. Ultrastructure of preimaginal stages of *Piophila megastigmata* McAlpine, 1978 (Diptera, Piophilidae): A fly of forensic importance. *Parasitol. Res.* **2013**, *112*, 3771–3788. [CrossRef]
14. Barros-Cordeiro, K.; Pujol-Luz, J.; Báo, S. A study of the pupal development of five forensically important flies (Diptera: Brachycera). *J. Med. Entomol.* **2021**, *58*, 1643–1653. [CrossRef]

15. Du, Z.; Hasegawa, H.; Cooley, J.R.; Simon, C.; Yoshimura, J.; Cai, W.; Sota, T.; Li, H. Mitochondrial genomics reveals shared phylogeographic patterns and demographic history among three periodical cicada species groups. *Mol. Biol. Evol.* **2019**, *36*, 1187–1200. [CrossRef] [PubMed]
16. Li, W.N.; Shao, R.; Zhang, Q.; Deng, W.; Xue, X.F. Mitochondrial genome reorganization characterizes various lineages of mesostigmatid mites (Acari: Parasitiformes). *Zool. Scr.* **2019**, *48*, 679–689. [CrossRef]
17. Xue, X.-F.; Dong, Y.; Deng, W.; Hong, X.-Y.; Shao, R. The phylogenetic position of eriophyoid mites (superfamily Eriophyoidea) in Acariformes inferred from the sequences of mitochondrial genomes and nuclear small subunit (18S) rRNA gene. *Mol. Phylogenetics Evol.* **2017**, *109*, 271–282. [CrossRef]
18. Su, X.; Fang, Y.; Xu, J.-Y.; Fang, W.-X.; Zhan, X.-B.; Fang, Y.; Chu, L.-M.; Feng, R.; Jin, Y.-L.; Sun, E.-T. The complete mitochondrial genome of the storage mite pest *Tyrophagus fanetzhangorum* (Acari: Acaridae). *Syst. Appl. Acarol.* **2020**, *25*, 1693–1701. [CrossRef]
19. Wang, T.; Wang, Z.; Bai, R.; Yu, Z.; Liu, J. The mitochondrial genome and phylogenetic analysis of the tick *Haemaphysalis qinghaiensis* Teng, 1980 (Acari: Ixodidae). *Syst. Appl. Acarol.* **2022**, *27*, 81–93. [CrossRef]
20. Lan, Y.-M.; Feng, S.-Q.; Xia, L.-Y.; Li, Z.-H.; Cao, Y.; Stejskal, V.; Aulicky, R.; Wu, Y. The first complete mitochondrial genome of *Cheyletus malaccensis* (Acari: Cheyletidae): Gene rearrangement. *Syst. Appl. Acarol.* **2020**, *25*, 1433–1443. [CrossRef]
21. Ye, S.; Zhang, H.-Y.; Song, Y.-F.; Yang, M.-F.; Li, L.-T.; Yu, L.-C.; Liu, J.-F. Complete mitochondrial genome of *Pyemotes zhonghuajia* (Acari: Pyemotidae). *Syst. Appl. Acarol.* **2022**, *27*, 1677–1686. [CrossRef]
22. Meng, G.; Li, Y.; Yang, C.; Liu, S. MitoZ: A toolkit for mitochondrial genome assembly, annotation and visualization. *Nucleic Acids Res.* **2019**, *47*, e63. [CrossRef] [PubMed]
23. Xie, Y.; Wu, G.; Tang, J.; Luo, R.; Patterson, J.; Liu, S.; Huang, W.; He, G.; Gu, S.; Li, S.; et al. SOAPdenovo-Trans: De novo transcriptome assembly with short RNA-Seq reads. *Bioinformatics* **2014**, *30*, 1660–1666. [CrossRef]
24. Gertz, E.M.; Yu, Y.-K.; Agarwala, R.; Schäffer, A.A.; Altschul, S.F. Composition-based statistics and translated nucleotide searches: Improving the TBLASTN module of BLAST. *BMC Biol.* **2006**, *4*, 41. [CrossRef]
25. Birney, E.; Clamp, M.; Durbin, R. GeneWise and Genomewise. *Genome Res.* **2004**, *14*, 988–995. [CrossRef]
26. Jühling, F.; Pütz, J.; Bernt, M.; Donath, A.; Middendorf, M.; Florentz, C.; Stadler, P.F. Improved systematic tRNA gene annotation allows new insights into the evolution of mitochondrial tRNA structures and into the mechanisms of mitochondrial genome rearrangements. *Nucleic Acids Res.* **2012**, *40*, 2833–2845. [CrossRef] [PubMed]
27. Nawrocki, E.P.; Eddy, S.R. Infernal 1.1: 100-fold faster RNA homology searches. *Bioinformatics* **2013**, *29*, 2933–2935. [CrossRef]
28. Bernt, M.; Donath, A.; Jühling, F.; Externbrink, F.; Florentz, C.; Fritzsch, G.; Pütz, J.; Middendorf, M.; Stadler, P.F. MITOS: Improved de novo metazoan mitochondrial genome annotation. *Mol. Phylogenet Evol.* **2013**, *69*, 313–319. [CrossRef] [PubMed]
29. Wilkinson, L. *The Grammar of Graphics. (Statistics and Computing)*; Springer: New York, NY, USA, 2005.
30. Rozas, J.; Ferrer-Mata, A.; Sánchez-DelBarrio, J.C.; Guirao-Rico, S.; Librado, P.; Ramos-Onsins, S.E.; Sánchez-Gracia, A. DnaSP 6: DNA sequence polymorphism analysis of large data sets. *Mol. Biol. Evol.* **2017**, *34*, 3299–3302. [CrossRef] [PubMed]
31. Xia, X. DAMBE7: New and improved tools for data analysis in molecular biology and evolution. *Mol. Biol. Evol.* **2018**, *35*, 1550–1552. [CrossRef]
32. Zhang, D.; Gao, F.; Jakovlić, I.; Zou, H.; Zhang, J.; Li, W.X.; Wang, G.T. PhyloSuite: An integrated and scalable desktop platform for streamlined molecular sequence data management and evolutionary phylogenetics studies. *Mol. Ecol. Resour.* **2020**, *20*, 348–355. [CrossRef]
33. Talavera, G.; Castresana, J. Improvement of phylogenies after removing divergent and ambiguously aligned blocks from protein sequence alignments. *Syst. Biol.* **2007**, *56*, 564–577. [CrossRef] [PubMed]
34. Ranwez, V.; Douzery, E.J.; Cambon, C.; Chantret, N.; Delsuc, F. MACSE v2: Toolkit for the alignment of coding sequences accounting for frameshifts and stop codons. *Mol. Biol. Evol.* **2018**, *35*, 2582–2584. [CrossRef]
35. Hofacker, I.L.; Stadler, P.F. Memory efficient folding algorithms for circular RNA secondary structures. *Bioinformatics* **2006**, *22*, 1172–1176. [CrossRef] [PubMed]
36. Kalyaanamoorthy, S.; Minh, B.Q.; Wong, T.K.; Von Haeseler, A.; Jermiin, L.S. ModelFinder: Fast model selection for accurate phylogenetic estimates. *Nat. Methods* **2017**, *14*, 587–589. [CrossRef] [PubMed]
37. Ronquist, F.; Teslenko, M.; Van Der Mark, P.; Ayres, D.L.; Darling, A.; Höhna, S.; Larget, B.; Liu, L.; Suchard, M.A.; Huelsenbeck, J.P. MrBayes 3.2: Efficient Bayesian phylogenetic inference and model choice across a large model space. *Syst. Biol.* **2012**, *61*, 539–542. [CrossRef]
38. Nguyen, L.-T.; Schmidt, H.A.; Von Haeseler, A.; Minh, B.Q. IQ-TREE: A fast and effective stochastic algorithm for estimating maximum-likelihood phylogenies. *Mol. Biol. Evol.* **2015**, *32*, 268–274. [CrossRef]
39. Drummond, A.J.; Suchard, M.A.; Xie, D.; Rambaut, A. Bayesian phylogenetics with BEAUti and the BEAST 1.7. *Mol. Biol. Evol.* **2012**, *29*, 1969–1973. [CrossRef]
40. Han, H.-Y.; Ro, K.-E. Molecular phylogeny of the superfamily Tephritoidea (Insecta: Diptera) reanalysed based on expanded taxon sampling and sequence data. *J. Zool. Syst. Evol. Res.* **2016**, *54*, 276–288. [CrossRef]
41. Rambaut, A.; Drummond, A.J.; Xie, D.; Baele, G.; Suchard, M.A. Posterior summarization in Bayesian phylogenetics using Tracer 1.7. *Syst. Biol.* **2018**, *67*, 901–904. [CrossRef]
42. Shang, J.; Xu, W.; Huang, X.; Zhang, D.; Yan, L.; Pape, T. Comparative Mitogenomics of Flesh Flies: Implications for Phylogeny. *Insects* **2022**, *13*, 718. [CrossRef]

43. Junqueira, A.C.M.; Lessinger, A.C.; Torres, T.T.; da Silva, F.R.; Vettore, A.L.; Arruda, P.; Azeredo Espin, A.M.L. The mitochondrial genome of the blowfly *Chrysomya chloropyga* (Diptera: Calliphoridae). *Gene* **2004**, *339*, 7–15. [CrossRef] [PubMed]
44. Ojala, D.; Montoya, J.; Attardi, G. tRNA punctuation model of RNA processing in human mitochondria. *Nature* **1981**, *290*, 470–474. [CrossRef] [PubMed]
45. Li, J.; Yan, B.; He, H.; Xu, X.; Ruan, Y.; Yang, M. Characterization of the Complete Mitochondrial Genome of a Flea Beetle *Luperomorpha xanthodera* (Coleoptera: Chrysomelidae: Galerucinae) and Phylogenetic Analysis. *Genes* **2023**, *14*, 414. [CrossRef]
46. Kumar, V.; Tyagi, K.; Chakraborty, R.; Prasad, P.; Kundu, S.; Tyagi, I.; Chandra, K. The Complete Mitochondrial Genome of endemic giant tarantula, *Lyrognathus crotalus* (Araneae: Theraphosidae) and comparative analysis. *Sci. Rep.* **2020**, *10*, 74. [CrossRef] [PubMed]
47. De Mandal, S.; Mazumder, T.H.; Panda, A.K.; Kumar, N.S.; Jin, F. Analysis of synonymous codon usage patterns of HPRT1 gene across twelve mammalian species. *Genomics* **2020**, *112*, 304–311. [CrossRef]
48. Pál, C.; Papp, B.; Lercher, M.J. An integrated view of protein evolution. *Nat. Rev. Genet.* **2006**, *7*, 337–348. [CrossRef]
49. Sharp, P.M.; Li, W.-H. An evolutionary perspective on synonymous codon usage in unicellular organisms. *J. Mol. Evol.* **1986**, *24*, 28–38. [CrossRef]
50. Ye, F.; Liu, T.; King, S.D.; You, P. Mitochondrial genomes of two phlebotomine sand flies, *Phlebotomus chinensis* and *Phlebotomus papatasi* (Diptera: Nematocera), the first representatives from the family Psychodidae. *Parasites Vectors* **2015**, *8*, 1–13. [CrossRef]
51. Gong, Y.-J.; Shi, B.-C.; Kang, Z.-J.; Zhang, F.; Wei, S.-J. The complete mitochondrial genome of the oriental fruit moth Grapholita molesta (Busck) (Lepidoptera: Tortricidae). *Mol. Biol. Rep.* **2012**, *39*, 2893–2900. [CrossRef]
52. Yuan, L.; Liu, H.; Ge, X.; Yang, G.; Xie, G.; Yang, Y. A Mitochondrial Genome Phylogeny of Cleridae (Coleoptera, Cleroidea). *Insects* **2022**, *13*, 118. [CrossRef]
53. Li, R.; Shu, X.; Li, X.; Meng, L.; Li, B. Comparative mitogenome analysis of three species and monophyletic inference of Catantopinae (Orthoptera: Acridoidea). *Genomics* **2019**, *111*, 1728–1735. [CrossRef] [PubMed]
54. Chen, Z.-T.; Zhao, M.-Y.; Xu, C.; Du, Y.-Z. Molecular phylogeny of *Systellognatha* (Plecoptera: Arctoperlaria) inferred from mitochondrial genome sequences. *Int. J. Biol. Macromol.* **2018**, *111*, 542–547. [CrossRef]
55. Huang, X.; Chen, B.; Wei, Z.; Shi, A. First Report of Complete Mitochondrial Genome in the Tribes Coomaniellini and Dicercini (Coleoptera: Buprestidae) and Phylogenetic Implications. *Genes* **2022**, *13*, 1074. [CrossRef] [PubMed]
56. Yong, H.-S.; Chua, K.-O.; Song, S.-L.; Liew, Y.J.-M.; Eamsobhana, P.; Chan, K.-G. Complete mitochondrial genome of *Dacus vijaysegarani* and phylogenetic relationships with congeners and other tephritid fruit flies (Insecta: Diptera). *Mol. Biol. Rep.* **2021**, *48*, 6047–6056. [CrossRef]
57. Sugimoto, N.; Takahashi, A.; Ihara, R.; Itoh, Y.; Jouraku, A.; Van Leeuwen, T.; Osakabe, M. QTL mapping using microsatellite linkage reveals target-site mutations associated with high levels of resistance against three mitochondrial complex II inhibitors in Tetranychus urticae. *Insect Biochem. Mol. Biol.* **2020**, *123*, 103410. [CrossRef]
58. Motyka, M.; Kusy, D.; Háva, J.; Jahodářová, E.; Bílková, R.; Vogler, A.P.; Bocak, L. Mitogenomic data elucidate the phylogeny and evolution of life strategies in Dermestidae (Coleoptera). *Syst. Entomol.* **2022**, *47*, 82–93. [CrossRef]
59. Sarkar, I.; Rathore, S.S.; Singh, G.D.; Singh, R.P. Whole-genome and mitogenome based in silico analysis of select *Plasmodium* and identification of a novel drug molecule against the malaria parasite. *bioRxiv* **2022**. [CrossRef]

genes

Article

The Transmission Patterns of the Endosymbiont *Wolbachia* within the Hawaiian Drosophilidae Adaptive Radiation

Renée L. Corpuz [1], M. Renee Bellinger [1,2,*], Anne Veillet [1], Karl N. Magnacca [3] and Donald K. Price [1,4]

1 Department of Biology, Tropical Conservation Biology and Environmental Science, University of Hawaii at Hilo, 200 West Kāwili Street, Hilo, HI 96720, USA; donald.price@unlv.edu (D.K.P.)
2 U.S. Geological Survey, Pacific Island Ecosystems Research Center, P.O. Box 44, Hawaii National Park, HI 96718, USA
3 Department of Land and Natural Resources, Division of Forestry & Wildlife, Native Ecosystem Protection and Management, Hawaii Invertebrate Program, 1151 Punchbowl Street Rm. 325, Honolulu, HI 96813, USA; knm956@gmail.com
4 School of Life Sciences, University of Nevada, Las Vegas, NV 89557, USA
* Correspondence: mbellinger@usgs.gov

Abstract: The evolution of endosymbionts and their hosts can lead to highly dynamic interactions with varying fitness effects for both the endosymbiont and host species. *Wolbachia*, a ubiquitous endosymbiont of arthropods and nematodes, can have both beneficial and detrimental effects on host fitness. We documented the occurrence and patterns of transmission of *Wolbachia* within the Hawaiian Drosophilidae and examined the potential contributions of *Wolbachia* to the rapid diversification of their hosts. Screens for *Wolbachia* infections across a minimum of 140 species of Hawaiian *Drosophila* and *Scaptomyza* revealed species-level infections of 20.0%, and across all 399 samples, a general infection rate of 10.3%. Among the 44 *Wolbachia* strains we identified using a modified *Wolbachia* multi-locus strain typing scheme, 30 (68.18%) belonged to supergroup B, five (11.36%) belonged to supergroup A, and nine (20.45%) had alleles with conflicting supergroup assignments. Co-phylogenetic reconciliation analysis indicated that *Wolbachia* strain diversity within their endemic Hawaiian Drosophilidae hosts can be explained by vertical (e.g., co-speciation) and horizontal (e.g., host switch) modes of transmission. Results from stochastic character trait mapping suggest that horizontal transmission is associated with the preferred oviposition substrate of the host, but not the host's plant family or island of occurrence. For Hawaiian Drosophilid species of conservation concern, with 13 species listed as endangered and 1 listed as threatened, knowledge of *Wolbachia* strain types, infection status, and potential for superinfection could assist with conservation breeding programs designed to bolster population sizes, especially when wild populations are supplemented with laboratory-reared, translocated individuals. Future research aimed at improving the understanding of the mechanisms of *Wolbachia* transmission in nature, their impact on the host, and their role in host species formation may shed light on the influence of *Wolbachia* as an evolutionary driver, especially in Hawaiian ecosystems.

Keywords: co-phylogenetic reconciliation; co-speciation; evolution; stochastic character trait mapping; horizontal transfer; vertical transfer

Citation: Corpuz, R.L.; Bellinger, M.R.; Veillet, A.; Magnacca, K.N.; Price, D.K. The Transmission Patterns of the Endosymbiont *Wolbachia* within the Hawaiian Drosophilidae Adaptive Radiation. *Genes* **2023**, *14*, 1545. https://doi.org/10.3390/genes14081545

Academic Editor: Zhiteng Chen

Received: 16 June 2023
Revised: 10 July 2023
Accepted: 11 July 2023
Published: 27 July 2023

1. Introduction

The Hawaiian Drosophilidae, long recognized as a striking example of adaptive radiation, are of considerable interest as model systems for understanding the underlying mechanisms of insular speciation [1]. Comprised of up to 1000 species in two major genera (*Scaptomyza* and *Drosophila*), which are believed to have diverged within the Hawaiian archipelago approximately 23.4 million years ago, this taxonomic grouping represents approximately 10% of the insect fauna endemic to the Hawaiian Islands [2,3] and one third of the world's *Drosophila* species [4]. Numerous mechanisms have been proposed to explain the explosive lineage diversification of Hawaiian Drosophilidae, including isolation, niche availability [5], sexual selection [6], and host plant and substrate shifts [1,3]; however, data are lacking on the potential role of symbiont pressures, despite recognition that symbionts, especially those associated with reproduction, could be a major contributor to insect species formation [7]. In particular, a growing body of empirical evidence suggests that the reproductive endosymbiont *Wolbachia* may play a role in the speciation process of some arthropods [8–10], including *Drosophila* [11].

Wolbachia is a widespread and common α-proteobacterium (order Rickettsiales) that infects arthropods and nematodes [12]. The relationship between *Wolbachia* and its host can span from parasitism to facultative or obligate mutualism to ultimate mutualism, and in some cases, beneficial and detrimental effects can be simultaneously conferred [13]. *Wolbachia* strains possess a remarkable ability to significantly alter the reproductive functions of its host in ways that serve to enhance the rate of *Wolbachia*'s transmission, be it through cytoplasmic incompatibility, male-killing, feminization of genetic males, increased fecundity of host, and parthenogenesis [13,14]. Thus, through multiple mechanisms, *Wolbachia* possess the means to give rise to reproductive isolation barriers, which could contribute to the divergence of populations into new species [15]. Consistent with that notion, cytoplasmic incompatibility is known to have a direct effect on gene flow and can serve as a mechanism of reproductive isolation between populations [11,16,17].

The primary mode of *Wolbachia* infection is vertical transmission to the host's progeny through the cytoplasm of the egg [14]. Horizontal transmission is believed to occur as well, especially in arthropods, as evidenced by the widespread distribution of *Wolbachia* and its potential to infect new host species [8,18], phylogenetic incongruence between hosts and endosymbionts [12,19]), and evidence for species sweeps [20,21]. In contrast, within filarial nematodes hosts, strict vertical inheritance of *Wolbachia* endosymbionts is evidenced by high levels of co-phylogenetic concordance for certain clades [22,23]. At present, the community-level interactions required for *Wolbachia* strains to be successfully transmitted horizontally and become stable within a new host species remain largely unknown, but in some cases, they are believed to involve transfer through plant tissues or parasitoids of insects [24,25].

Molecular methods have been invaluable for the study of *Wolbachia* because of an inability to culture it outside of its host or host cells, owing to its obligate intracellular status [14]. Based on molecular diversity analysis, the genus *Wolbachia* is subdivided into at least 17 possible supergroups [26,27], with terrestrial arthropods most commonly infected by *Wolbachia* belonging to supergroups A and B [28]. Estimates for the incidence of *Wolbachia* in terrestrial arthropod species worldwide range between 40–76% [13,29,30], whereas within-species estimates for *Wolbachia* incidence indicate that infection rates tend to be either exceedingly high (>90%) or considerably low (<10%), depending on the surveyed insect system [13,30]. In native Hawaiian insects, the overall incidence of *Wolbachia* infection at the species level was estimated to be ~14%, and for native Dipteran species (e.g., Drosophilidae and Calliphoridae), 12% [2].

Although many mechanisms have been proposed to explain the rapid and extensive diversification of the Hawaiian Drosophilidae, the potential contribution of *Wolbachia* as a driver of speciation and patterns of *Wolbachia* transmission have yet to be examined. Using a single gene marker, *Wolbachia surface protein* (*wsp*), Bennett et al. [2] found the incidence of infection within Hawaiian Drosophilidae, including genera *Drosophila* and *Scaptomyza*, was ~18%. *Wolbachia*'s presence in the Hawaiian Islands, and the knowledge of the potential impacts that it can have on host reproductive strategies, give rise to the question: could *Wolbachia* have played a role in the diversification of the native Hawaiian insects? To begin to address this larger question, in this study we conducted genetic analyses of *Wolbachia* and its Hawaiian Drosophilidae hosts to examine: (1) the *Wolbachia* strain diversity and phylogenetic affiliations; (2) the co-phylogenetic diversification patterns of *Wolbachia* and hosts; and (3) *Wolbachia* host-switching mechanisms through stochastic character trait mapping to construct host ancestral traits.

2. Methods

2.1. Biological Specimens Screened for Wolbachia Endosymbionts

The Hawaiian *Drosophila*, many of which are primarily single island endemics that have high levels of host plant specificity, can be subdivided into 4 main groups: *modified mouthparts*, *haleakalae*, *picture wing*, and the AMC clade (comprising the groups *antopocerus*, *modified tarsus*, and *ciliated tarsus*) [31]. The genus *Scaptomyza* is divided into 21 subgenera, 10 of which contain native Hawaiian species [31]. A total of 399 Hawaiian Drosophilidae specimens representing a minimum of 136 species of *Drosophila* and 14 species of *Scaptomyza* collected from Kaua'i (n = 50), Lāna'i (n = 1), Maui (n = 68), Moloka'i (n = 17), O'ahu (n = 29), and the Island of Hawai'i (n = 234) were screened for *Wolbachia* infections (Supplementary S1). A number of undescribed morphospecies in the *Scaptomyza*, *modified tarsus* and *modified mouthparts* groups of *Drosophila* are included. These Drosophilidae specimens were components of biological collections described in Magnacca and Price [3]. Additional screens for *Wolbachia* infections were conducted from DNA extracts of three species of insects that have invaded the Hawaiian archipelago: *D. suzukii* (n = 68 specimens from Kaua'i, O'ahu, and the Island of Hawai'i [32], Supplementary S1), *Aedes albopictus* (n = 1, collected on the Island of Hawai'i), and *Culex quinquefasciatus* (collected on the Island of Hawai'i, sample 6771, [33]. The *Wolbachia* DNA was sourced from whole-body soaks or digests of individual body parts (e.g., genitalia or abdomen) and DNA extractions were performed using Qiagen DNeasy Blood and Tissue Extraction Kits. For notation purposes, *Wolbachia* strains having published lineage assignments are denoted by their host following established practices, e.g., a *Wolbachia* endosymbiont of *Drosophila recens* is written as *w*Rec, or in the case of this study, sample number followed by host species name.

Using seven *Wolbachia* amplification targets (see below) and Sanger sequencing, individual specimens were classified as testing positive for a *Wolbachia* infection if any single amplification target was visible by gel electrophoresis and the sequenced amplicon matched to a *Wolbachia* sequence contained in the National Center of Biotechnology Information (NCBI) GenBank nucleotide sequence repository (approximate search dates: February 2018 to March 2019).

2.2. Wolbachia Gene Sequencing

2.2.1. Amplicon Sequencing and Primer Redesign

We aimed to characterize *Wolbachia* allele diversity and determine the phylogeny of *Wolbachia* by sequencing seven gene targets, five of which are components of the widely-accepted universal Multi-Locus Sequence Typing (MLST) system that assigns *Wolbachia* to a strain type using five housekeeping genes: *coxA* [cytochrome C oxidase subunit A], *fbpA* [fructose-bisphosphate aldolase], *hcpA* [hypothetical conserved protein], *ftsZ* [cell division protein], and *gatB* [aspartyl/glutamyl-tRNA aminotransferase subunit B] [34]. The sixth and seventh gene target, *wsp* [*Wolbachia* surface protein] [34], is duplicated in *Wolbachia* endosymbionts of *Drosophila*, with the paralogous gene denoted *wspB* [*Wolbachia* surface protein (duplicate)] [35]. The gene targets were amplified from DNA extracts using polymerase chain reaction (PCR), visualized using electrophoresis with 1.5% agarose gels, and amplification products purified in preparation for Sanger sequencing on an Applied Biosystems 3500 Genetic Analyzer (see Supplementary Information for details). The chromatograms were viewed and edited using Sequencher version 5.2.4 (Gene Codes Corporation). Based on chromatogram visualization, samples that showed evidence of a double *Wolbachia* infection were sequenced from clones generated with a TOPO-TA Cloning Kit using One Shot Chemically Competent TOP 10 *Escherichia coli* cells.

Preliminary amplification results showed high rates of amplification failures; therefore, to increase primer specificity, we redesigned primers for supergroups A and B in insect hosts. Primer re-design efforts utilized a combination of sequence data obtained from: (a) the n = 31 sequences generated in this study using original primer pairs, (b) *w*Drosophila gene sequences (n = 195) downloaded from the National Center for Biotechnology Information (NCBI), and (c) nucleotide sequences in silico extracted from five *w*Drosophila reference genomes belonging to supergroups A and B (Table 1). Those included: *Wolbachia* endosymbionts of *D. recens* (*w*Rec), *D. melanogaster* (*w*Mel), *D. simulans* (*w*No), *D. suzukii* (*w*Suzi), and *D. ananassae* (*w*Ana). Target regions within genomes were identified by BLASTn (v 2.2.30) using the 231 available sequences as queries, with the per-gene number of query sequences ranging from three (*fbpA*) to 141 (*wsp*) (accessions available from Supplementary S2). The BLASTn hits were filtered using a threshold *e*-value <0.001, and gene target regions were excised in silico along with 200 base pair regions flanking the 5′ and 3′ reading frames. Next, multiple sequence alignment was conducted for each gene in MEGA7 [36] and candidate primers were designed across sites internal or external to the MLST gene targets. Finally, all pairwise combinations of redesigned and original primers were tested for improved amplification and sequencing efficiency (see Supplementary Information). These efforts increased data for *hcpA*, *fbpA*, and *ftsZ* by 54 sequences obtained from 93 additional amplifications, yet the re-designed primers failed to improve amplifications for genes *coxA* and *gatB*. The overall poor amplification success for *wsp* and *wspB* (consistent with findings by Wu et al. [35]), led to the exclusion of those two genes for phylogenetic and strain typing analyses, while poor amplification of *gatB* led to the exclusion of that gene from phylogenetic analysis. The primer design strategy, PCR conditions for the original and modified primers, primer sequences, and those re-designed for this study are available from Supplementary Information.

2.2.2. *Wolbachia* Sequence Datasets

The final *Wolbachia* dataset included MLST genes amplified and sequenced from DNA extracts of native *Drosophila* spp., *Scaptomyza* spp. and invasive species *D. suzukii*, *C. quinquefasciatus*, and *A. albopictus* hosts as described above, plus published *Wolbachia* nucleotide sequences downloaded from the MLST database or extracted from genomes (Table 1). The published sequences were used as references for assigning *Wolbachia* alleles to supergroups and used as outgroups in phylogenetic reconstructions, plus represent *Wolbachia* endosymbionts of *Drosophila* hosts and mosquitoes sampled from around the world. After aligning sequences in MEGA7 using the ClustalW algorithm [36], the sequences were manually adjusted to ensure that all codons were in the correct reading frame and trimmed so that each sequence began and ended with a codon. The *Wolbachia* sequence data generated for this study are available from Supplementary Datafile S1.

Table 1. Data for *Wolbachia* genetic sequences used for the purpose (Purpose) of in silico extraction of sequence from genomes for primer redesign (PR) or *Wolbachia* allele strain typing and/or phylogenetic analysis (A/P). Shown are *Wolbachia* host species names, *Wolbachia* strain abbreviations, host collection locations or laboratory sources if known, National Center for Biotechnology Information (NCBI) accessions, genome references, and *Wolbachia* supergroup designations.

Wolbachia Host	*Wolbachia* Strain	Host Collect Location	Genome Accession	Citation	Supergroup	Purpose
Drosophila recens	*w*Rec	Rochester, New York, USA	GCF_000742435.1	[37]	A	PR
D. melanogaster	*w*Mel	Stock Center *D. melanogaster* strain yw67c23	GCF_000008025.1	[35]	A	PR, A/P
D. simulans	*w*No	Noumea, New Caledonia	GCF_000376585.1	[38]	B	PR, A/P
D. simulans	*w*Ha	Hawai'i, USA	GCF_000376605.1	[38]	A	A/P
D. ananassae	*w*Ana	Tucson Strain Center [strain 14024-0371.13]	GCF_000167475.1	[39]	A	PR
D. suzukii	*w*Suzi, strain valsugana	Trentino Alto Adige, Italy	GCF_000333795.1	[40]	A	PR, A/P
Culex quinquefasciatus	*w*Pip strain Pel	Sri Lanka	AM999887.1	[41]	B	A/P
C. quinquefasciatus	*w*Pip strain JHB	Johannesburg, Africa	ABZA0100000	[42]	B	A/P
Aedes albopictus	*w*AlbA	Unknown	[1]	[34]	A	A/P
Brugia malayi	*w*Bm	TRS Lab colony (Athens, GA, USA)	NC_006833.1	[43]	D	A/P
Cimex lectularius strain JESC	*w*Cle	Japan	AP013028.1	[44]	F	A/P

[1] "isolate 12", typed using MLST by Baldo et al. [34], NCBI accessions DQ842268.1, DQ842342.1, DQ842379.1, DQ842305.1.

2.2.3. *Wolbachia* Supergroup Designation

Previous studies have shown that phylogenetic clustering of individual MLST genes is sufficient for the classification of *Wolbachia* alleles into supergroups A and B [45]. To evaluate if sequence data from re-designed MSLT primers performed similarly well, we reconstructed single-gene phylogenies using our sequence data and eight published reference sequences. These included the following: supergroup A, *w*Mel, *w*Suzi (strain valsugana), endosymbionts of *D. simulans* (*w*Ha) and *A. albopictus* (*w*AlbA); supergroup B, *w*No, endosymbionts of *C. quinquefasciatus* *w*Pip (sample 6771, [33]); and supergroup D and F outgroup sequences from *Wolbachia* endosymbionts of *B. malayi* (nematode, *w*Bm) and *C. lectularius* (bed bug, *w*Cle) (Table 1). Phylogenetic patterns for individual gene trees were inferred using a Bayesian methodology implemented in MrBayes (v3.2.5) [46] and the Maximum-Likelihood methodology implemented in RAxML (v1.5b2) [47].

2.2.4. *Wolbachia* Strain Typing

The MLST strain typing protocol established by Baldo et al. [34] defines an 'allele' as a nucleotide sequence that differs by at least 1 nucleotide base, and it classifies a 'strain' as unique if any individual possesses at least one different allele across any of the five loci, with data at all five loci required for strain assignment. We were unable to apply established MLST conventions (http://pubmlst.org/wolbachia; accessed on 1 July 2017; [34]) for allele and strain categorizations for two reasons: the universal MLST primer sets failed to produce amplifications at all five loci across the majority of our samples, and the amplicon products produced with redesigned primers did not span the full-length of MLST gene sequences. Therefore, we categorized each allele by supergroup affiliation based on single-gene trees and assigned each allele an arbitrary numeric code, which permitted comparison of allele variability and supergroup designations across species.

2.2.5. Phylogenetic Reconstructions

Evolutionary relationships and genetic similarity of *Wolbachia* strains can be inferred through phylogenetic analyses, and phylogenetic concordance between host and symbiont phylogenies can indicate co-speciation or horizontal transfer events between the two groups. We performed phylogenetic reconstruction for *Wolbachia* strains and their hosts, including Hawaiian Drosophilidae, invasive *Drosophila* flies and mosquitoes, and outgroup taxa, using Bayesian methodologies implemented in MrBayes (v3.2.5) [46] and the Maximum-Likelihood methodology implemented in RAxML (v1.5b2) [47]. Model selection and procedures are available from Supplementary Information, and the final set of trees were visualized and edited in FigTree v1.4.3 [48].

2.2.6. *Wolbachia* Phylogenetic Signals

The five *Wolbachia* MLST gene targets were not successfully amplified in all samples. Therefore, to assess the impact of missing sequences on phylogenetic reconstructions, we examined concordance of *Wolbachia* supergroup designation based on single and concatenated gene trees. Phylogenetic reconstructions for 5-, 4-, and 3-gene MLST data sets revealed that strain assignments and tree topologies were consistent in nearly all cases (see Supplementary Information); therefore, we applied the 3-gene MLST dataset for co-phylogenetic reconciliation analysis and stochastic character trait mapping.

2.2.7. Host Sequence Data Set

Phylogenetic reconstruction for Hawaiian *Drosophila* and *Scaptomyza* was inferred using a sequence data set previously shown to produce a well-resolved Hawaiian Drosophilidae phylogeny [3]. However, we used only four of the five genes published in that study (*EF1g* [elongation factor 1-γ], *Gpdh* [glycerol-3-phosphate dehydrogenase], *Pgi* [phosphoglucose isomerase], *Yp2* [yolk protein 2]). The gene *Fz4* (*frizzled* 4) was excluded because of high levels of missing data in the original published dataset, which had negligible effects on the tree topology (compared to [3]). Only Hawaiian Drosophilids having confirmed

Wolbachia infections with three or more sequences were utilized for phylogenetic reconstructions, along with host sequences obtained by a BLASTn search of genome contents for *D. suzukii*, *D. melanogaster*, *D. simulans*, and two mosquito species, *A. albopictus* and *C. quinquefasciatus* (accessions available from Table S1). Searches for genes in mosquitoes recovered genes *EF1g*, *Gpdh*, and *Pgi* but not *Yp2* (or *Fz4*). The concatenated host sequence data set totaled to 1812 bp across the 4 genes (*EF1g* [507 bp], *Gpdh* [363 bp], *Pgi* [306 bp], *Yp2* [636 bp]).

2.2.8. Co-phylogenetic Assessment of Host Species and *Wolbachia* Strains

To evaluate biological events that might influence associations between host and symbiont phylogenies, we conducted co-phylogenetic reconstruction analyses for *Wolbachia* and the Hawaiian Drosophilidae, as well as *Wolbachia* and the 2 mosquito host species collected on the Island of Hawai'i. By considering five possible biological events (co-speciation, duplication, duplication and host switch, loss, and failure to diverge) and applying each a cost, JANE [49] used a heuristic approach to evaluate and find minimal cost solutions that best explain associations between host and endosymbiont phylogenies [49]. Two models were considered by setting the co-speciation cost parameter to 0 or 1, while keeping all other parameters fixed as follows: loss, failure to diverge, and duplication were each set to a cost of 1, and the parameter duplication and host switch was set to a cost of 2 [49,50]. The genetic algorithm parameters were set to a population size of 23 and the number of generations set to 45, as suggested by Conow et al. [49]. Additional statistical parameters included selecting the random tip mapping procedure with 1000 replicates. Data inputs included host and endosymbiont trees based on Bayesian inference using the codon position data set for the host species and the 3-gene, gene + codon position data set (*coxA*, *hcpA*, and *ftsZ*) for *Wolbachia* (see Supplementary Information for justification). Additionally, a co-phylogenetic tanglegram was produced using the cophylo function in the phytools v0.6-44 package in R [51].

2.2.9. Stochastic Character Mapping

Potential host-switching mechanisms were evaluated using stochastic character trait mapping [52], which characterizes associations between *Wolbachia* phylogenies and host species characteristics. When co-speciation can be explained by a particular host trait, evolutionarily conserved characters of the hosts are reflected in the phylogenetic reconstruction of their endosymbionts. Data inputs included three host species traits, island of collection, host plant families, and preferred ovipositional substrate [3], with analyses conducted using the *Wolbachia* 3-gene and gene + codon position data set (*coxA*, *hcpA*, and *ftsZ*) (see Supplementary Information for justification). The contemporary host character traits are depicted on branch tips as a pie chart, with a priori known character traits indicated by 1.0 probability (i.e., 100%) and unknown character traits depicted as equal probability across all possible categories (e.g., $0 \leq x \leq 1$), with the sum of all character state probabilities equaling 1. The internal nodes (also a pie chart) depict the posterior probability of each host character trait being the ancestral state, which reflects the strength of the association between that host trait and the endosymbiont phylogeny. This analysis was performed using phytools v0.6–44 package in R [51]. A total of 225 stochastic character maps were constructed using a model of even rates, as it was indicated to be the best model based on the computed Akaike Information Criterion (AIC) values using the phytools fitMK function (Table S3).

3. Results

3.1. *Incidence of Wolbachia Infection*

Among the 150 species of Hawaiian Drosophilidae screened (including undescribed morphospecies), *Wolbachia* infections were confirmed for 30 species (20.0% species infection rate), and across the entire data set, infections were confirmed for 41 of 399 specimens (10.3% overall specimen infection rate) (Table S2). At a genus level, infection frequencies were higher in *Scaptomyza* (seven of 14 species screened, 50.0%) than *Drosophila* (23 of 136 species screened, 16.9%). An additional 24 Hawaiian Drosophilidae specimens belonging to 17 species (including five undescribed) showed evidence of infection by presence of PCR bands, but infection by *Wolbachia* could not be confirmed owing to the amplicons failing Sanger sequencing. Had those samples been included in the *Wolbachia* infection tally (65/399), the overall infection rate would increase to 16.3%. Some insights into the variability of infection status by species (and sequencing success) can be gleaned from species having data from multiple samples. For example, among 13 species with five or more samples screened (excluding the taxa resembling *D. basimacula*, a complex of undescribed species), the proportion of within-species infections ranged from 0% to 29% (Table 2). A caveat to these findings is that within-species infection rates are known to vary widely (i.e., 10–90%), and a sample size larger than what was available in our specimen collection is required for a robust assessment of infection rates. Screens of the invasive *D. suzukii* indicated that 8 of 68 (11.8%) individuals possessed a *Wolbachia* infection, and that 20 additional individuals may have been infected based on PCR amplification alone. A record of PCR amplicons and sequencing is provided in Table S2.

3.2. Wolbachia Strain Typing and Supergroup Designations

A complete MLST profile (5 genes: *coxA*, *fbpA*, *gatB*, *hcpA*, and *ftsZ*) was obtained for *Wolbachia* endosymbionts of only 9 individual Hawaiian Drosophilidae, all of which belonged to supergroup B, plus *w*Bm and *w*Cle outgroup taxa belonging to supergroups D and F. The *gatB* gene failed PCR amplification across the majority of individual Drosophilidae and was not recovered from endosymbiont genomes belonging to hosts *D. suzukii*, *A. albopictus*, and *C. quinquefasciatus*, leaving only genes *coxA*, *fbpA*, *hcpA*, and *ftsZ* available for analytical inferences across the majority of *Wolbachia* datasets. Individual-gene phylogenetic reconstructions of *Wolbachia* based on *coxA*, *fbpA*, *hcpA*, and *ftsZ* gene sequences (n = 46, 33, 44, and 28 sequences, respectively) showed strong support for the clustering of alleles by supergroup, although supergroup sister status and placement relative to the supergroups D and F outgroups was inconsistent across trees, and placement of some individuals within supergroup clusters varied slightly (Figures S1 and S2, Bayesian and Maximum-Likelihood trees).

Phylogenetic reconstructions of *Wolbachia* based on the concatenated data set comprised of *coxA*, *hcpA*, and *ftsZ* genes, and the 25 individuals with data available at all three genes (including outgroups), showed clear separation between supergroups A and B (Figure S3), consistent with the four-gene dataset (Figure S4). However, the three-gene dataset showed supergroup B placed interior to supergroup A (instead of sister), possibly driven by inclusion of the additional set of *Wolbachia* sequences (247*w*D. engyochracea, 266*w*D. Hawaiiensis, and the invasive *w*Alb collected on the Island of Hawai'i) that had conflicting supergroup assignments and were positioned intermediately between supergroups A and B. Given that the three-gene data set recovered a reasonable degree of phylogenetic structure, and allowed use of the maximum available data, we selected that dataset, using the Bayesian method and partition scheme 'gene and codon position', for co-phylogenetic reconciliation analyses and stochastic character trait mapping. The analysis method (Bayesian versus Maximum-Likelihood analyses (Figures S3 and S4) had little effect on tree topologies, and no significant statistical differences were detected between their top likelihood scores (See Supplementary Information for model selection justification).

Table 2. A comparison of numbers of individual Hawaiian Drosophildae (genus *Drosophila*) species with at least five specimens per species screened, per-species total numbers of individuals screened, number of individuals with confirmed *Wolbachia* infections, numbers of individuals having no confirmed infections but positive for PCR amplifications that failed sequencing, total number of individuals having zero amplifications across all loci, and the proportion of infected individuals by species.

Species	Screened	Confirmed Infected	PCR Product Only	Zero Amplifications	Proportion Infected
D. ciliaticrus	5	0	0	5	0.00
D. engyochracea	7	2	2	3	0.29
D. hawaiiensis	15	2	0	13	0.13
D. heteroneura	5	2	0	3	0.40
D. murphyi	5	2	0	3	0.40
D. ochracea	11	1	1	9	0.09
D. odontophallus	5	0	0	5	0.00
D. orphnopeza	7	0	1	6	0.00
D. primaeva	11	0	1	10	0.00
D. silvestris	6	0	0	6	0.00
D. sproati	114	0	5	109	0.00
D. tanythrix	10	0	0	10	0.00
D. yooni	10	2	1	7	0.20
totals	219	17	11	191	n/a

n/a = not applicable.

3.3. Strain Typing

A total of 41 Hawaiian Drosophilidae were confirmed as having *Wolbachia* infections, with four individuals (w16 *D. large spots*, w208 *D. apodasta*, w215 *D. nr. perissopoda #1*, w250 *D. engyochracea*) doubly infected (Table 3, Table S2). Among the 44 *Wolbachia* typed with MLST markers, a minimum of 27 unique strains were present based on *Wolbachia* allelic diversity analysis. This minimum number of strains is conservative because only nine *Wolbachia* (representing seven unique strains) could be sequenced across all five gene targets (Table 3). Patterns of infection varied by species, for example, one individual of *D. engyochracea* was doubly infected, one was single-infected, and one showed a PCR amplification, but the PCR product failed to sequence. The majority (30/44, 68%) of *Wolbachia* alleles belonged to supergroup B across all loci (Table 3), based on individual gene trees, while only five (5/44, 11%) belonged to supergroup A, including two from within the double-infected *D. engyochracea*. A modest amount (9/44, 20%) of *Wolbachia* strains were characterized as having supergroup A and B alleles that conflicted across individual gene trees, including two *Drosophila* spp. (of four) with double-infections. The *hcpA* allele 11 was responsible for seven of the nine observed A/B allelic conflicts, and one allele (allele 3) did not clearly assign to supergroup A or B in the single-gene phylogeny (Figures S1 and S2). Additional patterns of interest were that the *hcpA* allele 14 was shared by the *Wolbachia* endosymbiont of native *S. undulata* and invasive *D. suzukii* hosts, and that allele 13 was detected in *Wolbachia* of two distantly related invasive host flies sampled in Hawai'i: *D. suzukii* and *D. simulans*. For *C. quinquefasciatus* host specimens collected on the Island of Hawai'i, South Africa and Sri Lanka, only a single strain of *Wolbachia* was detected. Two alleles, at two genes (*coxA*, allele 13; *hcpA*, allele 11), were detected in *Culex* and also >10 Hawaiian Drosophilidae, but in no cases were those two alleles observed in the identical combination in flies as was observed in mosquitoes. Conversely, *w*Alb, isolated from the *A. albopictus* specimen collected on the Island of Hawai'i(sequenced for this study), had no alleles in common with the other *w*Alb sample [34] or even with any Hawaiian Drosophilidae. A limitation to our study is that we were unable to match allele names to those contained in the online MLST database curated by Baldo and colleagues (http://pubmlst.org/wolbachia/, [34])) because we had to use redesigned primers to successfully sequence the genes in Hawaiian *Drosophila*. Therefore, the gene sequences in our dataset are of different sequence lengths compared to the MLST database and we could not determine if the alleles that were sequenced in this study are "novel" to Hawai'i or to what parts of the world they are most similar.

Table 3. A list of Hawaiian Drosophilidae, invasive mosquito, and outgroup host species screened for *Wolbachia* infections using PCR amplification and verified by Sanger sequencing. The five gene targets were amplified using a modified version of the multi-locus strain typing (MLST) approach for strain assignment to supergroup (see text for details). For each gene, alleles were assigned to a supergroup based on single-gene phylogenetic reconstructions, and unique sequences were assigned an arbitrary allele number. In some cases, supergroup assignments were discordant across alleles, and alleles that could not be assigned to a supergroup are denoted as (?). *Wolbachia* endosymbionts of double-infected hosts are denoted by bold font. MLST genes that failed amplification and/or sequencing are denoted as '---'.

Wolbachia Sample Name	Host Species Name	Island of Collection	*coxA*	*fbpA*	*hcpA*	*ftsZ*	*gatB*	Strain Type
247w	*Drosophila engyochracea*	Hawai'i	**2 (A)**	---	**11 (B)**	**2 (A)**	---	A/B
250_1w	*D. engyochracea*	Hawai'i	2	---	---	2	---	**A**
250_2w	*D. engyochracea*	Hawai'i	5	---	---	10	---	**A**
264w	*D. hawaiiensis*	Hawai'i	2	---	---	---	---	A

Table 3. *Cont.*

	Wolbachia Sample Name	Host Species Name	Island of Collection	*coxA*	*fbpA*	*hcpA*	*ftsZ*	*gatB*	Strain Type
Native Hawaiian Drosophilidae	266w	*D. hawaiiensis*	Hawai'i	2 (A)	---	11 (B)	2 (A)	---	A/B
	252w	*D. heteroneura*	Hawai'i	2 (A)	---	11 (B)	---	---	A/B
	253w	*D. heteroneura*	Hawai'i	6 (A)	---	11 (B)	---	---	A/B
	16_1w	***D. large spots***	Hawai'i	2	7	6	2	---	**A**
	16_2w	***D. large spots***	Hawai'i	4	8	2	1	---	**B**
	171w	*D. murphyi*	Hawai'i	---	4 (A)	3 (?)	4 (B)	---	A/?/B
	MLL6w (415)	*D. murphyi*	Hawai'i	13	---	1	---	---	B
	244w	*D. nigrocirrus*	Hawai'i	2	6	5	11	---	A
	256w	*D. ochracea*	Hawai'i	2 (A)	3	11 (B)	---	---	A/B
	123w	*D. prolaticilia*	Hawai'i	1	1	2	1	3	B
	197w	*D. prolaticilia*	Hawai'i	1	1	2	1	---	B
	221w	*D. seclusa*	Hawai'i	1	1	1	6	1	B
	291w	*D. yooni*	Hawai'i	13	---	---	---	---	B
	292w	*D. yooni*	Hawai'i	13	---	---	---	---	B
	20w	*Scaptomyza caliginosa*	Hawai'i	1	1	1	6	1	B
	152w	*S. cyrtandrae*	Hawai'i	2 (A)	---	11 (B)	---	---	A/B
	204w	*S. reducta*	Hawai'i	---	---	11	---	---	B
	205w	*S. reducta*	Hawai'i	3	2	11	---	---	B
	206w	*S. undulata*	Hawai'i	1	---	14	---	---	B
	185w	*D. ancyla*	Maui	13	12	1	7	1	B
	175w	*D. prostopalpis*	Maui	1	2	2	1	4	B
	145w	*D. quasiexpansa*	Maui	1	---	4	4	---	B
	216w	*D.* nr. *redunca*	Hawai'i	1	1	1	9	---	B
	200w	*S. crassifemur*	Maui	1	---	1	---	---	B
	201w	*S. crassifemur*	Maui	9	10	11	---	---	B
	202w	*S. nasalis*	Maui	1	---	---	---	---	B
	203w	*S. nasalis*	Maui	1	3	1	---	---	B
	208_1w	***D. apodasta***	Kaua'i	8 (A)	---	11 (B)	---	---	**A/B**
	208_2w	***D. apodasta***	Kaua'i	3	---	11	---	---	**B**
	187w	*D. atroscutellata*	Kaua'i	13	13	1	4	---	B
	41w	*D.* nr. *basimacula* #2	Kaua'i	13	1	1	5	1	B
	59w	*D.* nr. *basimacula* #1	Kaua'i	13	1	1	3	---	B
	209w	*D. basimacula*	Kaua'i	1	---	11	---	---	B
	212w	*D.* nr. *basimacula* #1	Kaua'i	13	1	1	3	1	B
	213w	*D.* nr. *basimacula* #2	Kaua'i	13	1	1	5	---	B
	5w	*D.* nr. *basimacula* #5	Kaua'i	13	1	1	3	1	B
	127w	*D. kikiko*	Kaua'i	7	---	8	---	---	B
	155w	*D. micromyia*	Kaua'i	10	11	7	8	2	B
	215_1w	***D.* nr. *perissopoda* #1**	Kaua'i	1 (B)	5 (A)	---	---	---	**A/B**
	215_2w	***D.* nr. *perissopoda* #1**	Kaua'i	13	9	---	---	---	**B**

Table 3. *Cont.*

	Wolbachia Sample Name	Host Species Name	Island of Collection	*coxA*	*fbpA*	*hcpA*	*ftsZ*	*gatB*	Strain Type
Non-native Drosophila and mosquitoes collected in Hawaii	3_A12w	*D. suzukii*	Kaua'i	---	---	12	---	---	A
	3_B11w	*D. suzukii*	Kaua'i	---	---	14	---	---	B
	3_C2w	*D. suzukii*	O'ahu	15	17	14	---	---	B
	3_C3w	*D. suzukii*	Kaua'i	16	---	13	---	---	A
	3_D5w	*D. suzukii*	Kaua'i	---	---	13	---	---	A
	3_E3w	*D. suzukii*	Kaua'i	---	---	13	---	---	A
	3_F6w	*D. suzukii*	Kaua'i	---	---	14	---	---	B
	3_H4w	*D. suzukii*	Kaua'i	---	---	13	---	---	A
	wHa [6]	*D. simulans wHa*	Hawai'i	16	20	13	15	---	A
	wAlb	*Aedes albopictus*	Hawai'i	11 (A)	14 (B)	9 (B)	12 (B)	---	A/B
	6771w	*Culex quinquefasciatus*	Hawai'i	13	16	11	14	---	B
Other Drosophila and Mosquitoes	wAlbA [1]	*A. albopictus*	Unknown	12	15	10	13	---	A
	*w*Pip [2]	*C. quinquefasciatus Pel*	Sri Lanka	13	16	11	14	---	B
	*w*Pip [3]	*C. quinquefasciatus JHB*	Johannesburg	13	16	11	14	---	B
	*w*Dmel [4]	*D. melanogaster*	Laboratory Stock	17	19	15	16	---	A
	*w*Dsuzi [5]	*D. suzukii*	Italy	14	18	12	15	---	A
	wNo [6]	*D. simulans*	New Caledonia	18	21	16	17	---	B
OG	wBm [7]	*Brugia malayi*	Unknown	19	22	17	18	5	D
	wCle [8]	*Cimex lectularius*	Unknown	20	23	18	19	6	F

[1] "isolate 12", typed using MLST by Baldo et al. [34]; [2] genome accession AM999887.1, Klasson et al. [41]; [3] genome accession ABZA01000002.1, Salzberg et al. [42]; [4] genome accession NC_002978.6, Wu et al. [35]; [5] *Wolbachia* endosymbiont str. *valsugana*, WGS project CAOU00000000, Siozios et al. [40]; [6] genome accession NC_021084.1, Ellegaard et al. [38]; [7] Wolbachia endosymbiont strain TRS, NC_006833.1, Foster et al. [43]; [8] Nikoh et al. [44].

Patterns of *Wolbachia* strain diversity corresponded to host relatedness in some, but not all cases. Two closely related, sympatric host species, *D. hawaiiensis* and *D. engyochracea* were possibly infected with the same, or if not the same, a similar *Wolbachia* strain (at 3-identical alleles, Table 3). Furthermore, within the same population, an additional *D. engyochracea* specimen was doubly infected with one *Wolbachia* strain identical to *D. engyochracea* and *D. hawaiiensis* (at two alleles), plus a second strain with two unique alleles, both belonging to the uncommon supergroup A. Evidence of infection by identical *Wolbachia* strains (at five loci) was found for the distantly related host species *S. caliginosa* and *D. seclusa*, both collected on the island of Hawaii. Interestingly, it was also found that the five members of the *D. basimacula/perissopoda* "bristle tarsus" complex were each infected by a different *Wolbachia* strain, while a sixth (*D.* nr. *perissopoda* #5) was not infected. Each is only represented by one or two individuals, but the strains appear to be the same within each taxon.

3.4. Phylogenetic Reconstruction Analysis

Phylograms for Hawaiian Drosophilidae host species showed nearly identical topologies between inferences made with Bayesian and Maximum Likelihood analyses (Figure S5) and were approximately concordant with the Hawaiian Drosophilidae phylogram previously published by Magnacca and Price [3]. The only discrepancy is the placement of the modified mouthparts group (represented here by *D. nigrocirrus* and *D.* "large spots") as sister to the *picture wing* group with the AMC clade basal, rather than with the *picture wing* group basal as they were found. However, Magnacca and Price [3] noted that the phylogenetic position of the modified mouthparts and AMC clade (outgroups) relative to the *picture wing* species group were not well supported, and in fact the arrangement found here is the same as in their analysis using BEAST. The addition of *A. albopictus* and *C. quinquefasciatus* had minimal effect on tree topology.

3.5. Co-Phylogenetic Reconciliation

The co-phylogenetic reconciliation analysis run in JANE [49] determined that the optimal solutions consisted of two main biological events: co-speciation and duplication with host switches (Table 4, Figure 1A). The co-phylogenetic reconstructions for the dataset consisting of only Hawaiian Drosophilidae and their *Wolbachia* endosymbionts resulted in identical optimal solutions, regardless of co-speciation being assigned a cost of 0 or 1, and the pattern of events was similar, differing only slightly by the projected timing of events (Figure S6, Panels A and B). When invasive mosquitoes collected in Hawaii and their *Wolbachia* endosymbionts were added to the data, the optimal solutions differed slightly by the number of each event, and the optimized cost between the two models differed significantly ($p < 0.01$), indicating that co-speciation has a significant effect on the overall model (Table 4, Figures 1A and S6). Lastly, a tanglegram illustrates that co-phylogenetic relationships between *Wolbachia* and its host show patterns consistent with both co-evolution (parallel connections) and horizontal transfer (crossed lines) (Figure 1B).

Table 4. Co-phylogenetic reconstructions implemented in JANE4 (see text for details) with cost-scheme parameters loss, failure to diverge, and duplication each set to 1, duplication and host switch set to 2, and varying the co-speciation cost (Cost) by 0 or 1.

	Drosophila and *Wolbachia*		Invasive Mosquitoes, *Drosophila*, and *Wolbachia*	
	Cost 0	Cost 1	Cost 0	Cost 1
Co-speciation	8	8	9	7
Duplication	0	0	0	1
Duplication and host switches	7	7	8	9
Loss	1	1	2	1

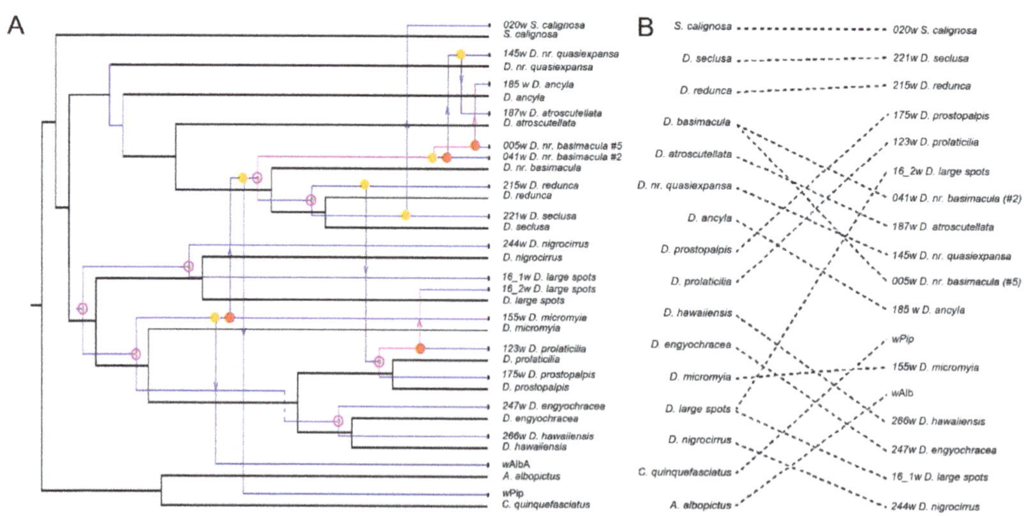

Figure 1. (**A**). Co-phylogenetic reconciliation analysis for Hawaiian Drosophilidae and two species of invasive mosquitoes and their *Wolbachia* endosymbionts based on the following cost scheme: co-speciation: 1; duplication: 1; duplication and host switch: 2; loss: 1; failure to diverge: 1. The estimated biological events that best describe the data are depicted on the phylogeny [open circle:

co-speciation; closed circle: duplication; closed circle with arrow: duplication and host switch; dashed line: loss]. Red indicates that the event is optimally placed, whereas yellow indicates that another placement exists that is equally valid. (**B**). A tanglegram depicting the co-phylogenetic relationship between the Hawaiian Drosophilidae and invasive mosquito phylogeny (**left**) and their *Wolbachia* endosymbiont phylogeny (**right**).

3.6. Stochastic Character Trait Mapping

The modeled ancestral state of host ovipositional substrate showed high posterior probabilities (depicted on the interior node) when mapped to the unrooted *Wolbachia* phylogeny, reflecting a phylogenetic signal for this character among similar *Wolbachia* strains (Figure 2). The bark and sap flux ancestral traits were, for the most part, conserved among supergroups A and A/B, while the bulk of supergroup B *Wolbachia* was affiliated with the trait leaf. In contrast, little support was evident for host trait associations to *Wolbachia* phylogenies for the host traits island of collection and host plant family (Figure S7).

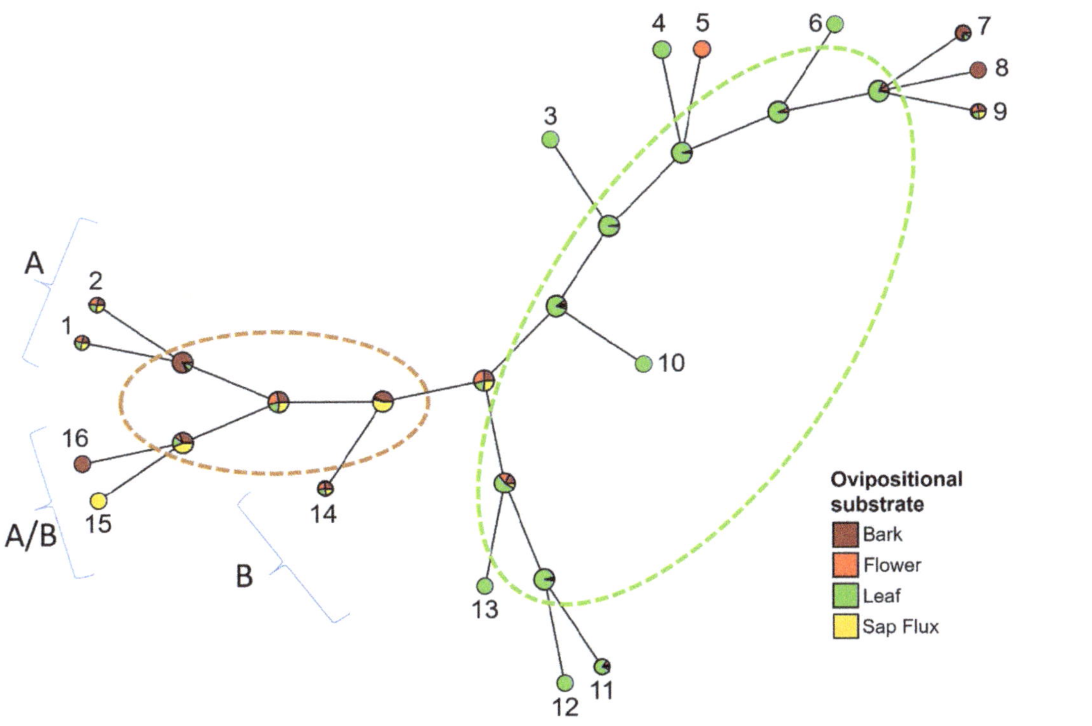

Figure 2. Stochastic character mapping of the Hawaiian Drosophilidae (*Drosophila* and *Scaptomyza*) host species' ancestral trait 'ovipositional substrate' (categories: bark, flower, leaf, and sap flux) mapped to an unrooted *Wolbachia* phylogeny. External nodes depict host trait assignments: solid = known, equal proportions = unknown. Interior nodes represent posterior probabilities that the host's character trait is ancestral and congruent with the phylogenetic signal of the strain of their *Wolbachia* endosymbiont. Key to *Wolbachia* found in host individual listed in Supplementary S1 (sample number, species name): (1) 244 w *D. nigrocirrus*, (2) 16_1w *D.* "large spots" (double-infected strain A), (3) 185w *D. ancyla*, (4) 221w *D. seclusa*, (5) 20w *S. caliginosa*, (6) 216w *D.* nr. *redunca*, (7) 175w *D. prostopalpis*, (8) 123w *D. prolaticilia*, (9) 16_2w *D.* "large spots" (double-infected strain B), (10) 5w *D.* nr. *basimacula #5*, (11) 187w *D. atroscutellata*, (12) 145 w *D. quasiexpansa*, (13) 41w *D.* nr. *basimacula #2*, (14) 155 w *D. micromyia*, (15) 266w *D. hawaiiensis*, and (16) 247w *D. engyochracea*. Two strains belong to supergroup A, two were intermediate A/B, and all other strains belong to supergroup B.

4. Discussion

Our assessment of *Wolbachia* within the Hawaiian Drosophilidae family contributes to the understanding of endosymbiont transmission and its potential role in speciation. Using a modified MLST strain typing protocol, and through phylogenetic analyses, we found evidence for both coevolution and horizontal transmission of *Wolbachia* within *Drosophila* sampled across the Hawaiian archipelago. Our study complements the singular previous broad-scale study of *Wolbachia* within natural populations of Hawaiian insect taxa by Bennett et al. [2], in which strain diversity was characterized using a single gene marker, *wsp*. These studies differed by taxonomic scope, in that our primary focus was to investigate *Wolbachia* strain diversity among members of native Hawaiian Drosophilidae (and select invasive insects), and we used a modified version of the MLST strain typing scheme developed by Baldo et al. [34]. Despite study design differences, findings across studies were largely concordant, with Bennett et al. [2] determining the species-level incidence of *Wolbachia* infection for native Hawaiian Drosophilidae to be 18.1%, compared to our finding of 20.0%. Across all samples screened, we found an infection rate of 10.3%, which is lower than Bennett et al.'s [2] incidence of infection at 18.1%. That difference in infection rate can be attributed to the sampling of different taxa, along with uneven sample numbers within individual species. We caution that many species considered in this study were represented by only a single individual; thus, infection status is not representative of the species as a whole. Indeed, we found strong differences in percent infection rate within individual species having data available for five or more individuals. Additionally, although our efforts to re-design *Wolbachia* MLST primers improved amplification efficiency and increased the number of confirmed infections, the amplification and sequencing of *Wolbachia* alleles still proved to be difficult and infection rates may thus be an underestimate. A few of the species (namely *D. claytonae* and *D. setosifrons*) are also represented only by older specimens with poor DNA extractions, which may not have yielded enough to detect *Wolbachia*. If specimens with PCR bands only (absent sequencing results) were to be counted as positive infections, the incidence of *Wolbachia* at both the species and individual level would increase to 28.1% and 16.3%, respectively.

Between supergroups A and B, the majority of *Wolbachia* strains in Hawaiian Drosophilidae were determined to belong to supergroup B (at 68%), consistent with previous screens in native Hawaiian insect taxa, using *wsp*, at ~75% [2]. Among the species included in Bennett et al.'s [2] study, and also screened here, the *Wolbachia* supergroup designations were concordant for endosymbionts of *D. basimacula*, *D.* nr. *basimacula*, *D. redunca*, and *D. ancyla*, which harbored *Wolbachia* from supergroup B, and *D. nigrocirrus*, which harbored *Wolbachia* from supergroup A. With regards to invasive *Drosophila*, Bennett et al. [2] found that *D. suzukii* was infected only by *Wolbachia* belonging to supergroup A, whereas we found individuals harboring infections belonging to supergroups A ($n = 5$) and B ($n = 3$). Interestingly, we observed that a *Wolbachia* infecting a *D. suzukii* individual collected from Hawai'i shared at least two identical alleles (*coxA* and *hcpA*) with the non-native species *D. simulans* that was also collected from Hawai'i by Ellegaard et al. [38]).

4.1. Mechanisms of Wolbachia Transmission

In the case of purely vertical transmission of *Wolbachia* within the Hawaiian Drosophilidae, the expectation is that *Wolbachia* strains would be most similar between closely related host species and that phylogenetic reconstructions of the host and endosymbiont would be fully congruent [18]. The alternative hypothesis is that host-switching may play a role in transmission, in which case host and endosymbiont phylogenies would be discordant. Using co-phylogenetic reconciliation analysis, we found that optimal solutions generated by JANE consistently showed co-speciation (i.e., vertical transmission) and duplication with host switching (i.e., horizontal transmission) events as significant parameters despite the costs associated with them. Further evidence for both scenarios—vertical and horizontal transmission—are evidenced through strain typing results. For example, the distantly related species *D. seclusa* and *S. caliginosa* possessed seemingly identical *Wolbachia* strains, and conversely, individual hosts belonging to the same species harbored differing *Wolbachia* strains (e.g., *D. engyochracea*). Mechanisms for horizontal transmission are suggested by stochastic character trait mapping results, which revealed a positive association between phylogenetic patterns of *Wolbachia* and their hosts' ancestral trait preferred host ovipositional substrate, a trait that is more evolutionarily conserved than affiliations with host plant families [3,31]. For preferred ovipositional substrate, in general, Hawaiian Drosophilidae from the genus *Scaptomyza* use flowers or rotting fruits (as well as many unusual substrates, such as living *Cyrtandra* leaves), the AMC clade (i.e., *antopocerus, modified-tarsus, ciliated-tarsus*) utilizes rotting leaves, the *picture wing* species group uses rotting bark or sap-flux, and the *modified mouthparts* clade (e.g., *D. nigrocirrus* and *D. large spots*) uses a range of ovipositional substrate types [31]. High posterior probabilities for ancestral states of host ovipositional substrate indicated associations between the trait 'bark' and 'sap flux' for supergroups A and A/B and the trait 'leaf' for supergroup B. This pattern was consistent even for the single *D. large spots* specimen doubly infected by *Wolbachia* strains belonging to supergroups A and B. Notably, the only other *Wolbachia* belonging to supergroup A isolated from Hawaiian *Drosophila* was isolated from *D. nigrocirrus*, also a member of the *modified mouthparts* sub-group. The host plant and substrate are unknown for both of these species. Bennett and colleagues [2] noted that phylogenetically, *wsp* alleles amplified from Hawaiian taxa tended to group closely together, and they found evidence for sharing of identical or similar *wsp* alleles between close and distantly related Hawaiian insect species. They postulated that this observation can be explained by *Wolbachia* infections persisting through speciation, as well as horizontal transmission occurring between host taxa.

An association of *Wolbachia* supergroup B with the decaying leaf substrate could play a role in one of the evolutionary puzzles of Hawaiian Drosophilidae, namely, why there are so many closely related, sympatric species utilizing the same host substrate. This is most readily seen in the *spoon tarsus* subgroup on Hawai'i and the *bristle tarsus* subgroup on Kaua'i. The latter is represented here by six members of the *D. basimacula–perissopoda* species complex, which can be distinguished by the number and arrangement of thickened bristles on the modified front tarsus of the male. Each was found to carry a different strain of *Wolbachia*, or none. Novel infection or loss of infection may initiate the localized equivalent of "founder events", leading to rapid speciation and maintenance of species boundaries when combined with the sexual selection for which Hawaiian *Drosophila* are well known [53].

Consistent with our findings, plants are thought to play key roles in the horizontal transmission of *Wolbachia* strains between infected and uninfected individuals, as well as between diverse insect species. For example, Sintupachee et al. [54]) found that distantly related species of arthropods found to co-occur on pumpkin leaves harbored *Wolbachia* with similar *wsp* sequences, and Li et al. [25] showed under a controlled experimental laboratory setting that a stable *Wolbachia* infection could be attained by uninfected whitefly individuals through feeding on the same leaf substrate previously exposed to *Wolbachia* infected individuals. In that study, *Wolbachia* was documented as dispersing to adjacent leaves within just a few days of the initial plant infection, where it remained within the phloem of the plant for a minimum of 50 days [25]. In Hawaiian insects, Bennett et al. [2] found that nearly identical *Wolbachia wsp* alleles were shared between some Diptera species (e.g., *Drosophila forficata*) and Hemiptera (*Nesophrosyne craterigena*), which they propose is explained by a reliance of both Drosophila and Nesophrosyne species on shared host plants across their ranges. Together, plant utilization and feeding habits may help explain why most native Drosophilidae species were infected with *Wolbachia* from supergroup B, why some members were infected with supergroup A (modified mouthparts group), and why identical alleles were shared between some distantly related taxa. Our findings are thus congruent with Bennet et al. [2], who proposed that horizontal transmission of *Wolbachia* occurs between Hawaiian taxa at multiple taxonomic scales.

Insects that possess piercing-sucking mouthparts may be more apt to transmitting *Wolbachia* to plants through feeding [19,54], and *Wolbachia* has been found to exist within insect salivary glands in addition to other somatic tissues [24,55]. Additionally, honeydew and infected leaves have been implicated in previous studies as a potential means of horizontal transmission [25,56]. Most non-native *Drosophila* included in this study were infected with supergroup A; however, infection by supergroup B *Wolbachia* within non-native *D. suzukii* individuals could be explained by their occasional use of native plants [31]. Full strain typing profiles, if available, could be used to test this idea. In other biological systems, although extremely rare, *Wolbachia* strains have been known to rapidly displace other strains, often in association with insect invasions. For example, the *Wolbachia* variant *w*Ri rapidly displaced *w*Au within their host *D. simulans* [57], and horizontal transmission occurred for *Wolbachia* endosymbionts and their host silverleaf whitefly (*Bemisia tabaci*), in which a host shift event occurred in China from indigenous members of the complex to the invader as well as from the invader to indigenous relatives [24]. An alternative explanation to plant-mediated horizontal transfer of *Wolbachia* is through non-lethal probing of infected nymphs and uninfected nymphs by parasitoid wasps ([24], reviewed by Sanaei et al. [58]). That mechanism for transmission is consistent with Bennett and colleagues [2] who postulated parasitoids to be a potential mechanism of horizontal transmission for *Wolbachia* in Hawaiian taxa, in addition to plant associations. They found that parasitoids, along with native and non-native Drosophila species, were grouped closely together based on the phylogenetic reconstruction of the *wsp* gene.

4.2. Discrepancy in Supergroup Designation of Loci

Whether supergroups can recombine has been the subject of debate. Ellegaard et al. [38] proposed that *Wolbachia* supergroups are irreversibly separated, and that barriers other than host-specialization are able to maintain distinct clades in recombining endosymbiont populations. Their conclusion was based on naturally occurring double-infections of *Wolbachia* strains *w*Ha and *w*No endosymbionts of *D. simulans*. Recent findings from a survey of 33 genome-sequences for *Wolbachia* strains belonging to supergroups A–F found that strains maintained a supergroup relationship across 210 conserved single-copy genes, yet an analysis of interclade recombination screening revealed that 14 inter-supergroup recombination events had occurred in six of the 210 core genes (6/210 = 2.9%) [59]. Consistent with recombination events, Baldo et al. [60]) found evidence for recombination between *gatB* and *fbpA* alleles, and intragenic re-combination was detected by comparing patterns of *gltA* to other housekeeping genes [60]. In this study, among the 44 *Wolbachia* strains isolated from Hawaiian Drosophilidae hosts, conflicting supergroup designations were observed for 20.4% of the strains (with data available at two or more genes), which in some cases resulted in an intermediate phylogenetic placement between supergroups A and B. In particular, *coxA* and *hcpA* alleles exhibited discordance between supergroup placement, congruent with discordance in supergroup designation for *coxA* and *hcpA* alleles observed within Lepidoptera species collected from West Siberia [61]. Although we cannot fully rule out that allelic discordance across strains may be a result of preferential amplification of certain alleles by primers in the presence of multiple infections—for example, double infections by strains belonging to supergroups A and B were observed to occur within w208 *D. apodasta* and w215 *D. nr. perissopoda*—the majority of individuals with conflicting alleles lacked evidence for the presence of a double infection. Therefore, the discrepancy in supergroup assignment between alleles may have resulted from a recombination event that occurred within a doubly infected host species and subsequent fixation of alleles. Further research could help to elucidate the complex interactions of endosymbionts and host taxa occurring within Hawaiian insect communities.

4.3. Conservation Implications

The rapid diversification of Hawaiian *Drosophila* results from a combination of evolutionary-time scale island isolation, rugged topography, and development of novel host plant associations that have persisted for millions of years [3]. Many species are single-island endemics with narrow ranges and are restricted to the natural distribution of their host plants, which makes populations especially vulnerable habitat degradation and climate change. At present the US Fish and Wildlife Service lists 13 Hawaiian Drosophilds as endangered (*D. aglaia*, *D. differens*, *D. digressa*, *D. hemipeza*, *D. heteroneura*, *D. montgomeryi*, *D. mulli*, *D. musaphilia*, *D. neoclavisetae*, *D. obatai*, *D. ochrobasis*, *D. sharpi*, *D. substenoptera*, and *D. tarphytrichia*) and one as threatened (*D. musaphilia*). These listed species represent 14.4% of all insects, and 4.8% of all listed invertebrates, within the USA (ECOS Environmental Conservation Online System https://ecos.fws.gov/ecp, accessed on 5 March 2023). Given *Wolbachia*'s impacts on reproduction, consideration of host–symbiont relationships and infection status might increase success of breeding programs and ensure that translocation efforts do not suffer from effects of cytoplasmic incompatibility. With regards to climate change, experimental data for Hawaiian *Drosophila* has demonstrated that species are locally adapted [62,63], thus, resilience to warming temperatures could perhaps be enhanced by manipulation of the host microbiomes, including *Wolbachia* endosymbionts. Endosymbiont-mediated responses to temperature stress are known to include transcription response and behavior [64,65].

5. Conclusions

This study sheds light on the infection status and coevolutionary history of *Wolbachia* endosymbionts within their Hawaiian Drosophilidae hosts. Co-phylogenetic reconciliations and comparative phylogenetic analyses indicate that the transmission patterns of *Wolbachia* is best explained by both co-speciation and host-switching events. Future studies that survey *Wolbachia* from a greater breadth of native Hawaiian arthropod taxa, as well as introduced arthropod invasive taxa, may help to improve our understanding of how *Wolbachia* transmission has occurred in Hawaiian ecosystems. Insights into *Wolbachia* infections and strain types could help guide conservation programs, possibly enhancing translocation efforts, impacting host behavioral response to temperatures, and conferring host thermal tolerance.

Supplementary Materials: The following supporting information can be downloaded at https://www.mdpi.com/article/10.3390/genes14081545/s1, Table S1. National Center for Biotechnology Information (NCBI) accession numbers for Hawaiian Drosophilidae gene sequences selected for phylogenetic reconstruction of individual species having a verified *Wolbachia* infection, and genome accessions for outgroup taxa (also infected) [3,66–70]. Table S2. Records for amplification and sequencing of *Wolbachia* endosymbionts of Hawaiian Drosophilids. Table S3. Record of Akaike information criterion (AIC) values obtained using the fitMK function in phytools v0.6-44 (Revell 2012) [51] package in R to determine the best rates model to apply to each data set for stochastic character mapping analyses. Figure S1. Phylogenetic reconstruction of *Wolbachia* housekeeping genes based on Bayesian inference analyses (see main text for details): (a) cytochrome C oxidase subunit A (*coxA*) [378 bp], with 46 sequences, (b) conserved hypothetical protein (*hcpA*) [381 bp], with 44 sequences; (c) fructose-bisphosphate aldolase (*fbpA*) [417 bp], with 30 sequences, and (d)) cell division protein (*ftsZ*) [354 bp], with 28 sequences. Individuals consistent in their supergroup designations across all genes considered are indicated as either pink for supergroup A or purple for supergroup B. Individuals that showed conflicting supergroup designation between genes are shown in grey. Outgroup taxa belonging to supergroups D and F are shown in green. A solid line indicates that supergroup designation was based on three or more genes, whereas a dotted line indicates that data for 2 of fewer genes were available for super group designation. The taxonomic standing is uncertain for *Wolbachia* endosymbiont host species *Drosophila basimacula* #5 and #2 (samples 5 and 41), *D. quasiexpansa* sample 145, *D. reduncа* sample 216 and *D. perrisopoda* sample 215 (see main text for details). Figure S2. Phylogenetic reconstruction of *Wolbachia* housekeeping genes based on Maximum Likelihood analyses (see main text for details]): (a) cytochrome C oxidase subunit A (*coxA*) [378 bp], with 46 sequences, (b) conserved hypothetical protein (*hcpA*) [381 bp], with 44 sequences; (c) fructose-bisphosphate aldolase (*fbpA*) [417 bp], with 30 sequences, and (d) cell division protein (*ftsZ*) [354 bp], with 28 sequences. Individuals consistent in their supergroup designations across all genes considered are indicated as either pink for supergroup A or purple for supergroup B. Individuals that showed conflicting supergroup designation between genes are shown in grey. Outgroup taxa belonging to supergroups D and F are shown in green. A solid line indicates that supergroup designation was based on three or more genes, whereas a dotted line indicates that data for two or fewer genes were available for super group designation. The taxonomic standing of *Wolbachia* endosymbiont host species *Drosophila basimacula* #5 and #2 (samples 5 and 41), *D. quasiexpansa* sample 145, *D. redunca* sample 216 and *D. perrisopoda* sample 215 is uncertain (see main text for details). Figure S3. Phylogenetic reconstruction of three concatenated *Wolbachia* MLST genes: cytochrome C oxidase subunit A (*coxA*), conserved hypothetical protein (*hcpA*), and cell division protein (*ftsZ*) [1113 bp] based on a by gene and codon partitioning scheme and 25 sequences analyzed using (A) Bayesian and (B) Maximum Likelihood approaches. Individuals consistent in their supergroup designation across all available MLST gene data are shown pink for supergroup A or purple for supergroup B. Individuals that showed conflicting supergroup designation between genes are shown in grey. Outgroup taxa (Supergroups D and F) are shown in green. The taxonomic standing of *Wolbachia* endosymbiont host species *Drosophila basimacula* #5 and #2 (samples 5 and 41), *D. quasiexpansa* sample 145, and *D. redunca* sample 216 is uncertain (see main text for details). Figure S4. Phylogenetic reconstruction of four concatenated *Wolbachia* MLST genes: cytochrome C oxidase subunit A (*coxA*), fructose-bisphosphate aldolase (*fbpA*), conserved hypothetical protein (*hcpA*), and cell division protein (*ftsZ*) [1530 bp] based on a by gene and codon partitioning scheme and (A) Bayesian and (B) Maximum Likelihood approaches. Individuals consistent in their supergroup designation across all available MLST gene data are shown pink for supergroup A or purple for supergroup B. Individuals that showed conflicting supergroup designation between genes are shown in grey. Outgroup taxa (Supergroups D and F) are shown in green. The taxonomic standing of *Wolbachia* endosymbiont host species *Drosophila basimacula* samples 5 and 41 and *D. redunca* sample 216 are uncertain (see main text for details). Figure S5. Phylogenetic reconstruction of Hawaiian Drosophilidae host species (*Drosophila* and *Scaptomyza*) and invasive mosquitoes *A. albopictus* and *C. quinquefasciatus* based on four concatenated genes, elongation factor 1-γ (*EF1g*), glycerol-3-phosphate dehydrogenase (*Gpdh*), phosphoglucose isomerase (*Pgi*), and yolk protein 2 (*Yp2*) [1812 bp] using a by codon partitioning scheme and Bayesian inference (A) or Maximum Likelihood (B). Phylogenetic analysis was restricted to Drosophilidae host species having *Wolbachia* sequence data available across three multilocus sequence typing genes (see main text for details). Figure S6. Implemented in JANE4 (Conow et al. 2010) [49], a possible solution for co-phylogenetic reconciliation analysis for Hawaiian Drosophilidae and their *Wolbachia* endosymbionts based on the cost scheme setting co-speciation assigned 0 (panel A) or 1 (panel B), with loss, failure to diverge, and duplication each set to 1 and duplication and host switch set to 2. Estimated biological events that best describe the data are depicted on the phylogeny [open circle: co-speciation; closed circle: duplication; closed circle with arrow: duplication and host switch; dashed line: loss]. Red indicates that the event is optimally placed, whereas yellow indicates that another placement exists that is equally valid. The taxonomic standings of *Wolbachia* endosymbiont host species *Drosophila basimacula* #5 and #2 (samples 5 and 41) and *D. perrisopoda* #1 (sample 215) are uncertain (see main text for details). Figure S7. Stochastic character mapping of the Hawaiian Drosophilidae host species' ancestral traits (A) island of collection and (B) host plant families mapped to an unrooted *Wolbachia* phylogeny. External nodes depict host trait assignments: solid = known, equal proportions = unknown. Interior nodes represent posterior probabilities that the host's character trait is ancestral and congruent with the phylogenetic signal of the *Wolbachia* strain. Key to *Wolbachia* found in host individual listed in Supplementary S1 (sample number, species name): (1) 244 w *D. nigrocirrus*, (2) 16_1w *D.* "large spots" (double-infected strain A), (3) 185w *D. ancyla*, (4) 221w *D. seclusa*, (5) 20w *S. caliginosa*, (6) 216w *D.* nr. *redunca*, (7) 175w *D. prostopalpis*, (8) 123w *D. prolaticilia*, (9) 16_2w *D.* "large spots" (double-infected strain B, (10) 5w *D.* nr. *basimacula #5*, (11) 187w *D. atroscutellata*, (12) 145 w *D. quasiexpansa*, (13) 41w *D.* nr. *basimacula #2*, (14) 155 w *D. micromyia*, (15) 266w *D. hawaiiensis*, and (16) 247w *D. engyochracea*. Two strains belong to supergroup A, two were intermediate A/B, and all other strains belong to supergroup B. Supplementary Information. Extended methods [34,35,71–76]. Supplementary Datafile S1. *Wolbachia* sequence data generated for this study and used for primer re-design and data analyses. Supplementary S1. Hawaiian Drosophilidae specimens (n = 399, genus *Drosophila* and *Scaptomyza*) and invasive *Drosophila* (n = 68) screened for the presence of endosymbiotic bacteria *Wolbachia* [3,32]. Supplementary S2. A list of *Wolbachia* host species and sequence accessions for five *Wolbachia* Multiple Locus Strain Typing sequences [cytochrome C oxidase subunit A (*coxA*), conserved hypothetical protein (*hcpA*), fructose-bisphophate adolase (*fbpA*), cell division protein (*ftsZ*) and aspartyl/glutamyl-tRNA aminotransferase subunit B (*gatB*)], and Wolbachia surface protein [*wsp* and paralog *wspB*] used for BLASTn queries as part of primer re-design efforts.

Author Contributions: Conceptualization, R.L.C. and D.K.P.; methodology, R.L.C., A.V., M.R.B. and D.K.P.; formal analysis, R.L.C., M.R.B. and D.K.P.; investigation, all authors; resources, D.K.P. and M.R.B.; data curation, R.L.C. and M.R.B.; writing—original draft preparation, R.L.C. and M.R.B.; writing—review and editing, D.K.P., A.V. and K.N.M.; visualization, R.L.C. and M.R.B.; supervision, D.K.P. and M.R.B.; project administration, D.K.P.; funding acquisition, D.K.P. All authors have read and agreed to the published version of the manuscript.

Funding: This work research was funded by the National Science Foundation (grant No. 1345247) and M.R.B. was partially funded by the U. S. Geological Survey, Pacific Island Ecosystems Research Center.

Institutional Review Board Statement: Listed endangered species were collected under USFWS permit TE-02997A-2.

Informed Consent Statement: Not applicable.

Data Availability Statement: The *Wolbachia* sequence data generated for this study (primer re-design and data analysis) are available from Supplementary Datafile S1; Drosophilidae sequences are available from the National Center for Biotechnology Information public repository.

Acknowledgments: Sanger sequencing was performed at the University of Hawai'i at Hilo's Evolutionary Genomics Core Facility. We thank J. Sutton for the *A. albopictus* sample and C. Atkinson for the *C. quinquefasciatus* sample. The technical support and advanced computing resources from University of Hawaii Information Technology Services—Cyberinfrastructure, funded in part by the National Science Foundation MRI award # 1920304, are gratefully acknowledged. Any opinions, findings, and conclusions or recommendations expressed in this material are those of the author(s) and do not necessarily reflect the views of the National Science Foundation. Any use of trade, firm, or product names is for descriptive purposes only and does not imply endorsement by the U.S. Government.

Conflicts of Interest: The authors have no conflict of interest to declare.

References

1. Kambysellis, M.P.; Ho, K.; Craddock, E.M.; Piano, F.; Parisi, M.; Cohen, J. Pattern of ecological shifts in the diversification of Hawaiian *Drosophila* inferred from a molecular phylogeny. *Curr. Biol.* **1995**, *5*, 1129–1139. [CrossRef]
2. Bennett, G.M.; Pantoja, N.A.; O'Grady, P.M. Diversity and phylogenetic relationships of *Wolbachia* in *Drosophila* and other native Hawaiian insects. *Fly* **2012**, *6*, 273–283.
3. Magnacca, K.N.; Price, D.K. Rapid adaptive radiation and host plant conservation in the Hawaiian picture wing *Drosophila* (Diptera: Drosophilidae). *Mol. Phylogenet. Evol.* **2015**, *92*, 226–242. [CrossRef] [PubMed]
4. O'Grady, P.M.; DeSalle, R. Phylogeny of the genus *Drosoph. Genet.* **2018**, *209*, 1–25. [CrossRef]
5. Zimmerman, E.C. *Insects of Hawaii: Volume 1*; University of Hawaii Press: Honolulu, HI, USA, 1948; 222p.
6. Kaneshiro, K.Y. Speciation in the Hawaiian "*Drosophila*": Sexual selection sppears to play an important role. *BioScience* **1988**, *38*, 258–263. [CrossRef]
7. Karvonen, A.; Seehausen, O. The role of parasitism in adaptive radiations-When might parasites promote and when might they constrain ecological speciation? *Int. J. Ecol.* **2012**, *2012*, 1–20. [CrossRef]
8. Werren, J.H.; Windsor, D.; Guo, L. Distribution of *Wolbachia* among neotropical arthropods. *Proc. R. Soc. Lond. B* **1995**, *262*, 197–204.
9. Stouthamer, R.; Breeuwer, J.A.J.; Hurst, G.D.D. *Wolbachia pipientis*: Microbial manipulator of arthropod reproduction. *Annu. Rev. Microbiol.* **1999**, *53*, 71–102. [CrossRef]
10. Bordenstein, S.R.; O'hara, F.P.; Werren, J.H. *Wolbachia*-induced incompatibility precedes other hybrid incompatibilities in *Nasonia*. *Nature* **2001**, *409*, 707–710. [CrossRef]
11. Jaenike, J.; Dyer, K.A.; Cornish, C.; Minhas, M.S. Asymmetrical reinforcement and *Wolbachia* infection in *Drosophila*. *Public Libr. Sci. (PLOS) Biol.* **2006**, *4*, 1852–1862. [CrossRef]
12. Werren, J.H.; Baldo, L.; Clark, M.E. *Wolbachia*: Master manipulators of invertebrate biology. *Nat. Rev. Microbiol.* **2008**, *6*, 741–751. [CrossRef] [PubMed]
13. Zug, R.; Hammerstein, P. Still a host of hosts for *Wolbachia:* Analysis of recent data suggests that 40% of terrestrial arthropod species are infected. *Public Libr. Sci. PLoS ONE* **2012**, *7*, e38544. [CrossRef] [PubMed]
14. Werren, J.H. Biology of *Wolbachia*. *Annu. Rev. Entomol.* **1997**, *42*, 587–609. [CrossRef] [PubMed]
15. Bourtzis, K.; O'Neill, S. *Wolbachia* infections and arthropod reproduction. *BioScience* **1998**, *48*, 287–293. [CrossRef]
16. Telschow, A.; Hammerstein, P.; Werren, J.H. Effects of *Wolbachia* on genetic divergence between populations: Mainland-island model. *Integr. Comp. Biol.* **2002**, *42*, 340–351. [CrossRef] [PubMed]
17. Telschow, A.; Hammerstein, P.; Werren, J.H. The effect of *Wolbachia* versus genetic incompatibilities on reinforcement and speciation. *Evolution* **2005**, *59*, 1607–1619. [PubMed]
18. Halfner, M.S.; Page, R.D.M. Molecular phylogenies and host-parasite cospeciation: Gophers and lice as a model system. *Philos. Trans. R. Soc. B* **1995**, *349*, 77–83.

19. Xu, X.; Ridland, P.M.; Umina, P.A.; Gill, A.; Ross, P.A.; Pirtle, E.; Hoffmann, A.A. High incidence of related *Wolbachia* across unrelated leaf-mining Diptera. *Insects* **2021**, *12*, 788. [CrossRef]
20. Riegler, M.; Sidhu, M.; Miller, W.J.; O'Neill, S.L. Evidence for a global *Wolbachia* replacement in *Drosophila melanogaster*. *Curr. Biol.* **2005**, *15*, 1428–1433. [CrossRef]
21. Turelli, M.; Cooper, B.S.; Richardson, K.M.; Ginsberg, P.S.; Peckenpaugh, B.; Antelope, C.X.; Kim, K.J.; May, M.R.; Abrieux, A.; Wilson, D.A.; et al. Rapid global spread of *w*Ri-like *Wolbachia* across multiple *Drosophila*. *Curr. Biol.* **2018**, *28*, 963–971. [CrossRef]
22. Casiraghi, M.; Anderson, T.J.C.; Bandi, C.; Bazzacchi, C.; Genchi, C. A phylogenetic analysis of filarial nematodes: Comparison with the phylogeny of *Wolbachia* endosymbionts. *Parasitology* **2001**, *122*, 93–103. [CrossRef] [PubMed]
23. Lefoulon, E.; Bain, O.; Makepeace, B.L.; d'Haese, C.; Uni, S.; Martin, C.; Gavotte, L. Breakdown of coevolution between symbiotic bacteria *Wolbachia* and their filarial hosts. *PeerJ* **2016**, *4*, e18401-30. [CrossRef] [PubMed]
24. Ahmed, M.Z.; Li, S.; Xue, X.; Yin, X.; Ren, S.; Jiggins, F.M.; Greeff, J.M.; Qiu, B. The intracellular bacterium *Wolbachia* uses parasitoid wasps as phoretic vectors for efficient horizontal *Transmission*. *Public Libr. Sci. PLoS Pathog.* **2015**, *11*, e1004672. [CrossRef]
25. Li, S.; Ahmed, M.Z.; Lv, N.; Shi, P.; Wang, X.; Huang, J.; Qiu, B. Plant-mediated horizontal transmission of *Wolbachia* between whiteflies. *Int. Soc. Microb. Ecol.* **2017**, *11*, 1019–1028.
26. Laidoudi, Y.; Levasseur, A.; Medkour, H.; Maaloum, M.; Ben Khedher, M.; Sambou, M.; Bassene, H.; Davoust, B.; Fenollar, F.; Raoult, D.; et al. An earliest endosymbiont, *Wolbachia massiliensis* sp. nov., strain PL13 from the bed bug (*Cimex hemipterus*), type strain of a new supergroup T. *Int. J. Mol. Sci.* **2020**, *21*, 8064. [CrossRef]
27. Kaur, R.; Shropshire, J.D.; Cross, K.L.; Leigh, B.; Mansueto, A.J.; Stewart, V.; Bordenstein, S.R.; Bordenstein, S.R. Living in the endosymbiotic world of *Wolbachia*: A centennial review. *Cell Host Microbe* **2021**, *29*, 879–893. [CrossRef] [PubMed]
28. Gerth, M.; Gansauge, M.T.; Weigert, A.; Bleidorn, C. Phylogenomic analyses uncover origin and spread of the *Wolbachia* pandemic. *Nat. Commun.* **2014**, *5*, 5117. [CrossRef]
29. Jeyaprakash, A.; Hoy, M.A. Long PCR improves *Wolbachia* DNA amplification: Wsp sequences found in 76% of sixty-three arthropod species. *Insect Mol. Biol.* **2000**, *9*, 393–405. [CrossRef]
30. Hilgenboecker, K.; Hammerstein, P.; Schlattmann, P.; Telschow, A.; Werren, J.H. How many species are infected with *Wolbachia*?—A statistical analysis of current data. *Fed. Eur. Microbiol. Soc. (FEMS) Microbiol. Lett.* **2008**, *281*, 215–220. [CrossRef]
31. Magnacca, K.N.; Foote, D.; O'Grady, P.M. A review of the endemic Hawaiian Drosophilidae and their host plants. *Zootaxa* **2008**, *1782*, 1–58. [CrossRef]
32. Koch, J.B.; Dupuis, J.R.; Jardeleza, M.K.; Ouedraogo, N.; Geib, S.M.; Follett, P.A.; Price, D.K. Population genomic and phenotype diversity of invasive *Drosophila suzukii* in Hawai'i. *Biol. Invasions* **2020**, *22*, 1753–1770. [CrossRef]
33. Atkinson, C.T.; Watcher-Weatherwax, W.; LaPointe, D.A. *Genetic Diversity of Wolbachia endosymbionts in Culex quinquefasciatus from Hawaii, Midway Atoll and American Samoa*; Technical Report HCSU-074; Hawaii Cooperative Studies Unit, University of Hawai'i at Hilo: Hilo, HI, USA, 2016; 33p.
34. Baldo, L.; Dunning Hotopp, J.C.; Jolley, K.A.; Bordenstein, S.R.; Biber, S.A.; Choudhury, R.R.; Hayashi, C.; Maiden, M.C.J.; Tettelin, H.; Werren, J.H. Multilocus sequence typing system for the endosymbiont *Wolbachia pipientis*. *Appl. Environ. Microbiol.* **2006**, *72*, 7098–7110. [CrossRef]
35. Wu, M.; Sun, L.V.; Vamathevan, J.; Riegler, M.; Deboy, R.; Brownlie, J.C.; McGraw, E.A.; Martin, W.; Esser, C.; Ahmadinejad, N.; et al. Phylogenomics of the reproductive parasite *Wolbachia pipientis w*Mel: A streamlined genome overrun by mobile genetic elements. *Public Libr. Sci. PLoS Biol.* **2004**, *2*, 327–341. [CrossRef] [PubMed]
36. Kumar, S.; Stecher, G.; Tamura, K. MEGA7: Molecular Evolutionary Genetics Analysis version 7.0. *Mol. Biol. Evol.* **2015**, *33*, 1870–1874. [CrossRef]
37. Metcalf, J.A.; Jo, M.; Bordenstein, S.R.; Jaenike, J.; Bordenstein, S.R. Recent genome reduction of *Wolbachia* in *Drosophila recens* targets phage WO and narrows candidates for reproductive parasitism. *PeerJ* **2014**, *2*, e529. [CrossRef]
38. Ellegaard, K.M.; Klasson, L.; Naslund, K.; Bourtzis, K.; Andersson, S.G.E. Comparative genomics of *Wolbachia* and the bacterial species concept. *Public Libr. Sci. (PLOS) Genet.* **2013**, *9*, e1003381. [CrossRef]
39. Salzberg, S.L.; Dunning Hotopp, J.C.; Delcher, A.L.; Pop, M.; Smith, D.R.; Eisen, M.B.; Nelson, W.C. Serendipitous discovery of *Wolbachia* genomes in multiple *Drosophila* species. *Genome Biol.* **2005**, *6*, R23.1–R23.8. [CrossRef]
40. Siozios, S.; Cestaro, A.; Kaur, R.; Pertot, I.; Rota-Stabelli, O.; Anfora, G. Draft genome sequence of the *Wolbachia* endosymbiont of *Drosophila suzukii*. *Genome Announc.* **2013**, *1*, e00032-13. [CrossRef]
41. Klasson, L.; Walker, T.; Sebaihia, M.; Sanders, M.J.; Quail, M.A.; Lord, A.; Sanders, S.; Earl, J.; O'Neill, S.L.; Thomson, N.; et al. Genome Evolution of *Wolbachia* strain *w*Pip from the *Culex pipiens* Group. *Mol. Biol. Evol.* **2008**, *25*, 1877–1887. [CrossRef]
42. Salzberg, S.L.; Puiu, D.; Sommer, D.D.; Nene, V.; Lee, N.H. Genome announcement: Genome sequence of the *Wolbachia* endosymbiont of *Culex quinquefasciatus* JHB. *J. Bacteriol.* **2009**, *191*, 1725. [CrossRef]
43. Foster, J.; Ganatra, M.; Kamal, I.; Ware, J.; Makarova, K.; Ivanova, N.; Bhattacharyya, A.; Kapatral, V.; Kumar, S.; Posfai, J.; et al. The *Wolbachia* genome of *Brugia malayi* endosymbiont evolution within a human pathogenic nematode. *Public Libr. Sci. PLoS Biol.* **2005**, *3*, 599–614. [CrossRef] [PubMed]
44. Nikoh, N.; Hosokawa, T.; Moriyama, M.; Oshima, K.; Hattori, M.; Fukatsu, T. Evolutionary origin of insect-*Wolbachia* nutritional mutualism. *Proc. Natl. Acad. Sci. USA* **2014**, *111*, 10257–10262. [CrossRef] [PubMed]

45. Baldo, L.; Werren, J.H. Revisiting *Wolbachia* supergroup typing based on wsp: Spurious lineages and discordance with MLST. *Curr. Microbiol.* **2007**, *55*, 81–87. [CrossRef] [PubMed]
46. Huelsenbeck, J.P.; Ronquist, F. MrBayes: Bayesian inference of phylogenetic trees. *Bioinform. Appl. Note* **2001**, *17*, 754–755. [CrossRef] [PubMed]
47. Stamatakis, A. RAxML version 8: A tool for phylogenetic analysis and post-analysis of large phylogenies. *Bioinformatics* **2014**, *30*, 1312–1313. [CrossRef] [PubMed]
48. Rambaut, A. FigTree v1. 3.1. 2009. Available online: http://tree.bio.ed.ac.uk/software/figtree/ (accessed on 1 January 2018).
49. Conow, C.; Fielder, D.; Ovadia, Y.; Libeskind-Hadas, R. Jane: A new tool for the cophylogeny reconstruction problem. *Algorithms Mol. Biol.* **2010**, *5*, 16. [CrossRef]
50. Viale, E.; Martinez-Sanudo, I.; Brown, J.M.; Simonato, M.; Girolami, V.; Squartini, A.; Bressan, A.; Faccoli, M.; Mazzon, L. Pattern of association between endemic Hawaiian fruit flies (*Diptera*: *Tephritidae*) and their symbiotic bacteria: Evidence of cospeciation events and proposal of "*Candidatus Stammerula trupaneae*". *Mol. Phylogenet. Evol.* **2015**, *90*, 67–79. [CrossRef]
51. Revell, L.J. Phytools: An R package for phylogenetic comparative biology (and other things). *Methods Ecol. Evol.* **2012**, *3*, 217–223. [CrossRef]
52. Huelsenbeck, J.P.; Nielsen, R.; Bollback, J.P. Stochastic mapping of morphological characters. *Syst. Biol.* **2003**, *52*, 131–158. [CrossRef]
53. Kaneshiro, K.Y. Dynamics of sexual selection in the *Hawaiian Drosophilidae*: A paradigm for evolutionary change. *Proc. Hawaii. Entomol. Soc.* **2006**, *38*, 1–19.
54. Sintupachee, S.; Milne, J.R.; Poonchaisri, S.; Baimai, V.; Kittayapong, P. Closely related *Wolbachia* strains within the pumpkin arthropod community and the potential for horizontal transmission via the plant. *Microb. Ecol.* **2006**, *51*, 294–301. [CrossRef] [PubMed]
55. Dobson, S.L.; Bourtzis, K.; Braig, H.R.; Jones, B.F.; Zhou, W.; Rousset, F.; O'Neill, S.L. *Wolbachia* infections are distributed throughout insect stomatic and germline tissues. *Insect Biochem. Mol. Biol.* **1999**, *29*, 153–160. [CrossRef] [PubMed]
56. Chrostek, E.; Pelz-Stelinski, K.; Hurst, G.D.D.; Hughes, G.L. Horizontal transmission of intracellular insect symbionts via Plants. *Front. Microbiol.* **2017**, *8*, 2237. [CrossRef] [PubMed]
57. Kriesner, P.; Hoffmann, A.A.; Lee, S.F.; Turelli, M.; Weeks, A.R. Rapid sequential spread of two *Wolbachia* variants in *Drosophila simulans*. *PLoS Pathog.* **2013**, *9*, e1003607. [CrossRef] [PubMed]
58. Sanaei, E.; Charlat, S.; Engelstädter, J. *Wolbachia* host shifts: Routes, mechanisms, constraints and evolutionary consequences. *Biol. Rev.* **2021**, *96*, 433–453. [CrossRef] [PubMed]
59. Wang, X.; Xiong, X.; Cao, W.; Zhang, C.; Werren, J.H.; Wang, X. Phylogenomic analysis of Wolbachia strains reveals patterns of genome evolution and recombination. *Genome Biol. Evol.* **2020**, *12*, 2508–2520. [CrossRef] [PubMed]
60. Baldo, L.; Bordenstein, S.; Wernegreen, J.J.; Werren, J.H. Widespread recombination throughout *Wolbachia* genomes. *Mol. Biol. Evol.* **2006**, *23*, 437–449. [CrossRef]
61. Ilinsky, Y.; Kosterin, O.E. Molecular diversity of *Wolbachia* in Lepidoptera: Prevalent allelic content and high recombination of MLST genes. *Mol. Phylogenetics Evol.* **2017**, *109*, 164–179. [CrossRef]
62. Uy, K.L.; LeDuc, R.; Ganote, C.; Price, D.K. Physiological effects of heat stress on Hawaiian picture-wing *Drosophila*: Genome-wide expression patterns and stress-related traits. *Conserv. Physiol.* **2015**, *3*, cou062. [CrossRef]
63. Eldon, J.; Bellinger, M.R.; Price, D.K. Hawaiian picture-winged *Drosophila* exhibit adaptive population divergence along a narrow climatic gradient on Hawaii Island. *Ecol. Evol.* **2019**, *9*, 2436. [CrossRef]
64. Hague, M.T.; Caldwell, C.N.; Cooper, B.S. Pervasive effects of *Wolbachia* on host temperature preference. *MBio* **2020**, *11*, e01768-20. [CrossRef] [PubMed]
65. Zhu, Y.X.; Song, Z.R.; Zhang, Y.Y.; Hoffmann, A.A.; Hong, X.Y. Spider mites singly infected with either *Wolbachia* or *Spiroplasma* have reduced thermal tolerance. *Front. Microbiol.* **2021**, *12*, 706321. [CrossRef] [PubMed]
66. Ometto, L.; Cestaro, A.; Ramasamy, S.; Grassi, A.; Revadi, S.; Siozios, S.; Moretto, M.; Fontana, P.; Varotto, C.; Pisani, D.; et al. Linking genomics and ecology to investigate the complex evolution of an invasive *Drosophila* pest. *Genome Biol. Evol.* **2013**, *5*, 745–757. [CrossRef] [PubMed]
67. dos Santos, G.; Schroeder, A.J.; Goodman, J.L.; Strelets, V.B.; Crosby, M.A.; Thurmond, J.; Emmeret, D.B.; Gelbart, W.M.; FlyBase Consortium. FlyBase: Introduction of the *Drosophila melanogaster* Release 6 reference genome assembly and large-scale migration of genome annotations. *Nucleic Acids Res.* **2015**, *43*, D690–D697. [CrossRef] [PubMed]
68. Hu, T.T.; Eisen, M.B.; Thornton, K.R.; Andolfatto, P. A second-generation assembly of the *Drosophila simulans* genome provides new insights into patterns of lineage-specific divergence. *Genome Res.* **2013**, *23*, 89–98. [CrossRef] [PubMed]
69. Arensburger, P.; Megy, K.; Waterhouse, R.M.; Abrudan, J.; Amedeo, P.; Antelo, B.; Bartholomay, L.; Bidwell, S.; Caler, E.; Atkinson, P.W.; et al. Sequencing of *Culex quinquefasciatus* establishes a platform for mosquito comparative genomics. *Science* **2010**, *330*, 86–88. [CrossRef]
70. Miller, J.R.; Koren, S.; Dilley, K.A.; Puri, V.; Brown, D.M.; Harkins, D.M.; Thibaud-Nissen, F.; Rosen, B.; Chen, X.G.; Shabman, R.S.; et al. Analysis of the *Aedes albopictus* C6/36 genome provides insight into cell line utility for viral propagation. *Gigascience* **2018**, *7*, 1–13. [CrossRef]
71. Ronquist, F.; Huelsenbeck, J.P. MrBayes3: Bayesian phylogenetic inference under mixed models. *Bioinformatics* **2003**, *19*, 1572–1574. [CrossRef] [PubMed]

72. Ronquist, F.; Deans, A.R. Bayesian phylogenetics and its influence on insect systematics. *Annu. Rev. Entomol.* **2010**, *55*, 189–206. [CrossRef]
73. Xie, W.; Lewis, P.O.; Fan, Y.; Kuo, L.; Chen, M.-H. Improving marginal likelihood estimation for Bayesian phylogenetic model selection. *Syst. Biol.* **2011**, *60*, 150–160. [CrossRef] [PubMed]
74. Ronquist, F.; Huelsenbeck, J.; Teslenko, M. Draft MrBayes Version 3.2 Manual: Tutorials and Model Summaries. 2011. Available online: https://pbil.univ-lyon1.fr/members/perriere/pdf/mb3.2_manual.pdf (accessed on 1 January 2018).
75. Kass, R.E.; Raftery, A.E. Bayes factors. *J. Am. Stat. Assoc.* **1995**, *90*, 773–795. [CrossRef]
76. Huelsenbeck, J.; Rannala, B. Frequentist properties of Bayesian posterior probabilities of phylogenetic trees under simple and complex substitution models. *Syst. Biol.* **2004**, *53*, 53904–53913. [CrossRef] [PubMed]

genes

MDPI

Article

Comparative Mitogenomic Analyses of Darkling Beetles (Coleoptera: Tenebrionidae) Provide Evolutionary Insights into tRNA-like Sequences

Su-Hao Wang [1,2,3,†], Shi-Yun Hu [1,3,4,†], Min Li [1,2,3], Min Liu [1,3,4], Hao Sun [1,3,4], Jia-Rui Zhao [1,2,3], Wen-Ting Chen [1,2,3] and Ming-Long Yuan [1,2,3,4,*]

1 State Key Laboratory of Herbage Improvement and Grassland Agro-Ecosystems, Lanzhou University, Lanzhou 730020, China; wangsh21@lzu.edu.cn (S.-H.W.); hushy20@lzu.edu.cn (S.-Y.H.); limin2015@lzu.edu.cn (M.L.); mliu19@lzu.edu.cn (M.L.); sunh2021@lzu.edu.cn (H.S.); zhaojr21@lzu.edu.cn (J.-R.Z.); chenwt18@lzu.edu.cn (W.-T.C.)
2 Key Laboratory of Grassland Livestock Industry Innovation, Ministry of Agriculture and Rural Affairs, Lanzhou University, Lanzhou 730020, China
3 College of Pastoral Agricultural Science and Technology, Lanzhou University, Lanzhou 730020, China
4 National Demonstration Center for Experimental Grassland Science Education, Lanzhou University, Lanzhou 730020, China
* Correspondence: yuanml@lzu.edu.cn
† These authors equally contributed to this study.

Abstract: Tenebrionidae is widely recognized owing to its species diversity and economic importance. Here, we determined the mitochondrial genomes (mitogenomes) of three Tenebrionidae species (*Melanesthes exilidentata*, *Anatolica potanini*, and *Myladina unguiculina*) and performed a comparative mitogenomic analysis to characterize the evolutionary characteristics of the family. The tenebrionid mitogenomes were highly conserved with respect to genome size, gene arrangement, base composition, and codon usage. All protein-coding genes evolved under purifying selection. The largest non-coding region (i.e., control region) showed several unusual features, including several conserved repetitive fragments (e.g., A+T-rich regions, G+C-rich regions, Poly-T tracts, TATA repeat units, and longer repetitive fragments) and tRNA-like structures. These tRNA-like structures can bind to the appropriate anticodon to form a cloverleaf structure, although base-pairing is not complete. We summarized the quantity, types, and conservation of tRNA-like sequences and performed functional and evolutionary analyses of tRNA-like sequences with various anticodons. Phylogenetic analyses based on three mitogenomic datasets and two tree inference methods largely supported the monophyly of each of the three subfamilies (Stenochiinae, Pimeliinae, and Lagriinae), whereas both Tenebrioninae and Diaperinae were consistently recovered as polyphyletic. We obtained a tenebrionid mitogenomic phylogeny: (Lagriinae, (Pimeliinae, ((Tenebrioninae + Diaperinae), Stenochiinae))). Our results provide insights into the evolution and function of tRNA-like sequences in tenebrionid mitogenomes and contribute to our general understanding of the evolution of Tenebrionidae.

Keywords: Tenebrionidae; mitochondrial genome; tRNA-like sequence; control region; phylogenetic analysis

Citation: Wang, S.-H.; Hu, S.-Y.; Li, M.; Liu, M.; Sun, H.; Zhao, J.-R.; Chen, W.-T.; Yuan, M.-L. Comparative Mitogenomic Analyses of Darkling Beetles (Coleoptera: Tenebrionidae) Provide Evolutionary Insights into tRNA-like Sequences. *Genes* **2023**, *14*, 1738. https://doi.org/10.3390/genes14091738

Academic Editor: Zhiteng Chen

Received: 16 June 2023
Revised: 25 August 2023
Accepted: 25 August 2023
Published: 30 August 2023

1. Introduction

Tenebrionidae is a large family of Coleoptera, with more than 20,000 extant species in eight to ten subfamilies [1–3]. Species of Tenebrionidae are widely distributed worldwide and are extremely well-adapted to their environments [4]. Complex feeding strategies and habitats lead to diverse morphological features and economic impacts of species in this family. Many tenebrionid species are agricultural pests; for example, *Tribolium* spp. are destructive cosmopolitan pests of stored flour, corn, peanuts, and other dried agricultural products [5]; however, its larvae are rich in protein and can be used as feeds for pets, such as reptiles and birds [6]. Some tenebrionid species are also valuable for ecological improvement. For example, *Tenebrio molitor* Linnaeus and *Uloma* spp. can biodegrade polystyrene (PS) by acting synergistically with intestinal microbiota [7,8].

During the past few decades, the phylogenetic relationships within Tenebrionidae have been investigated extensively based on both morphological [2,3,9,10] and molecular data [1,4,11–14]. However, the structure and relationships among many subfamilies of Tenebrionidae are not well-resolved [14]. Hunt et al. (2007) reconstructed a large-scale beetle phylogeny; however, Lagriinae and Pimeliinae did not group with the remaining subfamilies in Tenebrionidae [15]. Kergoat et al. (2014) utilized maximum likelihood to construct the Tenebrionidae phylogeny and found that each of the three subfamilies (Lagriinae, Alleculinae, and Stenochiinae) were monophyletic, whereas Pimeliinae was paraphyletic [16]. A recent phylogenetic analysis based on 37 species showed that Tenebrioninae and Diaperinae were polyphyletic, while the monophyly of four subfamilies (Lagriinae, Alleculinae, Stenochiinae, and Pimeliinae) could not be resolved using RAxML and MrBayes [17]. Further studies of phylogenetic relationships within Tenebrionidae are needed.

The tRNA-like structure was first identified in the RNA genome of *turnip yellow mosaic virus* [18]. Most tRNA-like structures have been found in plant viruses and their functions have been studied in depth [19–21]. However, some have been found in animals, such as a tRNA-like structure in human immunodeficiency virus type 1 (HIV-1) that can serve as a primer for reverse transcription of the HIV-1 genome [22]. In addition to viruses, tRNA-like structures have been detected in RNAs from many different species [23–25]. However, few tRNA-like sequences have been reported in insect mitochondrial genomes (mitogenomes) [26–28], and their functional and evolutionary significance are not well studied.

The insect mitogenome is a compact double-stranded circular molecule of 15–18 kb that encodes 37 genes, including 13 protein-coding genes (PCGs), two ribosomal RNA genes (rRNAs), and 22 transfer RNA genes (tRNAs) [29,30], as well as a large non-coding region (i.e., the control region, also termed the A+T-rich region in insects), which contains the origin of replication [31]. Owing to its simple genetic structure, matrilineal inheritance, and high copy number, insect mitogenomes are commonly used for phylogenetics, molecular evolutionary, and conservation genetics research [11,29,32]. Mitogenome sequences of 53 tenebrionid species have been deposited in GenBank (as of February 2023). However, several mitogenomes were not validated and lacked some genes or A+T-rich regions. Additional tenebrionid mitogenomes will be helpful for phylogenetic and comparative mitogenomic analyses of Tenebrionidae, especially for clarifying the evolutionary features of mitochondrial non-coding regions.

In this study, we sequenced and annotated the complete mitogenomes of three tenebrionid species, i.e., *Melanesthes exilidentata* Ren, *Anatolica potanini* Reitter, and *Myladina unguiculina* Reitter. We performed a comparative mitogenomic analysis, including analyses of the nucleotide composition, codon usage, tRNA secondary structure, and evolutionary rates of 13 PCGs. We also evaluated phylogenetic relationships among six subfamilies within Tenebrionidae based on three mitogenomic datasets and two tree inference methods. In addition, we comprehensively analyzed the structural features of A+T-rich regions, with a focus on the evolutionary characteristics of tRNA-like structures. The results of this study provide new insights into the phylogeny and evolution of Tenebrionidae from a mitogenomic perspective.

2. Materials and Methods

2.1. Sampling and DNA Extraction

Adult specimens of *M. exilidentata*, *A. potanini*, and *M. unguiculina* were collected from Lingwu City, Ningxia Hui Autonomous Region. Detailed sampling information is shown in Table 1. Samples and voucher specimens have been deposited in the College of Pastoral Agricultural Science and Technology, Lanzhou University, Lanzhou, China. All specimens were initially preserved in 100% ethanol at the sampling site and then transferred to −80 °C until used for DNA extraction. Total genomic DNA was extracted from a single specimen using a DNeasy Tissue Kit (Qiagen, Hilden, Germany) according to the manufacturer's protocols. The brief protocols were as follows: approximately 25 mg tissue were placed in a 1.5 mL microcentrifuge tube and ground with liquid nitrogen; 180 μL buffer ATL and 20 μL Proteinase K were added to the tube, and then incubated at 56 °C for 2 h; 200 μL buffer AL was added to the tube and mixed thoroughly by vortexing; the mixture was pipetted into the DNeasy Mini spin column and placed in a 2 mL collection tube and centrifuged at 8000× *g* for 1 min; the DNeasy Mini spin column was placed in a new 2 mL collection tube, 500 μL buffer AW1 was added, and centrifuged for 1 min at 8000× *g*; the DNeasy Mini spin column was placed in a new 2 mL collection tube, 500 μL buffer AW2 was added and then centrifuged for 3 min at 20,000× *g* to dry the DNeasy membrane; the DNeasy Mini spin column was placed in a clean 2 mL microcentrifuge tube, 200 μL buffer AE was pipetted directly onto the DNeasy membrane, then incubated at room temperature for 2 min, and finally centrifuged for 1 min at 8000× *g*. The quality of the extracted DNA was evaluated by 1.2% agarose gel electrophoresis and spectrophotometry with a NanoDrop ND-1000 (Thermo Scientific, Waltham, MA, USA).

Table 1. Sampling information of the three Tenebrionidae species that were newly sequenced in this study.

Subfamily	Species	Sampling Site	Voucher Specimen Code	Coordinate
Tenebrionidae	*M. exilidentata*	Lingwu County, Ningxia Province, China	YCLW-BJ1	106°34′ E, 38°10′ N
Tenebrionidae	*A. potanini*	Lingwu County, Ningxia Province, China	YCLW-BJ2	106°34′ E, 38°10′ N
Tenebrionidae	*M. unguiculina*	Lingwu County, Ningxia Province, China	YCLW-BJ3	106°34′ E, 38°10′ N

2.2. Mitogenome Sequencing, Assembly, and Annotation

The mitogenomes of the three species were sequenced using the Illumina NovaSeq 6000 (2 × 150 bp) platform by the Wuhan Benagen Tech Solutions Company Limited (Wuhan, China). Low-quality reads were removed using SOAPnuke (version: 2.1.0) [33] and high-quality reads were assembled using SPAdes (version: 3.13.0) [34]. The three assembled mitogenomes were annotated with the MITOS web server (http://mitos2.bioinf.uni-leipzig.de; accessed on 12 March 2022) [35] to identify 13 PCGs, 22 tRNAs, and two rRNAs using the mitogenomes from Tenebrionidae in GenBank as references. To avoid overlap between genes, all genes were manually verified and proofread after annotation. Then, the online Tandem Repeats Finder web tool (version: 4.09) (https://tandem.bu.edu/trf/trf.html; accessed on 20 May 2022) was used to predict tandem repeats in the control regions. The three newly obtained mitogenome sequences have been deposited in NCBI (GenBank accession numbers OQ534869, OQ511732, and OQ536315).

2.3. Comparative Mitogenomic Analysis of Tenebrionidae

Thirty-three Tenebrionidae mitogenomes were used to compare the full-length, size-differentiated variation and to calculate the genome-level base composition. MEGA 11 [36] was used to calculate the base composition of each PCG, including the overall composition and values at the first, second, and third loci. The number of codons and relative synonymous codon usage (RSCU) of the 13 PCGs of each species were calculated using MEGA 11, and results for each species were output individually and used to analyze differences in the most-used codons with respect to the A+T content. Asymmetry was evaluated using the following formula: AT-skew = (A − T)/(A + T), GC-skew = (G − C)/(G + C) [37]. The effective codon number (ENC) and codon bias index (CBI) were determined for each Tenebrionidae mitogenome using DnaSP [38]. To compare codon usage bias among these 34 species, the correlations between ENC, CBI, G+C content of all codons, and G+C content at the third codon position (GC3) were evaluated. Based on the ENC and GC3 for each species, a parabolic curve was fitted to ENC, providing a basis for evaluating whether codon usage was subject to natural selection [39]. We analyzed the control (AT-rich) and non-coding regions of the 33 tenebrionid species, focusing on their specific features (tRNA-like structures) and potential evolutionary significance. We also calculated the rates of synonymous substitutions (K_s) and non-synonymous substitutions (K_a) for each PCG using MEGA 11 [36].

2.4. Phylogenetic Analysis

Mitogenome sequences of 33 tenebrionid species and one outgroup (Meloidae) were used for a phylogenetic analysis. Details of the species included in this study are given in Table S1. All 13 PCGs were individually aligned by ClustalW (Codons) and each of the two rRNAs (*rrnL* and *rrnS*) were aligned by ClustalW, implemented in MEGA 11 [36]. Three mitogenome datasets were constructed for the phylogenetic analysis: (1) the P123 dataset, with nucleotide sequences for all codon positions of 13 PCGs, including 11,172 nucleotides; (2) the P123AA dataset, with inferred amino acid sequences of 13 PCGs, including 3724 amino acids; and (3) the P123RNA dataset, with P123 and nucleotide sequences of two rRNAs, comprising 13379 nucleotides. DAMBE 5.3.74 [40] was used for substitution saturation tests, revealing that none of the 13 PCGs had significant substitution saturation (Table S2). The IQ-TREE web server (http://iqtree.cibiv.univie.ac.at/; accessed on 2 July 2022) was used to determine the best partitioning schemes and corresponding nucleotide substitution model for each of the three datasets, and the results (Table S3) were used for the phylogenetic analysis.

We used RAxML-HPC2 [41] to construct maximum likelihood (ML) phylogenetic trees of Tenebrionidae, with the GTRGAMMAI model and 1000 bootstrap replicates (BS). Additionally, MrBayes 3.2.7 [42] was used for Bayesian inference (BI), employing four independent Markov chains, including three heated chains and one cold chain, and running 1×10^8 generations concurrently. When the estimated sample size was greater than 100 and the potential size reduction factor was close to 1.0, chain convergence was achieved [42].

3. Results

3.1. General Characteristics of Tenebrionidae Mitogenomes

The mitogenomes were 15542 bp in *M. unguiculina*, 15804 bp in *M. exilidentata*, and 15,809 bp in *A. potanini* (Figure 1), and all were typical closed-loop DNA molecules containing 37 genes (13 PCGs, 22 tRNAs, and 2 rRNAs) and containing a large non-coding region. The mitogenomes of all three species had a typical invertebrate mitochondrial gene arrangement. Fourteen genes (four PCGs, eight tRNAs, and two rRNAs) were transcribed from the minority strand (N strand), leaving 23 genes (nine PCGs and 14 tRNAs) transcribed from the majority strand (J strand). In all three Tenebrionidae mitogenomes, there was substantial gene overlap at *atp6*/*atp8* (7 bp) and *nad4*/*nad4L* (7 bp or 8 bp) (Table S4), and many intergeneric regions (generally ≤3 bp) were widely distributed. A large intergeneric space (16 bp to 20 bp) was found between *trnS2* (UCN) and *nad1* in all three Tenebrionidae mitogenomes. In addition, *M. exilidentata* and *A. potanini* had a large intergeneric space (47 bp) between *trnC* and *trnW*, while *M. unguiculina* had a 19 bp intergeneric space between *nad5* and *trnH* (Table S4). Except *trnS1*, which lacked the dihydrouridine (DHU) arm, the remaining 21 tRNAs possessed a canonical cloverleaf secondary structure, consisting of four arms. The anticodon arms and DHU stems of all tRNAs were highly conserved, with most substitutions and differences present in the DHU loop, the pseudouridine (TψC) arm, and the variable loop (Figure 2).

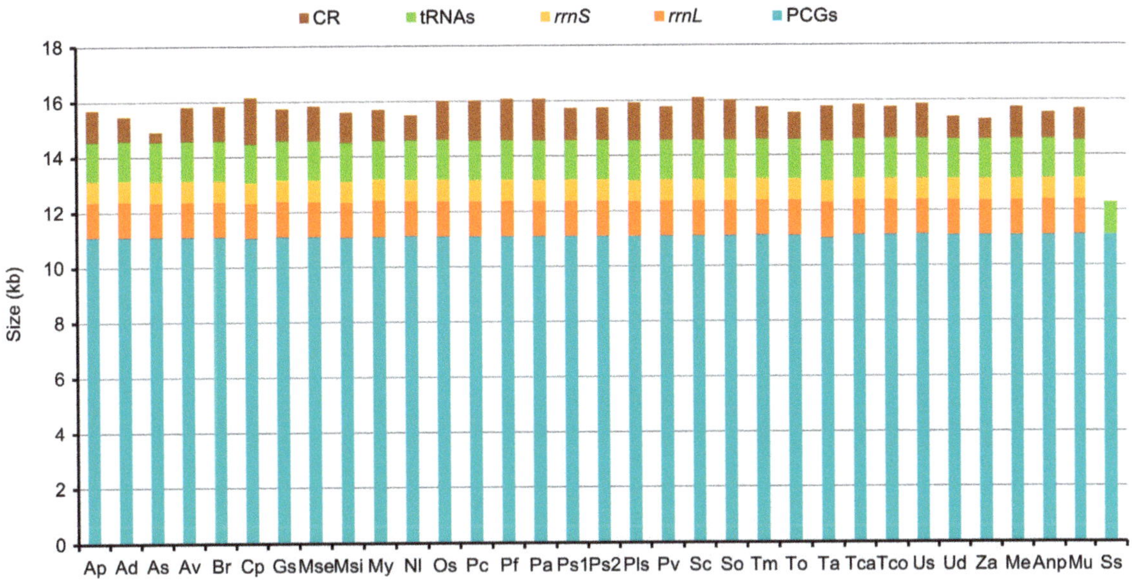

Figure 1. Sizes of PCGs, tRNAs, *rrnL*, *rrnS*, and CR in Tenebrioidae mitogenomes. Species are abbreviated as follows: *Adelium* sp, Ap; *Alphitobius diaperinus*, Ad; *Amarygmini* sp, As; *Asbolus verrucosus*, Av; *Blaps rhynchoptera*, Br; *Cerogria popularis*, Cp; *Gonocephalum* sp, Gs; *Machla setosa*, Mse; *Morphostenophanes sinicus*, Msi; *Morphostenophanes yunnanus*, My; *Nalassus laevioctostriatus*, Nl; *Opatrum sabulosum*, Os; *Pelecyphorus contortus*, Pc; *Pelecyphorus foveolatus*, Pf; *Philolithus aegrotus*, Pa; *Philolithus* sp1, Ps1; *Philolithus* sp2, Ps2; *Platydema* sp, Pls; *Promethis valgipes*, Pv; *Stenomorpha consobrina*, Sc; *Stenomorpha obovata*, So; *T. molitor*, Tm; *Tenebrio obscurus*, To; *Tribolium audax*, Ta; *Tribolium castaneum*, Tca; *Tribolium confusum*, Tco; *Uloma* sp, Us; *Ulomoides dermestoides*, Ud; *Zophobas atratus*, Za; *M. exilidentata*, Me; *A. potanini*, Anp; *M. unguiculina*, Mu; *Strongylium suspicax*, Ss.

Apart from *nad1* in three Tenebrionidae mitogenomes, all PCGs started with typical ATN codons, while all PCGs ended with complete termination codons (TAA/TAG) or truncated codons (TA or T) (Table S4). Among the 13 PCGs in the three mitogenomes, *cox1* had the lowest A+T content (65.43–65.70%) and *atp8* had the highest (79.49–81.41%). A component analysis of the 13 PCGs from 33 species showed that the third codon position had the highest A+T content, while the first and second codon positions had lower A+T contents (Figure 3). K_a/K_s values of 13 PCGs in the 33 tenebrionid species were less than 1.0 (Figure S1); *nad6* (0.37) and *atp8* (0.36) showed the highest K_a values, while *cox1* (0.05) had the lowest. There was a negative correlation between the K_a/K_s values and the G+C content of 13 PCGs (R^2 = 0.86, $p < 0.01$; Figure S2).

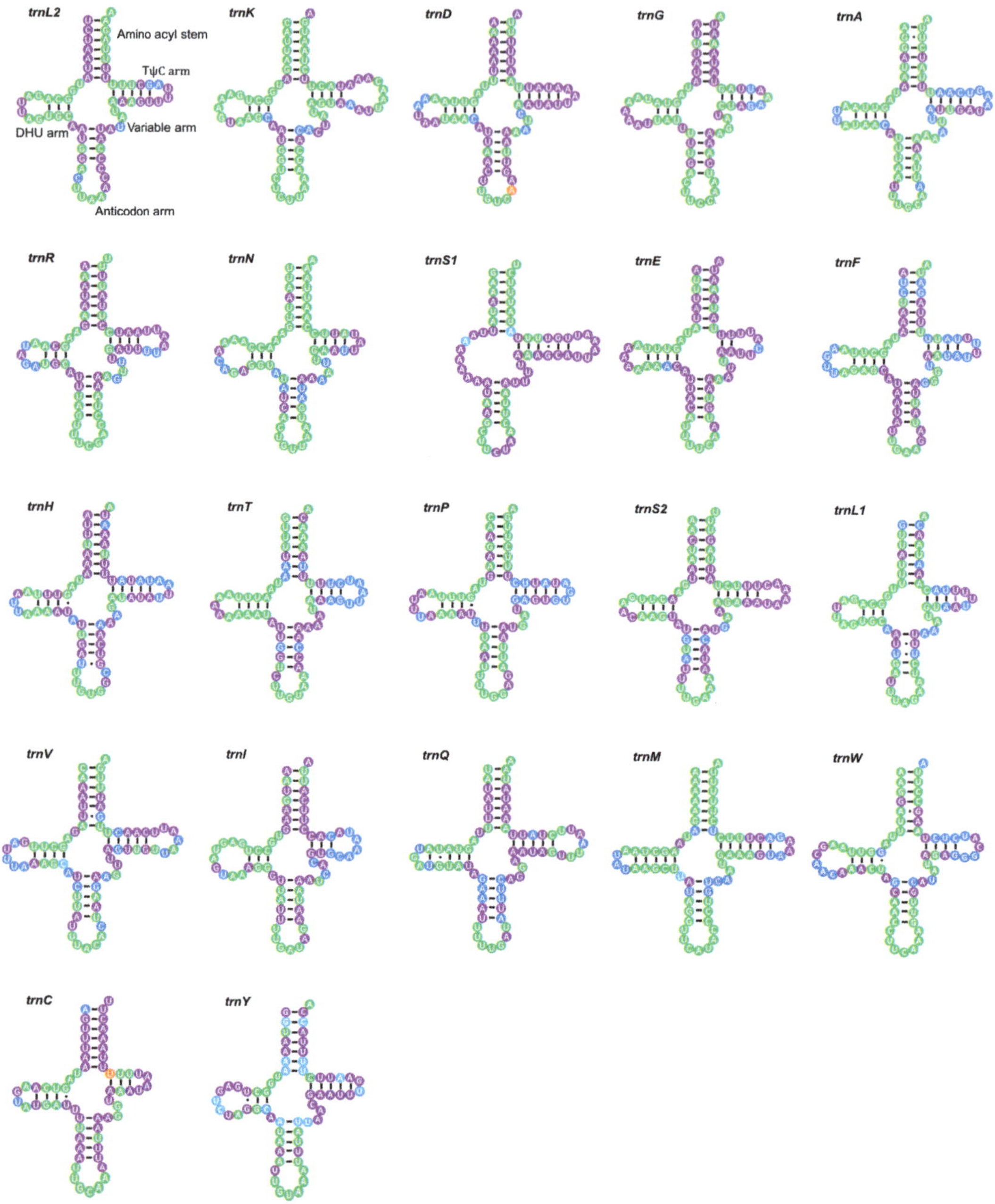

Figure 2. Putative secondary structures of 22 tRNA genes identified in the mitogenome of *M. unguiculina*. All tRNAs are shown in the order of occurrence in the mitogenome starting from *trnL2*. Bars indicate Waston-Crick base pairings, and dots between G and U pairs mark canonical base pairings in tRNA. Different colors represent different levels of base conservation. Green indicates conservation among all Tenebrionidae, orange indicates conservation among subfamilies, purple indicates conservation among the three newly sequenced species, and blue indicates a lack of conservation among the three newly sequenced species.

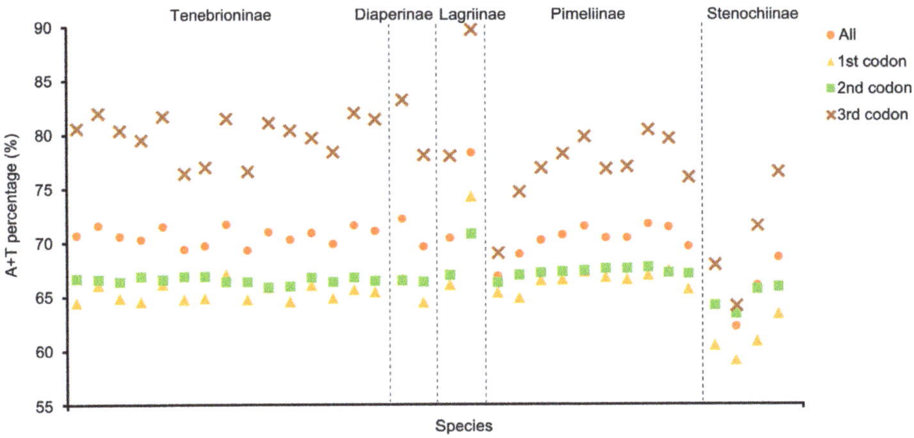

Figure 3. A+T% of the mitochondrial protein-encoding genes in the three groups within the 33 Tenebrionidae species. The dashed line separates the different subfamilies.

3.2. Nucleotide Composition and Codon Usage

The nucleotide compositions of the three new Tenebrionidae species were biased toward A and T, with A+T contents ranging from 71.0% to 71.5% (Figure 3). AT-skew ranged from 0.108 in *M. exilidentata* to 0.115 in *A. potanini*, while GC-skew ranged from −0.273 in *M. unguiculina* to −0.241 in *A. potanini* (Figure 4). The high A+T content and nucleotide bias in the mitogenomes were also reflected in the codon usage of PCGs. An RSCU analysis revealed that four AT-rich codons, i.e., AUU(I), UUU(F), UUA(L), and AUA(M), were used most frequently (Table S5). Except for the *M. unguiculina* mitogenome, which used 61 codons, both *M. exilidentata* and *A. potanini* used all 62 invertebrate codons.

To further investigate codon usage bias, the ENC and G+C content were analyzed for 13 PCGs of all 33 tenebrionid species. The mean ENC value for all species was 44.03, with a range of 33.89–54.51, indicating a strong codon bias. Positive correlations were found between the ENC of all codons and GC content of the third codon positions (GC3) ($R^2 = 0.97$, $p < 0.01$) (Figure 5A) and G+C content ($R^2 = 0.76$, $p < 0.01$) (Figure 5C). In addition, negative correlations were found between CBI and GC3 ($R^2 = 0.96$, $p < 0.01$) (Figure 5B), G+C content at all positions (GC) ($R^2 = 0.94$, $p < 0.01$) (Figure 5D), and ENC ($R^2 = 0.94$, $p < 0.01$) (Figure 5E). The ENC values for all Tenebrionidae species were below the ENC curve (Figure 6A) and no significant correlation ($R^2 = 0.61$, $p > 0.05$) was found between GC12 and GC3 (Figure 6B).

3.3. Non-Coding Region

The mitochondrial control region (A+T-rich region), the largest non-coding region, was located between *rrnS* and *trnI* (Figure 7), with a length of 930 bp in *M. unguiculina* and 1154 bp in both *M. exilidentata* and *A. potanini*. All 33 sequenced tenebrionid mitogenomes included a control region, with a maximum length difference of 1326 bp (Figure 7). The A+T content of control regions ranged from 80.59% in *B. rhynchoptera* Fairmaire to 94.34% in *T. molitor*. Several features were found in the control region of the 33 tenebrionid species, including A+T-rich regions (83.51–88.97%), G+C-rich regions, Poly-T stretches, TATA repeat units, as well as longer repeat fragments (1–4 repeats).

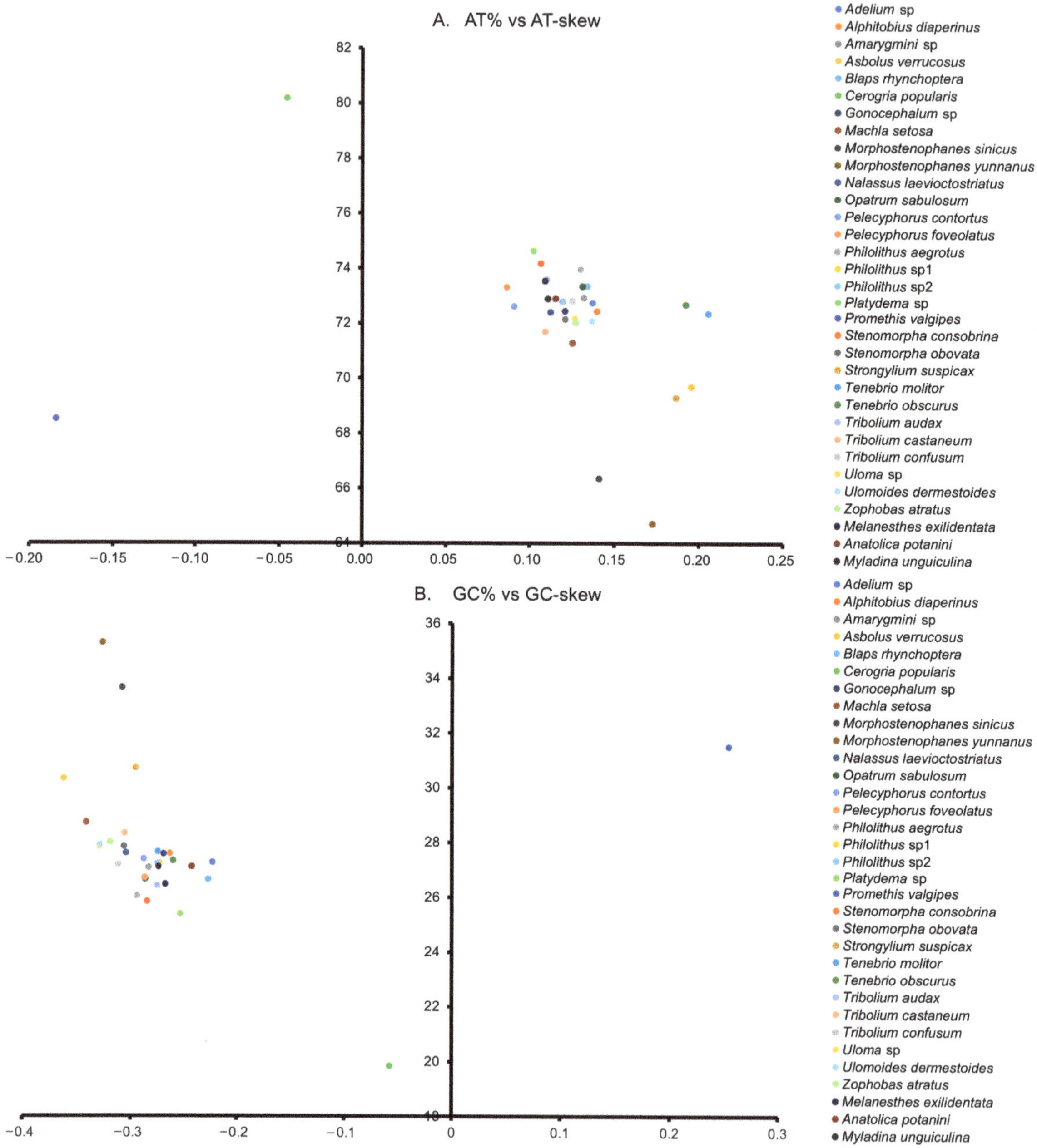

Figure 4. AT% vs. AT-skew (**A**) and GC% vs. GC-skew (**B**) in the 33 mitogenomes of the Tenebrionidae and one mitogenome of the outgroup. Measured in bp percentage (Y-axis) and level of nucleotide skew (X-axis). Values are calculated on J-strands for full-length mitochondrial genomes.

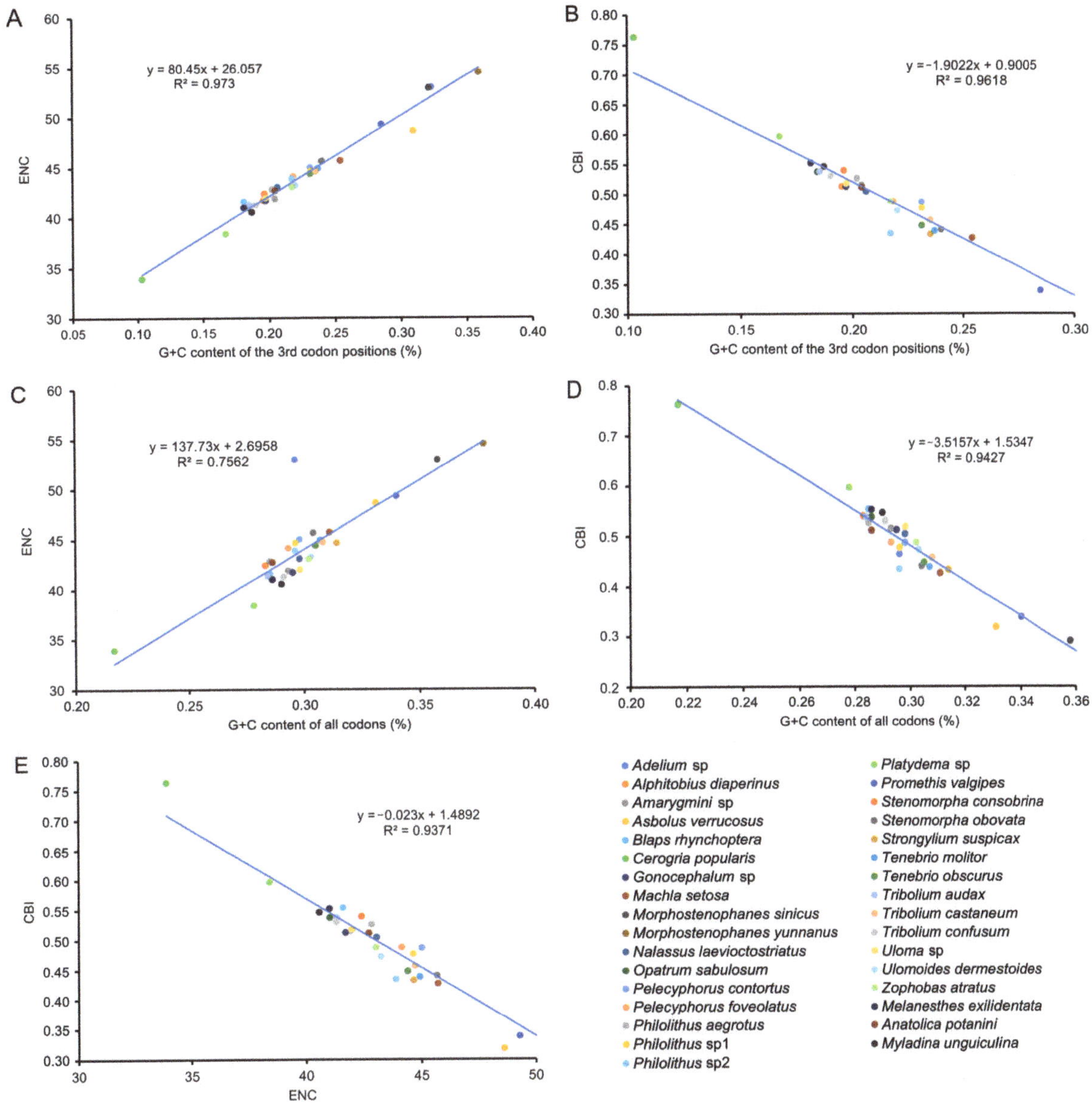

Figure 5. Evaluation of codon bias in the mitochondrial genomes of 33 Tenebrionidae species. (**A**) the correlation between ENC (effective number of codons) and the G + C content of the 3rd codon positions. (**B**) the correlation between CBI (codon bias index) and the 3rd codon positions. (**C**) the correlation between ENC and the G + C content of all codons. (**D**) the correlation between CBI and the G + C content of all codons. (**E**) the correlation between ENC and CBI.

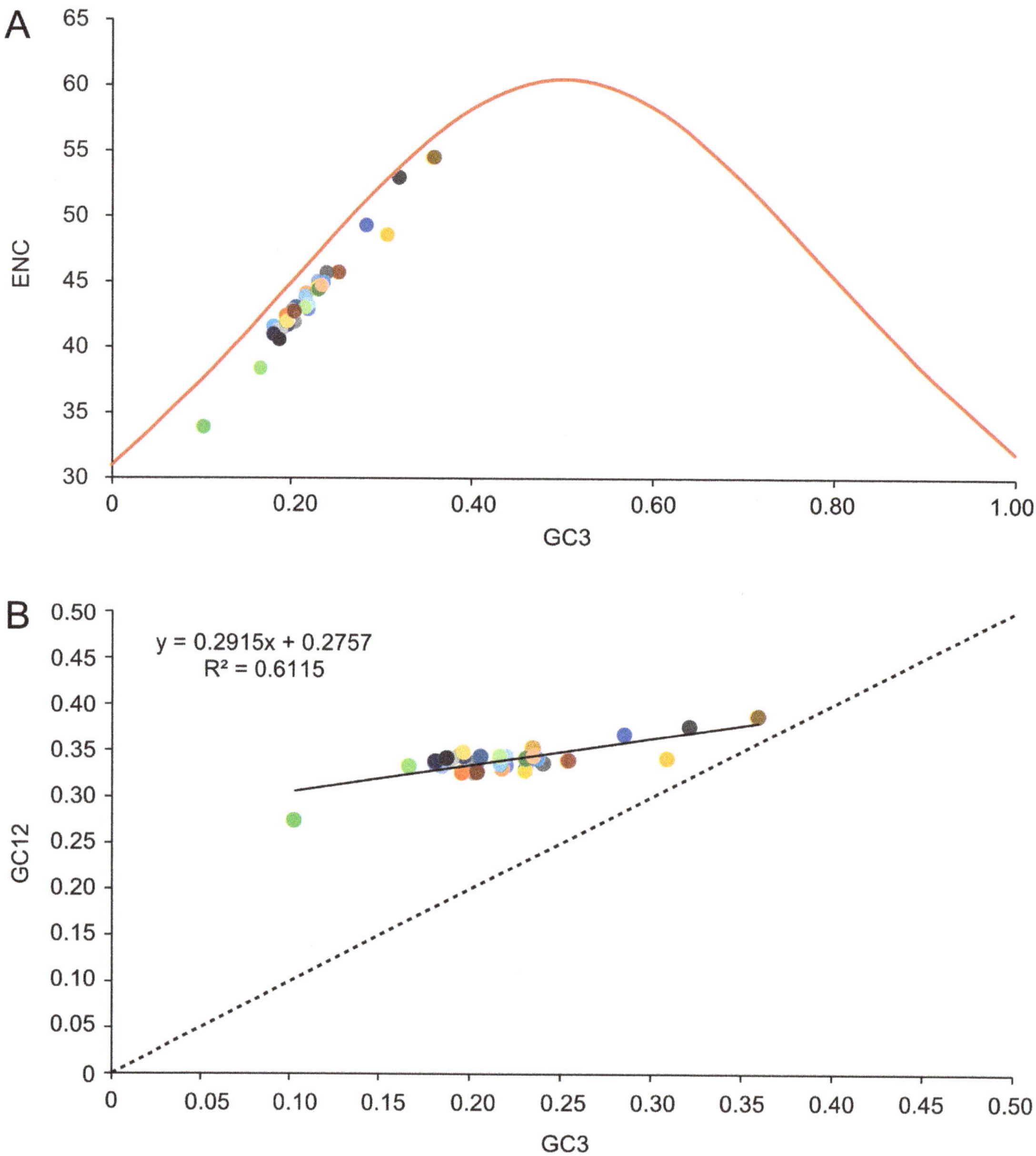

Figure 6. (**A**) Correlations between the effective number of codons (ENC) and G+C content of the third codon positions (GC3) for the 33 Tenebrionidae species. The solid line represents the relationship between the ENC* (2 + GC3 + (29/[(GC3)$^{-2}$ + (1 − GC3)$^{-2}$]) and the GC3. (**B**) The solid line represents the relationship between the GC3 and G+C content of the first and second positions (GC12), whereas the dotted line indicates y = x. Each color dot represents a tenebrionid species and colors match those in Figure 4.

Figure 7. Organization of the control regions in the Tenebrionidae mitogenomes. Location and copy number of tandem repeats are shown in orange and three colors (pink, purple, and grey) with Arabic numbers inside. Sky blue boxes represent interval sequences (positive numbers) or overlap (negative numbers) between two elements. Green boxes represent TA tandem repeat sequences. See Figure 1 for the full names of species.

We also found tRNA-like structures in the tenebrionid control regions. One, two, and three tRNA-like structures were detected in *M. exilidentata*, *M. unguiculina*, and *A. potanini*. Among the 33 tenebrionid species used for phylogenetic analyses, 22 species had at least one tRNA-like structure (Table S6), and the number of tRNA-like structures ranged from 1 to 9 (Table S7). The two most abundant tRNA-like structures were *trnL2*-like (nine) and *trnW*-like (seven), followed by the *trnV*-like and *trnE*-like structures, with five and four occurrences, respectively. The remaining tRNA-like sequences were only discovered in one to two cases (Table S7). These two tRNA-like structures (*trnL2*-like and *trnW*-like) were commonly found in tenebrionid species belonging to five subfamilies (Table S7).

Only three Pimeliinae species possessed both *trnL2*-like and *trnW*-like. These tRNA-like structures possessed the appropriate anticodons but generally lacked conserved base pairing arms, forming a cloverleaf secondary structure. We predicted several tRNA-like structures matches of greater than 65% with the corresponding conventional tRNAs (Figure 8). These had a relatively stable cloverleaf structure and more complete base pairing. Furthermore, tRNA-like sequences with less than a 65% match were aligned with the equivalent conventional tRNAs by tagging the same anticodon (Figure S3).

In addition to the control region, we identified several spacer regions of different lengths, including a highly conserved intergenic spacer sequence between *trnS2* and *nad1*, which was found in all 33 Tenebrionidae species and was 15–19 bp in size, with three highly conserved fragments (motif I, motif II, and motif III) (Figure S4).

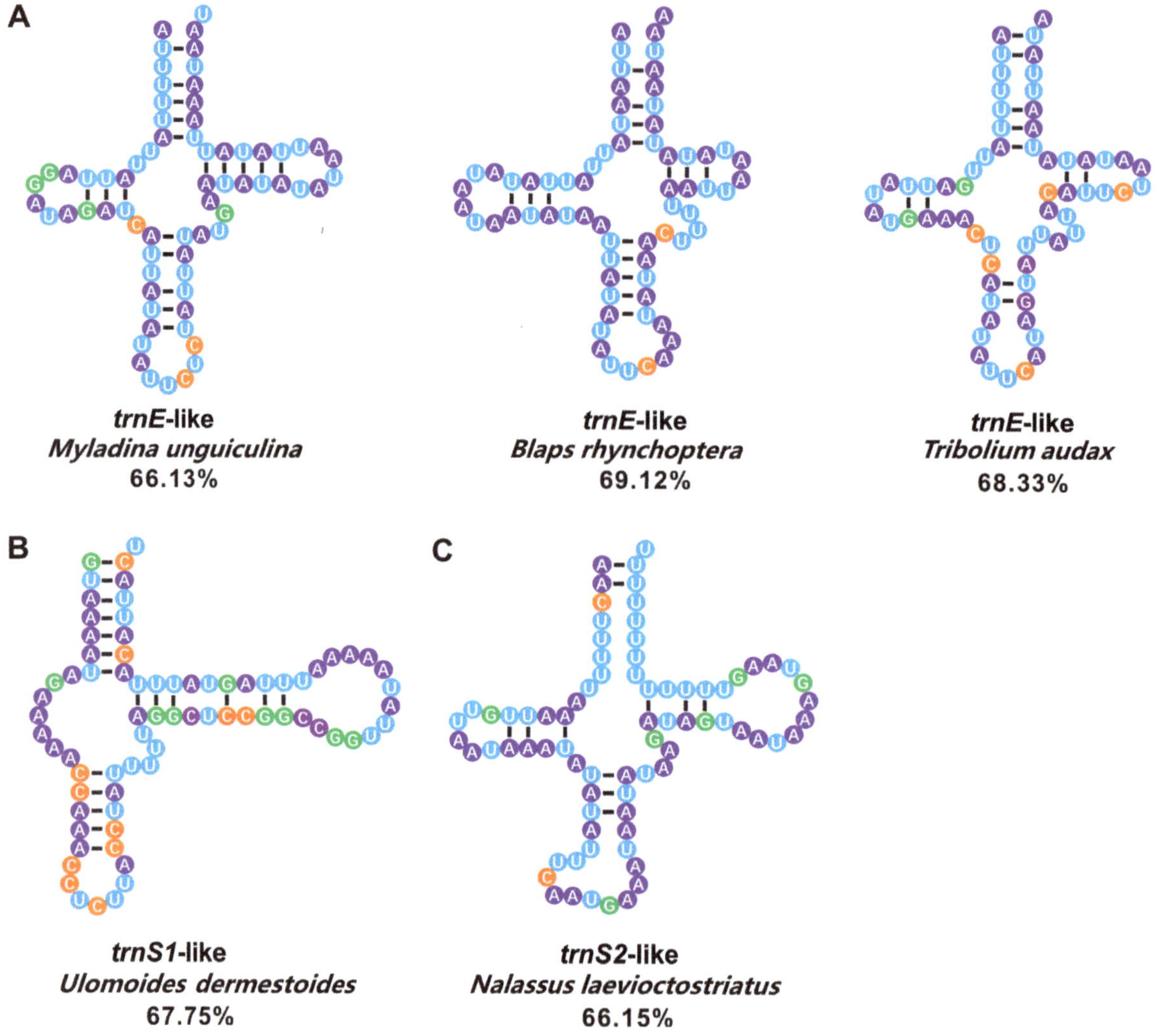

Figure 8. Five tRNA-like secondary structures with a match of 65% or more were selected from the A+T-rich regions of 33 Tenebrionidae species, which were involved in the construction of the phylogenetic tree. (**A**) Three *trnE*-like sequences. (**B**) one *trnS1*-like sequence. (**C**) one *trnS2*-like sequence. The tRNAs are labeled with the abbreviations of their corresponding amino acids. A dash (-) indicates Watson-Crick base-pairing. Blue, purple, green, and orange represent U, A, G, and C bases, respectively. Percentages represent the consensus sequences of tRNA-like sequences compared to each corresponding conventional tRNA.

3.4. *Mitochondrial Phylogeny of Tenebrionidae*

Three mitogenomic datasets (P123, P123AA, and P123RNA) and two analytical approaches (BI and ML) yielded six phylogenetic trees (Figures 9 and S5), involving 33 tenebrionid species in five subfamilies. Generally, BI trees had higher support values than those of ML trees (Figures 9 and S5). All analyses yielded similar phylogenetic relationships. However, seven species (*Uloma* sp., *U. dermestoides* Chevrolat, *Z. atratus*, *T. molitor*, *B. rhynchoptera*, and two Lagriinae species) showed unstable phylogenetic positions among trees (Figure S5).

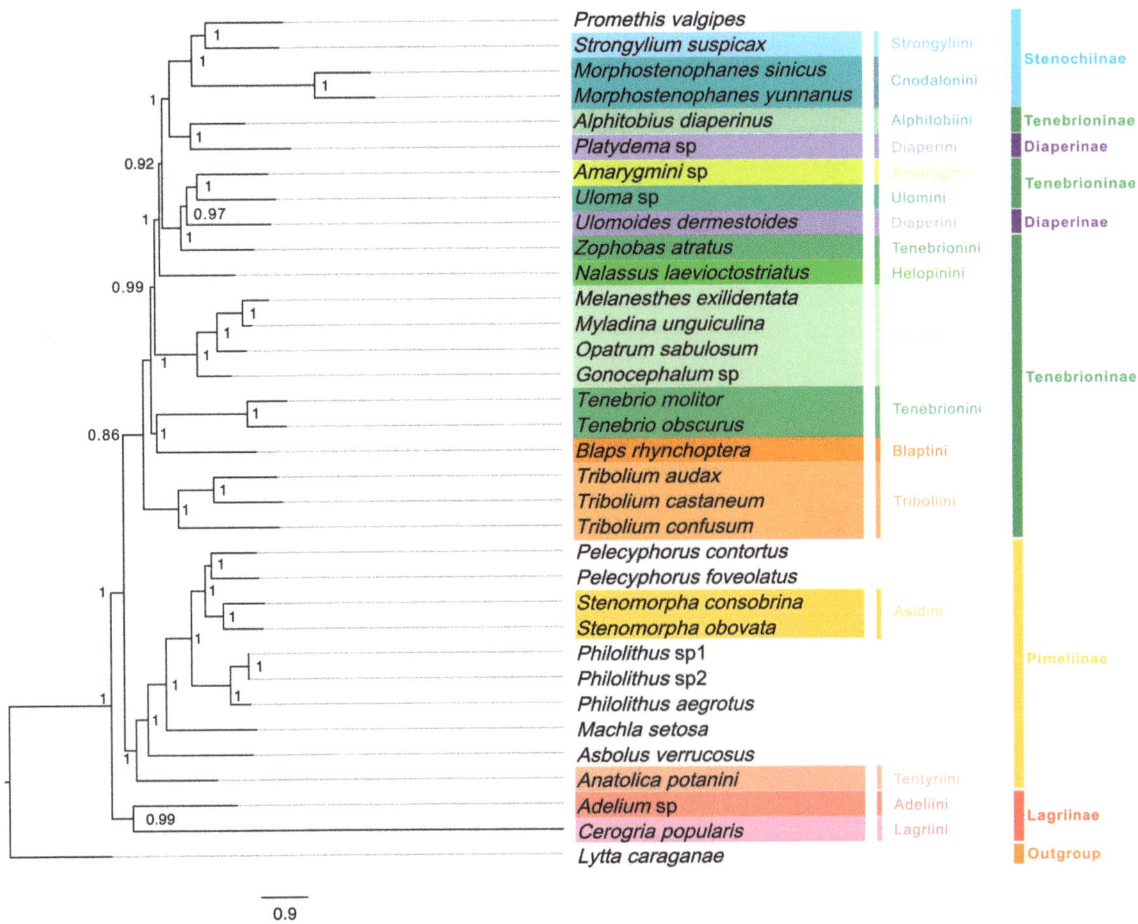

Figure 9. Mitogenome-based phylogenetic relationships among 33 Tenebrionidae species based on P123 datasets (protein-coding genes [PCGs] with all codon positions) using Bayesian inference (BI). Posterior probabilities for BI are shown on corresponding nodes in the topology of the BI tree. Different colors of species name blocks represent the different species, tribes, and subfamilies.

Among the five subfamilies, the phylogenies supported the monophyly of Lagriinae (PP = 1, BP > 57), Stenochiinae (PP = 1, BP ≥ 93), and Pimeliinae (PP = 1, BP ≥ 83), while Diaperinae and Tenebrioninae were polyphyletic (Figure 9). At the subfamily level, BI and ML trees based on dataset P123AA were concordant; however, the different positions of *Adelium* sp. and *C. popularis* Borchmann based on datasets P123 and P123RNA (Figure S5) did not support the monophyly of Lagriinae. For the eight tribes within Tenebrioninae, Tenebrionini was not monophyletic. Three phylogenetic groups were obtained for Tenebrioninae: the first included Alphitobiini (*A. diaperinus* Panzer), the second group included Amarygmi (*Amarygmini* sp.) and Ulomini (*Uloma* sp.), and the third group included five tribes was paraphyletic. At the tribe level, Cnodalonini of Stenochiinae (PP = 1, BP ≥ 93), Opatrini (PP = 1, BP = 100), and Triboliini (PP = 1, BP = 100) of Tenebrioninae, and Asidini of Pimiliinae (PP = 1, BP = 100) were monophyletic.

4. Discussion

4.1. Mitochondrial Genome Organization and Composition

The three newly sequenced mitogenomes of darkling beetles included the typical gene contents reported for other Tenebrionidae mitogenomes. Among the 33 Tenebrionidae species, *P. valgipes* (Marseul) and *C. popularis* both had aberrant AT-skews and GC-skews. The positive GC-skew and negative AT-skew in *P. valgipes* as well as the relatively high GC-skew in *C. popularis* indicated that there was a clear bias towards the utilization of G and C in both species. Ts also occurred more frequently in the backbone than As in both species. In most chains of other insects, there are fewer T than A and fewer G than C bases [43,44], suggesting that the chain asymmetry was reversed [45]. Furthermore, the AT-skew and GC-skew of *P. valgipes* were abnormally positive and negative, suggesting that the chain asymmetry of its mitogenome is reversed.

Several codons with high RSCU values ended in A/T, and codons ending in A/T were used more frequently than those ending in G/C, suggesting a codon bias, as determined by RSCU analyses of 33 Tenebrionidae species. Similar results have been obtained in previous studies [17,46,47]. On the basis of their GC content in relation to their synonyms, codons respond to genome composition according to Robin's theories of neutral mutations [48]. The A+T content at the third codon position was higher than those of the first and second codons in the mitogenomes of all Tenebrionidae species, indicating that the third codon was more vulnerable to AT alterations [49]. We discovered that species in the subfamily Stenochiinae had lower A+T contents than those of species in other subfamilies; this suggested that these species evolved relatively slowly, in turn suggesting that they showed high phylogenetic stability. This supports the theory that the Stenochiinae subfamily belongs to a stable monophyletic lineage (as described below).

All the 33 Tenebrionidae mitogenomes contained an intergeneric space (15 bp to 19 bp) between *trnS2* (UCN) and *nad1*. Previous studies of tenebrionid mitogenomes had also reported intergenic spacers at this location, with lengths of approximately 17 bp [14,46,50]. Additionally, other Coleoptera families had a similar spacer in this location [51,52]. Li et al. (2007) discovered that the spacer was absent from two members of the genus *Rhagophthalmus* [53]. Insertions and deletions near the end of *nad1* resulting from sequencing errors or correction by posttranslational modification were the cause of this difference [50]. Given that this position denotes the conclusion of the major-strand coding region, it has been hypothesized that this interval may correspond to the binding site of the transcription attenuation factor mtTERM [54]. These intergenic spacer sequences have been shown to be binding sites of a transcription termination factor (DmTTF) [55]. Both *M. exilidentata* and *A. potanini* had a 47 bp spacer between *trnC* and *trnY*; however, this was not a uniform finding, as other species belonging to the same tribe did not have the same spacer. A long intergenic space, measuring 35–376 bp, has been detected between *trnW* and *trnC*; these two tRNAs were transposed and rearranged. This lengthy intergenic region contains tandem repeat units, which are thought to be *trnW* remnants [56]. When comparing the 47 bp sequence to *trnW* of *A. potanini* and *M. exilidentata*, we discovered partial identity. As a result, we assumed that this intergenic region was a remnant of *trnW* post-replication.

With the exception of *trnS1*, the remaining 21 tRNAs had the normal cloverleaf form. Nematode *trnS1* lacks the DHU arm; however, nuclear magnetic resonance studies of its tertiary structure revealed that such an abnormal tRNA may fit the ribosome by altering its structural conformation and function [57]. The tRNA function is ensured by the structural conservation of tRNA in the anticodon and tRNA core region, where the D-loop and T-loop interact to stabilize the overall structure [58].

4.2. tRNA-like Structures

We uncovered several tRNA-like structures in the A+T-rich area of 23 darkling beetles, which have never been reported in Tenebrionidae but demonstrated in other Coleoptera [59], Lepidoptera [27], and Hymenoptera [26,28]. Twenty-three out of

33 (69.69%) reported Tenebrionidae mitogenomes had tRNA-like structures, indicating that tRNA-like structures were very common in A+T-rich regions of insects, consistent with a previous study of Lepidoptera [27]. Of note, *trnL2*-like sequences were most frequent, possibly due to the high frequency of usage of the TTA codon encoding leucine. *trnL2*-like sequences may have the highest turnover rate in cells, which may increase the chance of binding to cells [27]. Despite the relatively low usage, the codon that codes for tryptophan has a more conserved sequence, which might explain the high frequency of *trnW*-like sequences in Tenebrionidae mitogenomes. Remarkably, both *trnL2*-like and *trnW*-like sequences were found in three of 32 species. The phylogenetic tree showed that these three species belonged to the same subfamily, Pimeliinae; however, the shared *trnL2*-like and *trnW*-like sequences in these three taxa may not be informative for reconstructing relationships. We analyzed the *trnL2*-like and *trnW*-like sequences of *A. potanini*, *P. contortus* LeConte, and *Pelecyphorus foveolatus* Solier. We discovered that the *trnL2*-like similarity between *A. potanini* and *P. contortus* reached 59.10%, including highly conserved sequences (results not shown). Both *P. contortus* and *Stenomorpha obovate* LeConte, within the same monophyletic lineage, had *trnR*-like and *trnW*-like sequences. Additionally, the two *trnW*-like sequences collectively had multiple conserved sequences of 9–11 bp, which may also explain their grouping into monophyletic lineages. We also analyzed the noncoding regions of *Platydema* sp. and *A. diaperinus*. When the 22 tRNAs of *A. diaperinus* were compared with their A+T-rich regions, there was a highly repetitive fragment with eight tRNA sequences; however, they did not have the same anticodon and therefore we did not identify these as tRNA-like structures. We compared this sequence (74 bp) with the A+T-rich region of *Platydema* sp. and found a high sequence similarity, with 79.73% conservation. The most conserved regions in the A+T-rich regions of *Platydema* sp. and *A. diaperinus* matched the most tRNA-like sequences, which may be related to their evolutionary conservatism, as indicated by present and previous studies which consistently supported a sister relationship between *Platydema* sp. (Diaperinae) and *A. diaperinus* (Tenebrioninae).

In some previous studies, sequences with identical anticodons were regarded as tRNA-like sequences [27,60], while alternative anticodons can be utilized in other studies [26,61]; a consistent standard has not been established [62]. In the tRNA-like analysis in this study, the same anticodon was used as a criterion for identification. Certain appropriate anticodons and secondary structures made A+T-rich sequences partially functional. Some sequences with different anticodons can also form trilobal secondary structures, suggesting that they had unidentified functions. In this regard, Serine transfer between the three tRNA-Ser sequences had been investigated [26]. A *trnN*-like sequence with a different anticodon in *M. unguiculina* had the highest match to the conventional tRNA (72.3%) among tRNAs of all three newly sequenced species. The *trnN*-like sequence was located between in the major-strand *trnI* and minor-strand *rrnS* and overlapped with conventional *trnN* by 47 bp. It had a CCA-OH3′, a TψC arm, an anticodon arm, and a variable arm (Figure S6). We speculated that this *trnN*-like structure had the same function of receiving aspartic acid.

The *trnN*-like ATT anticodon was relatively rare in Tenebrionidae, although it was possible. Anticodon point mutations can lead to codon reassignment [63], and further studies are needed to explore the functional effects of anticodon point mutations on tRNA-like sequences. The pairwise distance between *trnN*-like and regular *trnN* sequences was 0.714 and between *trnN*-like and *trnN* of species in the same tribe (*M. exilidentata*, *O. sabulosum* Linnaeus, and *Gonocephalum* sp.) were 0.714, 0.786, and 0.786, respectively. The pairwise distance between *trnN*-like and *trnN* of the same subfamily was 0.133–0.200, indicating that the regular *trnN* was more closely related to *trnN* of other species (Figure S6). A phylogenetic analysis of *trnN*-like of darkling beetles in the same tribe revealed that regular *trnN* and those of other insects were distantly related to *trnN*-like, forming an independent group (Figure S6).

Possible sources of tRNA-like sequences include strand slippage and mispairing and incorrect replication initiation or termination [64], leading to copy duplication. Mistakes during gene rearrangement can potentially result in unexpected mutations. In studies of mammalian mtDNA heavy strand replication origins, nucleotide sequences that fold into tRNA-like structures may serve as primers for DNA synthesis [65]. For example, tRNAs can function as primers for mtDNA synthesis and integrate a tRNA gene into the mitogenome if the primer is not cleaved from the growing DNA strand [66]. The non-coding A+T-rich region will show increased tRNA-like sequence diversity after insertion. Gene rearrangements, which did not occurred recently in the Tenebrionidae mitogenome, may be involved. As the forward and reverse strands are transcribed separately and the start is in the control region, the tRNA at the intersection should theoretically be closer to the typical structure. Despite the discovery of multiple tRNA-like structures in plant and animal viruses and studies of their essential activities [62], few studies have evaluated their evolutionary characteristics and functions in insect mitogenomes. The conventional and tRNA-like structures differed in sequence homology by 47.14% to 69.12% in this study. All tRNA-like sequences had many stem mismatches. Many studies have found that these tRNA-like structures are not pseudogenes, since they comprise anticodon and clover-leaf structures, which must have similar functional values [26]. Invertebrates have a unique non-coding section called the A+T-rich region that functions similarly to the regulatory region of vertebrate mtDNA [67]. Hence, tRNA-like sequences are common in many insect species, suggesting that the non-coding region may be involved in transcription and termination. Does a better match indicate greater functional effects? Are they still functional? These questions need to be investigated.

4.3. Phylogeny of Tenebrionidae

Our mitochondrial phylogeny was generally consistent with traditional morphological classifications and recent molecular studies. However, there were contradictory results for some species depending on the dataset and method of inference, suggesting that mitogenome sequences are still useful genetic markers for resolving molecular phylogenetic relationships at various taxonomic levels in beetles.

At the subfamily level, phylogenies based on different datasets assigned *Adelium* sp. and *C. popularis* to different positions. The monophyly of Lagriinae has been argued in many previous molecular and morphological studies [9,14,16,68]. In this study, based on the dataset P123, Lagriinae was a sister group to the other subfamilies, followed by the subfamily Pimeliinae, in agreement with previous molecular studies [14]. Stenochiinae was part of a clade composed of partial Diaperini (Diaperinae) and Alphitobiini (Tenebrioninae); both Diaperinae and Tenebrioninae were clearly polyphyletic. These results were consistent with some previous molecular studies where Diaperinae was polyphyletic [1,14,17]; however, Diaperinae has been identified as monophyletic based on morphological characters [69]. Furthermore, conflicting results have been obtained based on molecular data; for example, studies have shown that the split between *Platydema* sp. and *U. dermestoides* is more recent than observed in our study [17]. Stenochiinae is a monophyletic group according to morphological studies based on 80 characters [69], consistent with our findings. However, other studies suggested that Stenochiinae belonged to a clade that included partial Diaperini (Diaperinae), Helopinini, and Alphitobini (Tenebrioninae) [14]. Our study supported the monophyly of Pimeliinae, which was in agreement with results of other molecular studies [14]; however, its monophyly was not supported by all previous studies. In morphological studies, Tenebrioninae has been identified as monophyletic, whereas molecular studies supports this group as polyphyly [1,14,16]. This inconsistency may be due to differences in taxa or analytical methods; the dataset is important to phylogenetic tree topology [70].

In the present study, the two species within Diaperini were in different positions on the phylogenetic tree; *U. dermestoides* was sister to *Z.atratus* and *Platydema* sp. was a sister of Alphitobiini in previous studies [71,72]; however, some studies suggested that the two were more distantly related [14,73]. The relationships within Tenebrioninae also varied among three trees constructed using Bayesian methods. In the P123-based tree, Blaptini was in the middle of Tenebrionini, while Diaperinini was more derived than Tenebrionini.

In the ML tree, Asidini based on P123AA was sister to *P. contortus*; in the P123-based and P123RNA-based trees, Asidini was a sister group to the monophyletic group formed by *P. contortus* and *P. foveolatus*.

In a morphological analysis based on 80 characters, each of Opatrini, Diaperini, and Tenebrionini within Tenebrioninae was supported as monophyly [69]. The monophyly of Opatrini in Tenebrioninae proposed in this study was well-supported by morphological and molecular data [1,14,16,69], while the monophyly of Diaperini and Tenebrionini is not well-supported by morphological data [69]. Triboliini was monophyletic in the Bayesian tree based on the P123 dataset but not in the ML tree, as reported in previous studies [1,19].

5. Conclusions

In this study, we characterized three Tenebrionidae mitogenomes and performed comparative mitogenomic analyses of darkling beetles. The genome size, gene arrangement, base composition, and codon usage were all conserved within the darkling beetles, while diversity in sequence length and structural features was detected in control regions. We presented a brief structural and functional analysis of tRNA-like structures, discussed the possible functions of tRNA-like structures in terms of how well they have been conserved during evolution, and argued that this may be part of the basis for the classification of Tenebrionidae species. This provides some data on the current inadequate tRNA-like studies in the insect mitogenomic field. This study recovered a tenebrionid mitogenomic phylogeny: (Lagriinae, (Pimeliinae, ((Tenebrioninae + Diaperinae), Stenochiinae))), indicating that mitogenomic data could provide useful information for resolving the Tenebrionidae phylogeny. However, additional morphological (more characters) and molecular analyses (more loci) with wider taxonomic sampling are needed for a better understanding of the evolution of darkling beetles.

Supplementary Materials: The following supporting information can be downloaded at: https://www.mdpi.com/article/10.3390/genes14091738/s1, Figure S1: Rates of evolution of 13 protein-coding genes in the mitogenomes of 33 Tenebrionidae species. The left Y-axis provides the substitution rate of the mitochondrial gene, while the right Y-axis provides the G+C content. Synonymous nucleotide substitutions per synonymous site (K_s) and nonsynonymous nucleotide substitutions per nonsynonymous site (K_a) are calculated using the Kumar method. The standard errors were obtained by a bootstrap procedure (1000 replicates). Figure S2: K_a/K_s values of the 13 PCGs were negatively correlated with the percentage of GC ($R^2 = 0.86$, $P < 0.01$). Figure S3: tRNA-like secondary structures in the A+T-riched regions of 33 Tenebrionidae species with a match of 65% or less. Comparisons with the corresponding conventional tRNA sequences are provided. Orange nucleotides indicate anticodons, which specify the corresponding tRNAs. *, consensus sequences in the alignment. Figure S4: Alignment of the gap region between tRNA-Ser (UCN) and *nad1* in the genomes of all Tenebrionidae species. Colored boxes indicate the conserved nucleotide regions in all beetles, with the highly conserved pentanucleotide region (motif I, motif II, and motif III) in the middle. Figure S5: Phylogenetic relationships among 33 beetles obtained from analyses of PCGs alone using MrBayes (A/C/E) and RAxML (B/D/F). Support values (PP > 0.90; BP > 70) are shown next to nodes in the trees. The scale bar represents substitutions/site. Figure S6: Relationship between *M. unguiculina trnN*, *M. unguiculina trnN*-like sequences and representative insect *trnN*. (A) Genetic distances between *M. unguiculina trnN*, *M. unguiculina trnN*-like sequences, and representative insect *trnN*. (B) Phylogenetic relationships among *M. unguiculina trnN*, *M. unguiculina trnN*-like sequences, and representative insect *trnN*. The number on each node indicates the bootstrap support based on 1000 replicates. Table S1: Characteristics of mitogenomes of 33 Tenebrionidae species. Table S2: Saturation test for the 13 protein-coding genes (PCGs, the P123 dataset), concentration of 13 PCGs and two ribosomal RNA genes (the P123RNA dataset), and each of the three positions of 13 PCGs as implemented in DAMBE. Table S3: The best partitioning schemes and substitution models selected by IQ-TREE for the three datasets. Table S4: Annotation and organization of three mitochondrial genomes newly sequenced in this study. Table S5: Codon usage for the 13 mitochondrial protein-coding genes of 33 Tenebrionidae species. n: number of occurrences; %: percentage of total amino acids; RSCU: Relative synonymous codon usage. Table S6: Distribution of tRNA-like sequences in all Tenebrionidae species in this study. Table S7: Number of occurrences and matching of tRNA-like structures in Tenebrionidae mitochondrial genomes.

Author Contributions: M.-L.Y. conceived and designed the experiments. S.-H.W., S.-Y.H., M.L. (Min Li), M.L. (Min Liu), H.S., J.-R.Z. and W.-T.C. sampled insect specimens and conducted experiments. S.-H.W., S.-Y.H., M.L. (Min Li) and M.-L.Y. performed data analyses. S.-H.W., S.-Y.H. and M.L. (Min Li) wrote the manuscript. M.-L.Y. revised the manuscript. All authors have read and agreed to the published version of the manuscript.

Funding: This research was funded by the National Science & Technology Fundamental Resources Investigation Program of China (2019FY100400/2019FY100404) and the Key Research & Development Program of Forestry and Grassland Administration of Ningxia Hui Autonomous Region, China.

Institutional Review Board Statement: Not applicable.

Informed Consent Statement: Not applicable.

Data Availability Statement: The data presented in this study are openly available in NCBI (GenBank accession numbers OQ534869, OQ511732, and OQ536315).

Conflicts of Interest: The authors declare no conflict of interest.

References

1. Gunter, N.L.; Levkaničová, Z.; Weir, T.H.; Ślipiński, A.; Cameron, S.L.; Bocak, L. Towards a phylogeny of the Tenebrionoidea (Coleoptera). *Mol. Phylogenet. Evol.* **2014**, *79*, 305–312. [CrossRef]
2. Aalbu, R.L.; Triplehorn, C.A.; Campbell, J.M.; Brown, K.W.; Somerby, R.E.; Thomas, D.B. Tenebrionidae Latreille 1802. In *Polyphaga: Scarabaeoidea through Curculionoidea*; Arnett, R.H., Jr., Thomas, M.C., Skelley, P.E., Frank, J.H., Eds.; CRC Press: Boca Raton, FL, USA, 2022; Volume 3, pp. 463–509.
3. Bouchard, P.; Lawrence, J.F.; Davies, A.E.; Newton, A.F. Synoptic classification of the world Tenebrionidae (Insecta: Coleoptera) with a review of family-group names. In *Annales Zoologici. Museum and Institute of Zoology*; Polish Academy of Sciences: Warsaw, Poland, 2005; Volume 55, pp. 499–530.
4. Levkaničová, Z. Molecular Phylogeny of the Superfamily Tenebrionoidea (Coleoptera: Cucujiformia). Ph.D. Thesis, Palaký University, Olomouc, Czech Republic, 2009.
5. Throne, J.E.; Hallman, G.J.; Johnson, J.A.; Follett, P.A. Post-harvest entomology research in the United States Department of Agriculture-Agricultural Research Service. *Pest Manag. Sci.* **2003**, *59*, 619–628. [CrossRef]
6. Selaledi, L.; Mbajiorgu, C.A.; Mabelebele, M. The use of yellow mealworm (*T. molitor*) as alternative source of protein in poultry diets: A review. *Trop. Anim. Health Prod.* **2020**, *52*, 7–16. [CrossRef]
7. Kundungal, H.; Synshiang, K.; Devipriya, S.P. Biodegradation of polystyrene wastes by a newly reported honey bee pest *Uloma* sp. larvae: An insight to the ability of polystyrene-fed larvae to complete its life cycle. *Environ. Chall.* **2021**, *4*, 100083. [CrossRef]
8. Pivato, A.F.; Miranda, G.M.; Prichula, J.; Lima, J.E.A.; Ligabue, R.A.; Seixas, A.; Trentin, D.S. Hydrocarbon-based plastics: Progress and perspectives on consumption and biodegradation by insect larvae. *Chemosphere* **2022**, *293*, 133600. [CrossRef]
9. Doyen, J.T.; Matthews, E.; Lawrence, J.F. Classification and annotated checklist of the Australian genera of Tenebrionidae (Coleoptera). *Invertebr. Syst.* **1989**, *3*, 229–260. [CrossRef]
10. Maeno, K.O.; Nakamura, S.; Babah, M.A.O. Sexing Live Adults of the Three Species of Darkling Beetle (Coleoptera: Tenebrionidae) and Morphological Characteristics. *Ann. Entomol. Soc. Am.* **2012**, *105*, 726–730. [CrossRef]
11. Song, N.; Zhang, H.; Li, H.; Cai, W. All 37 Mitochondrial Genes of Aphid *Aphis craccivora* Obtained from Transcriptome Sequencing: Implications for the Evolution of Aphids. *PLoS ONE* **2016**, *11*, e0157857. [CrossRef]
12. Angelini, D.R.; Jockusch, E.L. Relationships among pest flour beetles of the genus *Tribolium* (Tenebrionidae) inferred from multiple molecular markers. *Mol. Phylogenet. Evol.* **2008**, *46*, 127–141. [CrossRef]
13. Ge, X.; Yuan, L.; Kang, Y.; Liu, T.; Liu, H.; Yang, Y. Characterization of the First Complete Mitochondrial Genome of Cyphonocerinae (Coleoptera: Lampyridae) with Implications for Phylogeny and Evolution of Fireflies. *Insects* **2021**, *12*, 570. [CrossRef]
14. Wu, C.; Zhou, Y.; Tian, T.; Li, T.J.; Chen, B. First report of complete mitochondrial genome in the subfamily Alleculinae and mitochondrial genome-based phylogenetics in Tenebrionidae (Coleoptera: Tenebrionoidea). *Insect Sci.* **2022**, *29*, 1226–1238. [CrossRef] [PubMed]
15. Hunt, T.; Bergsten, J.; Levkanicova, Z.; Papadopoulou, A.; John, O.S.; Wild, R.; Hammond, P.M.; Ahrens, D.; Balke, M.; Caterino, M.S.; et al. A Comprehensive Phylogeny of Beetles Reveals the Evolutionary Origins of a Superradiation. *Science* **2007**, *318*, 1913–1916. [CrossRef] [PubMed]
16. Kergoat, G.J.; Soldati, L.; Clamens, A.L.; Jourdan, H.; Jabbour-Zahab, R.; Genson, G.; Bouchard, P.; Condamine, F.L. Higher level molecular phylogeny of darkling beetles (Coleoptera: Tenebrionidae). *Syst. Entomol.* **2014**, *39*, 486–499. [CrossRef]
17. Song, N.; Liu, H.-Y.; Yang, X.-J.; Zhao, X.-C.; Lin, A.-L. Complete mitochondrial genome of the darkling beetle *Gonocephalum outreyi* (Coleoptera: Tenebrionidae) with phylogenetic implications. *J. Asia-Pac. Entomol.* **2018**, *21*, 721–730. [CrossRef]

18. Yot, P.; Pinck, M.; Haenni, A.-L.; Duranton, H.M.; Chapeville, F. Valine-Specific tRNA-like Structure in Turnip Yellow Mosaic Virus RNA. *Proc. Natl. Acad. Sci. USA* **1970**, *67*, 1345–1352. [CrossRef] [PubMed]
19. Zhang, X.; Kim, C.H.; Sivakumaran, K.; Kao, C. Stable RNA structures can repress RNA synthesis in vitro by the brome mosaic virus replicase. *RNA* **2003**, *9*, 555–565. [CrossRef]
20. Vieweger, M.; Holmstrom, E.D.; Nesbitt, D.J. Single-Molecule FRET Reveals Three Conformations for the TLS Domain of Brome Mosaic Virus Genome. *Biophys. J.* **2015**, *109*, 2625–2636. [CrossRef]
21. Ni, P.; Vaughan, R.C.; Tragesser, B.; Hoover, H.; Kao, C.C. The Plant Host Can Affect the Encapsidation of Brome Mosaic Virus (BMV) RNA: BMV Virions Are Surprisingly Heterogeneous. *J. Mol. Biol.* **2014**, *426*, 1061–1076. [CrossRef]
22. Comandur, R.; Olson, E.D.; Musier-Forsyth, K. Conservation of tRNA mimicry in the 5′-untranslated region of distinct HIV-1 subtypes. *RNA* **2017**, *23*, 1850–1859. [CrossRef]
23. Johanson, K.; Hoang, T.; Sheth, M.; Hyman, L.E. GRS1, a Yeast tRNA Synthetase with a Role in mRNA 3′ End Formation. *J. Biol. Chem.* **2003**, *278*, 35923–35930. [CrossRef]
24. Plewka, P.; Thompson, A.; Szymanski, M.; Nuc, P.; Knop, K.; Rasinska, A.; Bialkowska, A.; Szweykowska-Kulinska, Z.; Karlowski, W.M.; Jarmolowski, A. A stable tRNA-like molecule is generated from the long noncoding RNA *GUT15* in *Arabidopsis*. *RNA Biol.* **2018**, *15*, 726–738. [CrossRef] [PubMed]
25. Yu, C.H.; Liao, J.Y.; Zhou, H.; Qu, L.H. The rat mitochondrial Ori L encodes a novel small RNA resembling an ancestral tRNA. *Biochem. Biophys. Res. Commun.* **2008**, *372*, 634–638. [CrossRef]
26. Cha, S.Y.; Yoon, H.J.; Lee, E.M.; Yoon, M.H.; Hwang, J.S.; Jin, B.R.; Han, Y.S.; Kim, I. The complete nucleotide sequence and gene organization of the mitochondrial genome of the bumblebee, *Bombus ignitus* (Hymenoptera: Apidae). *Gene* **2007**, *392*, 206–220. [CrossRef] [PubMed]
27. Kim, M.I.; Baek, J.Y.; Kim, M.J.; Jeong, H.C.; Kim, K.G.; Bae, C.H.; Han, Y.S.; Jin, B.R.; Kim, I. Complete Nucleotide Sequence and Organization of the Mitogenome of the Red-Spotted Apollo Butterfly, *Parnassius bremeri* (Lepidoptera: Papilionidae) and Comparison with Other Lepidopteran Insects. *Mol. Cells* **2009**, *28*, 347–363. [CrossRef] [PubMed]
28. Hong, M.Y.; Cha, S.Y.; Kim, D.Y.; Yoon, H.J.; Kim, S.R.; Hwang, J.S.; Kim, K.G.; Han, Y.S.; Kim, I.S. Presence of several tRNA-like sequences in the mitochondrial genome of the bumblebee, *Bombus hypocrita sapporoensis* (Hymenoptera: Apidae). *Genes Genom.* **2008**, *30*, 307–318.
29. Boore, J.L. Animal mitochondrial genomes. *Nucleic Acids Res.* **1999**, *27*, 1767–1780. [CrossRef]
30. Cameron, S.L. Insect Mitochondrial Genomics: Implications for evolution and phylogeny. *Annu. Rev. Entomol.* **2014**, *59*, 95–117. [CrossRef]
31. Zhang, D.X.; Hewitt, G.M. Insect mitochondrial control region: A review of its structure, evolution and usefulness in evolutionary studies. *Biochem. Syst. Ecol.* **1997**, *25*, 99–120. [CrossRef]
32. Moritz, C.; Dowling, T.E.; Brown, W.M. Evolution of Animal Mitochondrial DNA: Relevance for Population Biology and Systematics. *Annu. Rev. Ecol. Evol. S* **1987**, *18*, 269–292. [CrossRef]
33. Chen, Y.; Chen, Y.; Shi, C.; Huang, Z.; Zhang, Y.; Li, S.; Li, Y.; Ye, J.; Yu, C.; Li, Z.; et al. SOAPnuke: A MapReduce acceleration-supported software for integrated quality control and preprocessing of high-throughput sequencing data. *Gigascience* **2018**, *7*, 1–6. [CrossRef]
34. Dierckxsens, N.; Mardulyn, P.; Smits, G. NOVOPlasty: De novo assembly of organelle genomes from whole genome data. *Nucleic Acids Res.* **2017**, *45*, e18. [PubMed]
35. Bernt, M.; Donath, A.; Juhling, F.; Externbrink, F.; Florentz, C.; Fritzsch, G.; Putz, J.; Middendorf, M.; Stadler, P.F. MITOS: Improved *de novo* metazoan mitochondrial genome annotation. *Mol. Phylogenet. Evol.* **2013**, *69*, 313–319. [CrossRef]
36. Tamura, K.; Stecher, G.; Kumar, S. MEGA11: Molecular Evolutionary Genetics Analysis Version 11. *Mol. Biol. Evol.* **2021**, *38*, 3022–3027. [CrossRef] [PubMed]
37. Perna, N.T.; Kocher, T.D. Patterns of Nueleotide Composition at Fourfold Degenerate Sites of Animal Mitochondrial Genomes. *J. Mol. Evol.* **1995**, *41*, 353–358. [CrossRef]
38. Rozas, J.; Ferrer-Mata, A.; Sanchez-DelBarrio, J.C.; Guirao-Rico, S.; Librado, P.; Ramos-Onsins, S.E.; Sanchez-Gracia, A. DnaSP 6: DNA Sequence Polymorphism Analysis of Large Data Sets. *Mol. Biol. Evol.* **2017**, *34*, 3299–3302. [CrossRef]
39. Zang, M.; He, W.; Du, F.; Wu, G.; Wu, B.; Zhou, Z. Analysis of the codon usage of the ORF2 gene of feline calicivirus. *Infect. Genet. Evol.* **2017**, *54*, 54–59. [CrossRef] [PubMed]
40. Xia, X. DAMBE5: A comprehensive software package for data analysis in molecular biology and evolution. *Mol. Biol. Evol.* **2013**, *30*, 1720–1728. [CrossRef]
41. Stamatakis, A. RAxML version 8: A tool for phylogenetic analysis and post-analysis of large phylogenies. *Bioinformatics* **2014**, *30*, 1312–1313. [CrossRef]
42. Ronquist, F.; Teslenko, M.; van der Mark, P.; Ayres, D.L.; Darling, A.; Hohna, S.; Larget, B.; Liu, L.; Suchard, M.A.; Huelsenbeck, J.P. MrBayes 3.2: Efficient Bayesian phylogenetic inference and model choice across a large model space. *Syst. Biol.* **2012**, *61*, 539–542. [CrossRef]
43. Wei, S.J.; Shi, M.; Chen, X.X.; Sharkey, M.J.; van Achterberg, C.; Ye, G.Y.; He, J.H. New views on strand asymmetry in insect mitochondrial genomes. *PLoS ONE* **2010**, *5*, e12708. [CrossRef]

44. Wei, S.J.; Shi, M.; Sharkey, M.J.; van Achterberg, C.; Chen, X.X. Comparative mitogenomics of Braconidae (Insecta: Hymenoptera) and the phylogenetic utility of mitochondrial genomes with special reference to Holometabolous insects. *BMC Genomics* **2010**, *11*, 371. [CrossRef] [PubMed]
45. Yu, P.; Zhou, L.; Zhou, X.Y.; Yang, W.T.; Zhang, J.; Zhang, X.J.; Wang, Y.; Gui, J.F. Unusual AT-skew of *Sinorhodeus microlepis* mitogenome provides new insights into mitogenome features and phylogenetic implications of bitterling fishes. *Int. J. Biol. Macromol.* **2019**, *129*, 339–350. [CrossRef]
46. Ou, J.; Liu, J.B.; Yao, F.J.; Wang, X.G.; Wei, Z.M. The complete mitochondrial genome of the American black flour beetle *Tribolium audax* (Coleoptera: Tenebrionidae). *Mitochondrial. DNA A* **2016**, *27*, 958–959. [CrossRef] [PubMed]
47. Liu, L.N.; Wang, C.Y. Complete mitochondrial genome of yellow meal worm (*Tenebrio molitor*). *Zool Res.* **2014**, *35*, 537–545.
48. Knight, R.D.; Freeland, S.J.; Landweber, L.F. A simple model based on mutation and selection explains trends in codon and amino-acid usage and GC composition within and across genomes. *Genome Biol.* **2001**, *2*, RESEARCH0010. [PubMed]
49. Eyre-Walker, A. Differentiating between selection and mutation bias. *Genetics* **1997**, *147*, 1983–1987. [CrossRef]
50. Sheffield, N.C.; Song, H.; Cameron, S.L.; Whiting, M.F. A Comparative Analysis of Mitochondrial Genomes in Coleoptera (Arthropoda: Insecta) and Genome Descriptions of Six New Beetles. *Mol. Biol. Evol.* **2008**, *25*, 2499–2509. [CrossRef]
51. Du, C.; He, S.; Song, X.; Liao, Q.; Zhang, X.; Yue, B. The complete mitochondrial genome of *Epicauta chinensis* (Coleoptera: Meloidae) and phylogenetic analysis among Coleopteran insects. *Gene* **2016**, *578*, 274–280. [CrossRef]
52. Liao, F. The complete mitochondrial genome of the fall webworm, *Hyphantria cunea* (Lepidoptera: Arctiidae). *Int. J. Biol. Sci.* **2010**, *6*, 172–186. [CrossRef]
53. Li, X.; Ogoh, K.; Ohba, N.; Liang, X.; Ohmiya, Y. Mitochondrial genomes of two luminous beetles, *Rhagophthalmus lufengensis* and *R. ohbai* (Arthropoda, Insecta, Coleoptera). *Gene* **2007**, *392*, 196–205. [CrossRef]
54. Taanman, J.W. The mitochondrial genome: Structure, transcription, translation and replication. *Biochim. Biophys. Acta* **1999**, *1410*, 103–123. [CrossRef] [PubMed]
55. Roberti, M.; Polosa, P.L.; Bruni, F.; Musicco, C.; Gadaleta, M.N.; Cantatore, P. DmTTF, a novel mitochondrial transcription termination factor that recognises two sequences of *Drosophila melanogaster* mitochondrial DNA. *Nucleic Acids Res.* **2003**, *31*, 1597–1604. [CrossRef] [PubMed]
56. Ge, X.; Yuan, L.; Kang, Y.; Liu, T.; Liu, H.; Yang, Y. First complete mitochondrial genomes of Ototretinae (Coleoptera, Lampyridae) with evolutionary insights into the gene rearrangement. *Genomics* **2022**, *114*, 110305. [CrossRef] [PubMed]
57. Ohtsuki, T.; Kawai, G.; Watanabe, K. The minimal tRNA: Unique structure of *Ascaris suum* mitochondrial tRNA(Ser)(UCU) having a short T arm and lacking the entire D arm. *FEBS Lett.* **2002**, *514*, 37–43. [CrossRef]
58. Lorenz, C.; Lunse, C.E.; Morl, M. tRNA Modifications: Impact on Structure and Thermal Adaptation. *Biomolecules* **2017**, *7*, 35. [CrossRef]
59. Hong, M.Y.; Jeong, H.C.; Kim, M.J.; Jeong, H.U.; Lee, S.H.; Kim, I. Complete mitogenome sequence of the jewel beetle, *Chrysochroa fulgidissima* (Coleoptera: Buprestidae). *Mitochondrial. DNA* **2009**, *20*, 46–60. [CrossRef]
60. Kim, M.J.; Jeong, H.C.; Kim, S.R.; Kim, I. Complete mitochondrial genome of the nerippe fritillary butterfly, *Argynnis nerippe* (Lepidoptera: Nymphalidae). *Mitochondrial. DNA* **2011**, *22*, 86–88. [CrossRef]
61. Zhang, B.; Ma, C.; Edwards, O.; Fuller, S.; Kang, L. The mitochondrial genome of the Russian wheat aphid *Diuraphis noxia*: Large repetitive sequences between *trnE* and *trnF* in aphids. *Gene* **2014**, *533*, 253–260. [CrossRef]
62. Wu, S.; Li, X.; Wang, G. tRNA-like structures and their functions. *FEBS J.* **2022**, *289*, 5089–5099. [CrossRef]
63. Li, M.; Chen, W.T.; Zhang, Q.L.; Liu, M.; Xing, C.W.; Cao, Y.; Luo, F.Z.; Yuan, M.L. Mitochondrial phylogenomics provides insights into the phylogeny and evolution of spiders (Arthropoda: Araneae). *Zool Res.* **2022**, *43*, 566–584. [CrossRef]
64. Macey, J.R.; Larson, A.; Ananjeva, N.B.; Fang, Z.; Papenfuss, T.J. Two Novel Gene Orders and the Role of Light-Strand Replication in Rearrangement of the Vertebrate Mitochondrial Genome. *Mol. Biol. Evol.* **1997**, *14*, 91–104. [CrossRef]
65. Brown, C.R. Cliff swallow colonies as information centers. *Science* **1986**, *234*, 83–85. [CrossRef]
66. Cantatore, P.; Gadaleta, M.N.; Roberti, M.; Saccone, C.; Wilson, A.C. Duplication and remoulding of tRNA genes during the evolutionary rearrangement of mitochondrial genomes. *Nature* **1987**, *329*, 853–855. [CrossRef] [PubMed]
67. Goddard, J.M.; Wolstenholme, D.R. Origin and direction of replication in mitochondrial DNA molecules from the genus Drosophila. *Nucleic Acids Res.* **1980**, *8*, 741–757. [PubMed]
68. Timmermans, M.J.; Barton, C.; Haran, J.; Ahrens, D.; Culverwell, C.L.; Ollikainen, A.; Dodsworth, S.; Foster, P.G.; Bocak, L.; Vogler, A.P. Family-Level Sampling of Mitochondrial Genomes in Coleoptera: Compositional Heterogeneity and Phylogenetics. *Genome Biol. Evol.* **2015**, *8*, 161–175. [CrossRef]
69. Doyen, J.T.; TSCHINKEL, W.R. Phenetic and cladistic relationships among tenebrionid beetles (*Coleoptera*). *Syst. Entomol.* **1982**, *7*, 127–183. [CrossRef]
70. Zhang, H.L.; Liu, B.B.; Wang, X.Y.; Han, Z.P.; Zhang, D.X.; Su, C.N. Comparative Mitogenomic Analysis of Species Representing Six Subfamilies in the Family Tenebrionidae. *Int. J. Mol. Sci.* **2016**, *17*, 841. [CrossRef] [PubMed]
71. Bai, Y.; Chen, J.; Li, G.; Luo, J.; Wang, H.; Yang, Y.; Liang, S.; Ouyang, B. Complete mitochondrial genome of *Promethis valgipes valgipes* (Marseul) (Insecta: Coleoptera: Tenebrionidae). *Mitochondrial. DNA B* **2021**, *6*, 538–539. [CrossRef] [PubMed]

72. Bai, Y.; Gao, X.; Yu, Y.; Long, X.; Zeng, X.; Wei, D.; Ye, L. Complete mitochondrial genome of *Morphostenophanes sinicus* (Zhou, 2020) (Insecta: Coleoptera: Tenebrionidae). *Mitochondrial. DNA B* **2021**, *6*, 2946–2948. [CrossRef]
73. Hong, K.J.; Ki, W.; Park, D.S.; Yang, B.K.; Lee, H.; Park, J.; Lee, W. The complete mitochondrial genome of *Alphitobius diaperinus* Panzer, 1797 (Coleoptera: Tenebrionidae). *Mitochondrial. DNA B* **2020**, *5*, 2291–2293. [CrossRef]

genes

MDPI

Article

The Genetic Diversity of White-Backed Planthoppers (*Sogatella furcifera*) between Myanmar and Yunnan Province of China

Yue Liu [1,2,†], Khin Nyein Chan [1,3,†], Xiangyong Li [1], Xueqing Zhao [1], Dong Chu [4], Yanqiong Yin [1], Ying Liu [1,*] and Aidong Chen [1]

1 Key Laboratory of Green Prevention and Control of Agricultural Transboundary Pests of Yunnan Province/Agricultural Environment and Resource Research Institute, Yunnan Academy of Agricultural Sciences, Kunming 650205, China; yinyq1977@sina.com
2 State Key Laboratory of Biocatalysis and Enzyme Engineering, School of Life Sciences, Hubei University, Wuhan 430062, China
3 Biotechnology Research Department, Ministry of Education, Mandalay 05151, Myanmar
4 Key Lab of Integrated Crop Pest Management of Shandong Province, College of Plant Health and Medicine, Qingdao Agricultural University, Qingdao 266109, China
* Correspondence: liuying@yaas.org.cn
† These authors contributed equally to this work.

Abstract: In order to clarify the migration route and the source of white-backed planthopper (WBPH) (*Sogatella furcifera*) between Myanmar and Yunnan Province, China, we collected six populations throughout Myanmar and five populations around the border areas in Yunnan Province, China. A total of 790 base pairs in the mtDNA COI genes from 416 individuals were obtained. A total of 43 haplotypes were identified, among which 37 were unique haplotypes, and the remaining 6 were shared among different populations. Two common shared haplotypes (H_1 and H_2) had a widespread distribution in all populations and accounted for 88.8% of the total haplotype frequency, suggesting a high-level gene flow among the Myanmar and Yunnan populations. Bayesian skyline plot (BSP) analysis results indicated that the effective population size of WBPH expanded between about 10,000 and 7000 years ago, and *S. furcifera* might follow the post-LGM (Last Glacial Maximum) expansion pattern. Based on the total migrant (N_em) value, it can be deduced that north and northeast Myanmar were the primary migration sources for WBPH populations in the southwest and south Yunnan regions. This study aims to contribute to the sustainable regional management of this important rice pest and provide new insights into the genetic diversity of WBPH in Southeast Asia.

Keywords: mitochondrial DNA; population structure; gene flow; *Sogatella furcifera*

Citation: Liu, Y.; Chan, K.N.; Li, X.; Zhao, X.; Chu, D.; Yin, Y.; Liu, Y.; Chen, A. The Genetic Diversity of White-Backed Planthoppers (*Sogatella furcifera*) between Myanmar and Yunnan Province of China. *Genes* **2023**, *14*, 2164. https://doi.org/10.3390/genes14122164

Academic Editor: Brenda Oppert

Received: 13 October 2023
Revised: 17 November 2023
Accepted: 22 November 2023
Published: 30 November 2023

1. Introduction

The white-backed planthopper, *Sogatella furcifera* (Horváth), belongs to Hemiptera: Delphacidae and is one of the most serious insect pests in rice production [1,2]. It is widely distributed in South, Southeast, and East Asia, as well as the South Pacific islands and Australia [3]. The white-backed planthopper (WBPH) possess formidable migratory abilities, enabling them to cover vast distances between different seasons and regions, spanning thousands of kilometers [4–7]. This migratory capability allows them to swiftly move from one area to another in a relatively short period, posing a significant threat to crops. The threat to rice primarily manifests during the process of feeding on plant sap, causing abnormal plant growth and development that leads to reduced yields or even plant death. Additionally, by transmitting plant viruses, such as the *Southern rice black-streaked dwarf virus* [7–10], the WBPH further exacerbates its harm to crops. This can result in significant economic losses, with estimated reductions in crop yield ranging from 10% to 30% [11], posing a serious threat to agricultural production [12]. Therefore, understanding the migration patterns of the WBPH is crucial for the development of effective regional management and prevention strategies.

In recent years, researchers have also made some progress in exploring the migration origins of the WBPH through trajectory analysis [13–15]. According to research by Chaoxing Hu et al., WBPH are unable to overwinter in temperate regions. Annually, the primary WBPH originate from warmer areas such as the Indochinese Peninsula, and their typical migration pattern usually begins around March annually [16]. This migration marks the peak of their movement from tropical regions into China. Regarding the migration of WBPH in Yunnan Province, it is speculated that from April to early May, these planthoppers primarily choose Myanmar as their migration source. However, by mid-May, it is confirmed that the main source of these migrating insects is the northern regions of Vietnam. This migration process is, to some extent, influenced by the seasons and climate, forming an organized and sustained migration chain [15,17]. Despite some evidence based on molecular marker analysis [18,19], our understanding of the genetic flow and migration patterns of the WBPH between Myanmar and Yunnan Province, China remains limited.

In our previous research, we found that the main migration sources of the WBPH in Yunnan Province, China include Myanmar, Vietnam, Laos, and Cambodia [13,20–22]. However, it is important to note that our previous studies in Myanmar primarily focused on samples from the southern region of the country. To comprehensively understand the genetic diversity of WBPH populations between Myanmar and Yunnan Province, China, we collected a total of 416 individual WBPH specimens from six locations in Myanmar and five locations in Yunnan Province, China. We chose the mitochondrial DNA cytochrome c oxidase subunit I (mtCOI) gene for our study because, compared to other mitochondrial genes, the mtCOI gene provides broader phylogenetic information and is considered a reliable molecular marker for evolution [23–25]. To gain a deeper understanding of the relationships between these individuals, we amplified the mtCOI gene from these specimens and conducted analyses on both phylogenetic relationships and population genetic structure. By delving into the genetic relationships and population structure among individual WBPH, we can gain a more comprehensive understanding of their genetic diversity and migration patterns. This study provides a scientific basis for developing more targeted cross-border monitoring and source control measures. The findings not only contribute to optimizing agricultural pest management strategies, mitigating the damage caused by WBPH to crops, but also lay a solid foundation for sustainable cross-border regional management between Myanmar and Yunnan Province, China in the future.

2. Materials and Methods

2.1. Sampling and DNA Extraction

In 2018, we collected adult *S. furcifera* samples from six locations in Myanmar and subsequently, in 2019, collected adult *S. furcifera* samples from five locations in Yunnan Province, China. The sample collection sites and population data of *S. furcifera* in Myanmar and Yunnan, China are shown in Figure 1 and Table 1. These samples were brought back to the laboratory and under controlled environmental conditions with a constant temperature of 27 ± 1 °C, relative humidity maintained at 70% ± 10%, and a light–dark cycle alternating between 14 h of light and 10 h of darkness, the WBPH population was sustained using the TN1 rice variety. During the sample preparation process, all insect samples were washed three times in 75% ethanol for 5 min each, followed by two washes in sterile ddH$_2$O for 3 min each, and air-dried at room temperature [26]. Finally, genomic DNA was extracted from each individual insect using the Cwbiotech kit from Beijing, China, and the extracted DNA was stored at −20 °C.

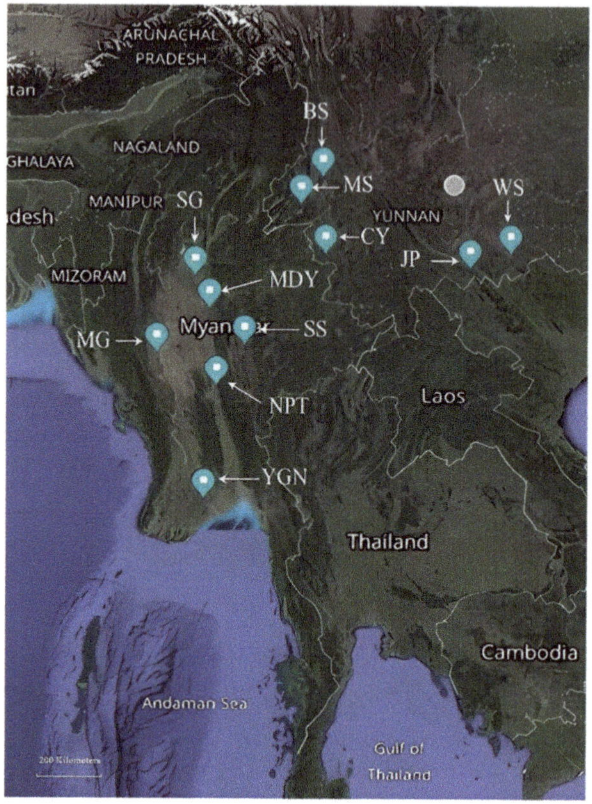

Figure 1. Sample collection sites of *S. furcifera* populations in Myanmar and Yunnan Province, China. The map was generated using Google Maps application (https://www.google.com/maps/, accessed on 1 November 2020).

Table 1. Sampling information of *S. furcifera* collected from various locations in Myanmar and Yunnan Province, China.

Code	Location	Longitude	Latitude	Date
YGN	Tontay, Yangon Division, Myanmar	16.71	95.94	5 September 2018
NPT	Pyinmana, Naypyitaw, Myanmar	19.71	96.28	10 May 2018
MG	Salin, Magway Division, Myanmar	20.58	94.65	8 June 2018
SG	Shwebo, Sagaing Division, Myanmar	22.58	95.70	15 May 2018
MDY	Sintkaing, Mandalay Division, Myanmar	21.74	96.09	11 April 2018
SS	Taunggyi, Shan State, Myanmar	20.91	96.94	10 June 2019
BS	Baoshan, Yunnan, China	25.13	99.17	10 May 2019
CY	Cangyuan, Yunnan, China	23.15	99.25	17 April 2019
JP	Jinping, Yunnan, China	22.78	103.23	15 May 2019
MS	Mangshi, Yunnan, China	24.44	98.58	10 May 2019
WS	Wenshan, Yunnan, China	23.16	104.35	5 May 2019

2.2. Mitochondrial COI Gene Amplification and Sequencing

The *mitochondrial cytochrome oxidase subunit I* gene (mtCOI) was used to determine the genetically distinct populations of *S. furcifera*. All samples were first amplified using the mtCOI primers CI-J-2183 (5′-CAACATTTATTTGATTTTTTGG-3′) and L2-N-3014 (5′-TCCAATGCACTAATC TGCCATATTA-3′), followed by Sanger sequencing [27]. Polymerase chain reaction (PCR) assays were performed in 20 μL of buffer containing 2 μL of

10× buffer, 1.5 mM $MgCl_2$, 0.2 μM dNTPs, 1 unit of Taq DNA polymerase, 2 μL of template DNA, and 0.2 μM of each primer. PCR amplification was carried out as follows: initial denaturation at 94 °C for 5 min, followed by 35 cycles each of 30 s at 94 °C, 30 s at 55 °C, and 60 s at 72 °C, and a final elongation step at 72 °C for 10 min.

2.3. Population Genetic Diversity and Structure Analysis

The mtCOI sequences were aligned using Clustal W [28]. Genetic diversity indices for each population were analyzed based on mtCOI using DnaSP v. 5.0 [29]. This analysis included the number of polymorphic (segregating) sites (S), total number of mutations (η) [30], average number of nucleotide differences (K) [31], number of haplotypes (H), haplotype diversity (Hd), nucleotide diversity (π) [32], and nucleotide diversity with Jukes and Cantor correction π (JC) [33]. Tajima's D (D) [34] and Fu's F test [35] were performed to detect deviation from neutrality. The dispersal of different *S. furcifera* populations in Myanmar was determined by calculating the effective numbers of migrants per generation N_em using mitochondrial COI data. N_em is θM ($\theta = N_e\mu$, where μ is the mutation rate per site per generation; $M = m/\mu$, where m is the migration rate) calculated using Bayesian search strategies in MIGRATE v. 3.2.16 software [36]. Pairwise *Fst* values and analysis of molecular variance (AMOVA) among populations were implemented with the Tamura–Nei model in Arlequin v. 3.5 [37].

Phylogenetic analyses were performed by the Bayesian inference (BI) in MrBayes v. 3.1.1 [38] and maximum likelihood (ML) methods in RAxML v. 8.2 [39]. Clustering of individuals was performed for the mtCOI dataset with BAPS v. 6.0 [40] based on the spatial clustering of groups of individual models. Twenty runs (K = 20) were made to ensure consistency and convergence of the results. The haplotype network of mtCOI genes was inferred using the median-joining algorithm [30]. In order to elucidate the relationship between haplotypes, network constructions were created using the median-joining method [41] with Network version 4.6.1.0.

2.4. Demographic History Analysis

To estimate the population expansion time, we utilized the Bayesian skyline plot based on the mtCOI gene in BEAST version 1.6.1 [42]. In the analysis, we used a substitution rate of 1.77% per million years [43] and the GTR + G model. The Markov Chain length was set to 300 million generations, and an uncorrelated lognormal relaxed clock model was employed, allowing rate variation among branches. During the analysis, we sampled every 1000 steps. We employed the piecewise linear skyline model with Bayesian skyline coalescent tree priors, and default values were used for other parameters. To ensure the accuracy of the results, a 10% burn-in was applied, and the results of the Bayesian skyline plot were thoroughly examined and analyzed using Tracer version 1.7 [42].

3. Results

3.1. Genetic Diversity Analysis

This study identified 43 mtCOI haplotypes (designated as H_1 to H_43), as shown in Figure 2. Among these, 20 unique haplotypes were discovered in Myanmar, and 17 unique haplotypes were found in Yunnan Province, China. Additionally, six haplotypes were shared among different populations from both Myanmar and Yunnan Province as follows. (1) Two common haplotypes (H_1 and H_2) were shared by all populations in both Myanmar and Yunnan Province, China. These haplotypes were the dominant shared types in every population, with H_1 comprising approximately 62.5%, and H_2 comprising approximately 25.7% in each population. (2) H_16 was shared by the SG (Myanmar) and MS (Yunnan Province, China) populations. (3) H_20 was shared by the SG (Myanmar) and BS (Yunnan Province, China) populations. (4) The other two shared haplotypes were found in only Myanmar or Yunnan, China: H_3 was shared by the YGN (Myanmar) and SS (Myanmar) populations, while H_27 was shared by the BS (Yunnan Province, China) and WS (Yunnan Province, China) populations.

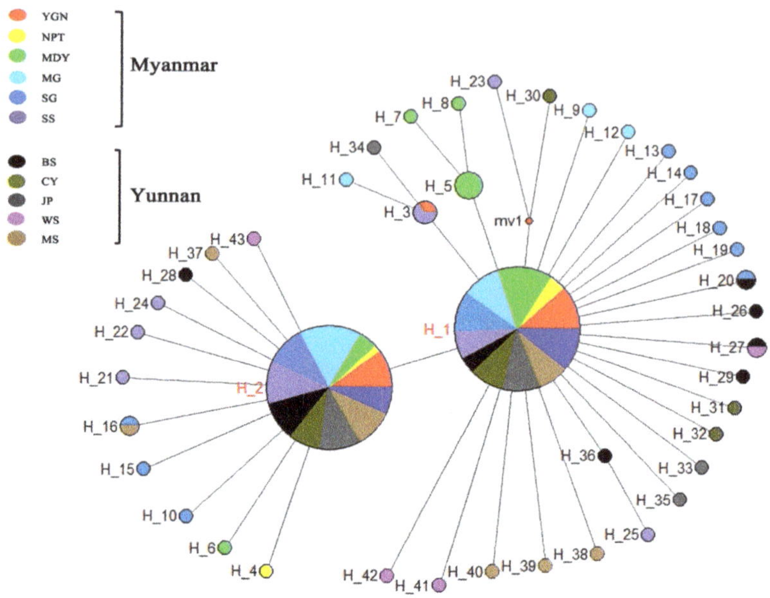

Figure 2. Haplotype network of the mtCOI gene. The haplotype network of COI gene sequences was inferred using the median-joining algorithm and the software Network v. 4.6.1.0. The calculated MP was used to identify and remove unnecessary median vectors and links. Each color represents a specific population.

3.2. Population Genetic Structure and Haplotype Distribution

As shown in Table 2, we conducted an analysis of genetic diversity among 11 different populations based on the mtCOI gene. The results revealed that in Yunnan Province, China, and northern Myanmar, the number of haplotypes (*H*) was more abundant compared to the central part of Myanmar (NPT) and the southern part of Myanmar (YGN). The haplotype diversity (*Hd*) among populations in Myanmar ranged from 0.362 (NPT) to 0.662 (SS), while in Yunnan Province, China, it ranged from 0.488 (WS) to 0.662 (BS). The average *Hd* value in Myanmar (0.589) was not significantly different from that in Yunnan Province, China (0.571). Nine of the eleven populations, except for those from the Yangon and Naypyitaw regions in Myanmar, had negative Fu's *F* and Tajima's *D* indices ($p < 0.05$). These results indicated a recent population expansion following a bottleneck event.

Table 2. Genetic diversity indices of *S. furcifera* populations in Myanmar and Yunnan Province, China.

Population Code (No. of Individuals Tested)	*S*	*η*	*H*	*D* (*p*)	*Fs* (*p*)	*Hd* (SD)	*π* (SD)	*K*	*π* (JC)
YGN (43)	2	2	3	−0.10389 ($p > 0.10$)	0.011 (0.280)	0.424 (0.068)	0.00055 (0.00010)	0.436	0.00055
NPT(15)	2	2	3	−0.59419 ($p > 0.10$)	0.518 (0.251)	0.362 (0.145)	0.00060 (0.00027)	0.476	0.00060
MDY (48)	5	5	6	−1.20378 ($p > 0.10$)	−2.582 (0.039)	0.427 (0.085)	0.00072 (0.00017)	0.572	0.00073
MG (47)	6	6	6	−1.24044 ($p > 0.10$)	−2.187 (0.067)	0.574 (0.047)	0.00089 (0.00014)	0.705	0.00090
SG (47)	10	10	10	−1.88055 ($p < 0.05$)	−7.005 (0.001)	0.599 (0.068)	0.00100 (0.00020)	0.792	0.00100
SS (38)	10	10	8	−1.72907 ($p > 0.10$)	−3.424 (0.022)	0.662 (0.060)	0.00104 (0.00017)	0.825	0.00129

Table 2. *Cont.*

Population Code (No. of Individuals Tested)	*S*	*η*	*H*	*D* (*p*)	*Fs* (*p*)	*Hd* (SD)	*π* (SD)	*K*	*π* (JC)
BS (26)	5	5	6	−1.05038 ($p > 0.10$)	−2.447 (0.056)	0.662 (0.060)	0.00104 (0.00017)	0.825	0.00105
CY (36)	5	5	5	−1.29300 ($p > 0.05$)	−1.767 (0.098)	0.505 (0.078)	0.00077 (0.00017)	0.608	0.00077
JP (39)	5	5	5	−1.21593 ($p > 0.10$)	−1.632 (0.107)	0.521 (0.068)	0.00079 (0.00015)	0.621	0.00079
MS (38)	7	7	8	−1.52318 ($p > 0.05$)	−4.725 (0.007)	0.623 (0.065)	0.00096 (0.00015)	0.760	0.00096
WS (40)	6	6	7	−1.51888 ($p > 0.10$)	−4.204 (0.011)	0.488 (0.087)	0.00077 (0.00017)	0.608	0.00077

S, number of polymorphic (segregating) sites; *η*, total number of mutations; *H*, number of haplotypes; *Hd*, haplotype diversity; *π*, nucleotide diversity; *K*, average number of nucleotide differences; *π* (JC), nucleotide diversity with Jukes and Cantor correction; *D*, Tajima's *D* statistic; *Fs*, Fu's *F* test statistic; *p*, significance values of the parameters evaluated using 10,000 simulations.

Nine of fifty-five pairwise *Fst* values based on the mitochondrial gene were significant ($p < 0.05$), such as the differentiation between YGN and BS (*Fst* = 0.02709, $p < 0.05$); SS and WS (*Fst* = 0.06724, $p < 0.05$); and MDY and YGN, MG, SG, SS, BS, JP, and MS (*Fst* = 0.03923, 0.00080, 0.02784, 0.00056, 0.00051, 0.01922, 0.00908, respectively, $p < 0.05$). The highest *Fst* value (*Fst* = 0.98360, $p > 0.05$) was observed between the YGN population and SG population. Most pairwise *Fst* values were not significant ($p > 0.05$), which suggested high-level gene flow among the populations in Myanmar and Yunnan Province, China (Table 3).

Table 3. Pairwise *Fst* values (*p* value) for *S. furcifera* populations in Myanmar and Yunnan Province, China.

	YGN	NPT	MDY	MG	SG	SS	BS	CY	JP	MS	WS
YGN	---										
NPT	0.33603	---									
MDY	0.03923 *	0.62196	---								
MG	0.27267	0.28552	0.00080 *	---							
SG	0.98360	0.83533	0.02784 *	0.25266	---						
SS	0.29515	0.44171	0.00056 *	0.23759	0.30503	---					
BS	0.02709 *	0.09100	0.00051 *	0.25605	0.23489	0.50366	---				
CY	0.55549	0.56302	0.06487	0.41749	0.98038	0.27200	0.06905	---			
JP	0.60843	0.45765	0.01922 *	0.54649	0.99198	0.30037	0.12561	0.85094	---		
MS	0.31108	0.73535	0.00908 *	0.43340	0.87622	0.58144	0.39849	0.68066	0.64490	---	
WS	0.38822	0.88951	0.11603	0.08123	0.83152	0.06724 *	0.04464	0.68255	0.38785	0.26380	---

* Significant values at 0.05 level.

The analysis of molecular variance (AMOVA) revealed that 98.7% of the variation was attributed to differences within populations ($p < 0.05$), 1.9% of the variation was due to differences among populations within the same region ($p < 0.05$), and −0.55% of the variation was attributed to differences among different regions ($p > 0.05$). On the contrary, ΦCT was not significant, indicating no difference in the genetic structure among the *S. furcifera* populations in Myanmar and Yunnan Province of China (Table 4). The maximum likelihood (ML) tree topology showed that there were no significant genealogical branches for all *S. furcifera* populations in Myanmar and Yunnan Province (Figure 3).

Table 4. AMOVA analysis of *S. furcifera* populations in Myanmar and Yunnan Province, China.

Source of Variation	Sum of Squares	Variance Components	Percentage Variation (*p* Value)
Among areas (ΦCT)	0.203	−0.00189	−0.55% (0.90029)
Among population within areas (ΦSC)	5.208	0.00638	1.85% (0.02639)
Within population (ΦST)	137.766	0.33933	98.69% (0.03519)

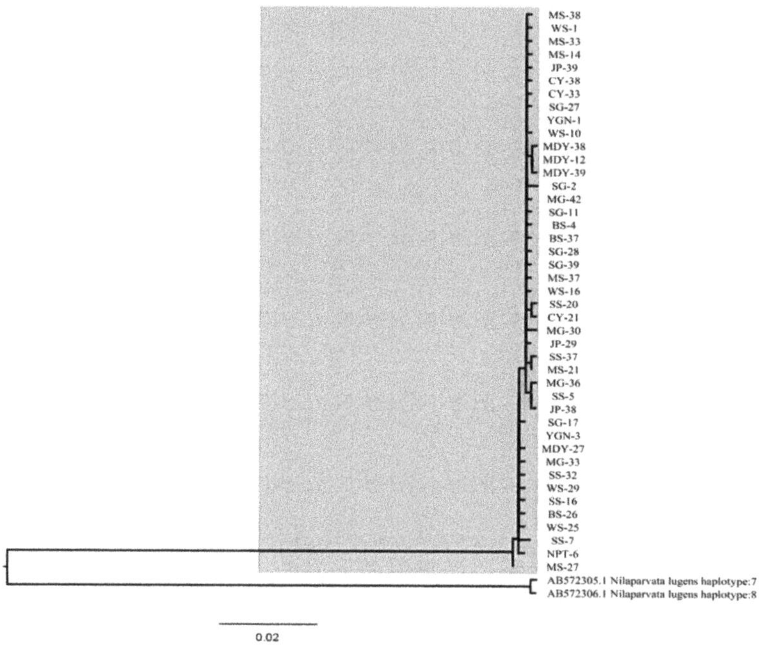

Figure 3. Maximum likelihood tree based on mitochondrial haplotypes of *S. furcifera*. The gray background represents the white-backed planthopper, while the other two sequences represent the brown planthopper.

3.3. Gene Flow Based on Mitochondrial DNA

The unidirectional estimates of total migrants (N_em) ranged from 0.0254 (from YGN to SG, and YGN to BS) to 51.3476 (from SS to JP). The highest numbers of total migrants ($N_em > 200$) were found in some populations of Myanmar (SS 255.15, MG 235.52, and SG 202.46), while moderate numbers of total migrants ($N_em > 100$) were found in some populations of Yunnan Province, China (CY 178.71, WS 169.78, MS 129.30, and JP 108.04). The lowest numbers of total migrants were found in Mandalay Division, Myanmar (MDY 21.80) and Baoshan, Yunnan Province, China (BS 50.29). The total number of migrants (N_em) was less in southern Myanmar, e.g., from YGN to CY (0.035), JP (0.029), MS (0.034), and WS (0.050) (Yunnan Province, China) compared with that in north and northeast Myanmar, which was from MG to CY (47.96), from SG to CY (35.77), from SS to JP (51.35), and from SS to WS (37.11), indicating that intensive gene flow existed among north and northeast Myanmar and Yunnan Province, China (Figure 4, Table 5).

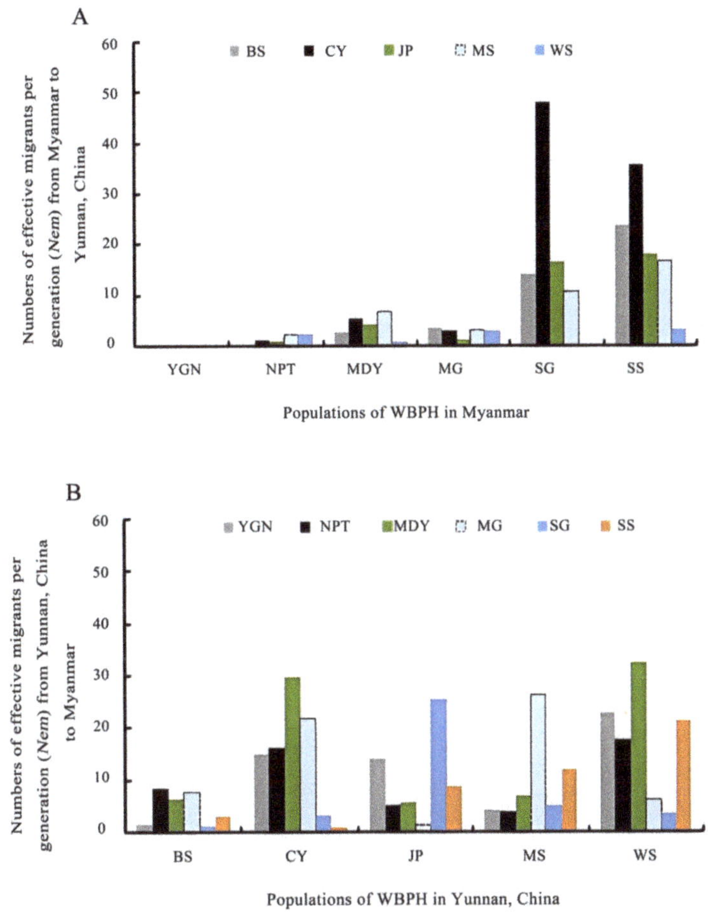

Figure 4. Comparison of the effective migrants per generation (N_em) of *S. furcifera* populations in Myanmar (**A**) and Yunnan Province, China (**B**). The eleven populations were grouped into six populations in Myanmar and five populations in Yunnan Province.

Table 5. Numbers of effective migrants per generation (N_em) in the *S. furcifera* populations found in Myanmar and Yunnan Province, China.

Population Code (i)	θi	YGN→i	NPT→i	MDY→i	MG→i	SG→i	SS→i	BS→i	CY→i	JP→i	MS→i	WS→i	∑(immigration)
YGN	0.0769	---	4.8837	2.0503	50.8836	35.6291	49.9904	1.4941	14.9216	14.2943	4.3621	22.7011	201.2103
NPT	0.0070	19.7556	---	3.1556	0.0284	25.2077	5.8986	8.5400	16.3450	5.3087	4.0499	17.8145	106.104
MDY	0.0033	2.0240	1.7057	---	8.6284	35.7985	13.5892	6.5054	29.7385	5.7505	7.1291	32.4422	143.3115
MG	0.0860	4.0226	4.2957	1.9391	---	6.5406	23.6466	7.6845	21.8516	1.5264	26.3409	6.1254	103.9734
SG	0.0363	0.0254	0.3617	1.4617	73.4156	---	28.8666	1.1993	3.1616	25.4497	5.1209	3.6489	142.7114
SS	0.0522	2.9464	2.3777	0.0534	23.5356	11.2074	---	3.2281	1.0616	8.8223	12.0602	21.2152	86.5079
BS	0.0087	0.0254	0.4643	2.6978	3.8124	14.1333	23.7510	---	22.1785	12.7008	25.2923	7.5781	112.6339
CY	0.0350	1.4090	5.5603	3.1468	47.9596	35.7742	2.6972	8.2047	---	5.8271	7.1291	19.4985	137.2065
JP	0.0288	0.8456	4.0950	1.1783	16.4836	18.1713	51.3476	1.1011	33.6350	---	13.2876	19.9606	160.1057
MS	0.0335	2.5882	6.8810	3.2079	10.7500	16.7448	18.2528	8.5573	33.0050	0.0095	---	18.8051	118.8016
WS	0.0495	2.4852	0.9777	2.9136	0.0284	3.2523	37.1142	3.7773	2.8116	28.3487	24.5335	---	106.2425
∑(emigration)	---	36.1274	31.6028	21.8045	235.5256	202.4592	255.1542	50.2918	178.7100	108.0380	129.3056	169.7896	---

In Myanmar, where six populations (YGN, NPT, MDY, MG, SG, SS) were identified, and in Yunnan Province, China, where five populations (BS, CY, JP, MS, WS) were studied, it was observed that the effective migration values in the northern and northeastern regions of Myanmar (SG, SS) were higher than those in the central and southern parts of Myanmar (YGN, NPT). Simultaneously, in the southwestern (BS, CY) and southern (JP) regions of

Yunnan Province, the effective immigrant values (immigration from Myanmar) were higher than the effective emigrant values (emigration to Myanmar) (Figure 4). By comparing the migration values in Myanmar and Yunnan Province, China, it can be inferred that north and northeast Myanmar were the main migration sources of WBPH populations in southwest and south Yunnan Province.

3.4. Demographic History Analysis

Five out of the eleven populations showed negative Fu's F indices with statistical significance ($p < 0.05$), indicating recent demographic expansion in *S. furcifera* (Table 2). These findings were consistent with the results obtained from the Bayesian skyline plot (BSP) analysis. The BSP analysis, conducted with a substitution rate of 1.77% per million years based on the mtCOI gene, revealed that the effective population size of the WBPH expanded approximately 1000 to 7000 years ago (Figure 5).

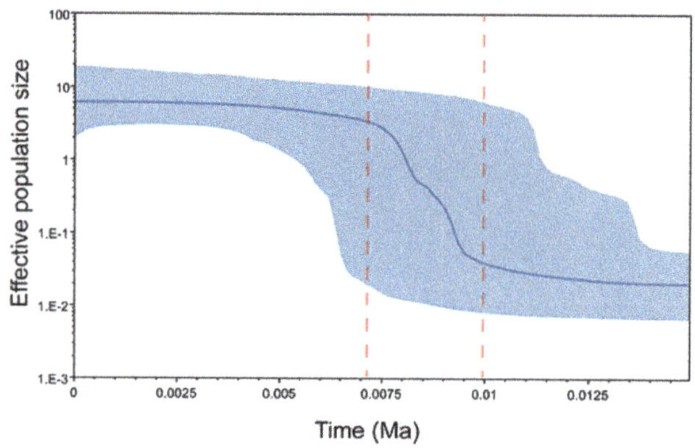

Figure 5. The demographic history of *S. furcifera* was reconstructed through Bayesian skyline plots utilizing the mtCOI gene with a substitution rate of 0.0177. The X-axis represents the timescale in the past, while the Y-axis displays the estimated effective population size. Solid curves indicate the median effective population size, and the shaded range represents the 95% highest posterior density (HPD) intervals.

4. Discussion

Mitochondrial genes are extremely suitable for tracing the history of populations, estimation of migration, and gene flow [44,45]. Since the mitochondria of most insects are haploid and inherited maternally, the mitochondrial genome is transmitted clonally [46]. In contrast, the nuclear genome is diploid with biparental inheritance. Therefore, mitochondrial DNA variations allow for more precise tracking of recent population histories compared to nuclear loci [47]. In this study, 417 individuals were sampled from six Provinces in Myanmar and five sites in Yunnan Province, China. The mtCOI was used to estimate genetic variations among geographic populations of *S. furcifera* in Myanmar and Yunnan Province, China. Among the confirmed 43 mtCOI haplotypes, haplotypes H_1 and H_2 were found to be widely distributed in all populations of *S. furcifera* in Myanmar and Yunnan Province, China. This suggests extensive gene flow among these populations. In addition, H_16 was shared by the SG (Myanmar) and MS (Yunnan Province, China) populations and H_20 was shared by the SG (Myanmar) and BS (Yunnan Province, China) populations, indicating a close relationship between WBPH populations from north Myanmar and southwest–south Yunnan Province, China. Matsumoto et al. identified 20 WBPH haplotypes based on the COI fragment [19], 23 haplotypes less than present study, possibly due to fewer samples being tested (20 individuals per population) and lacking samples collected from Myanmar and Yunnan Province.

Genetic variations were found in migratory insects; the population's genetic structure can be affected by evolutionary factors, such as mating system, gene flow, mode of reproduction, and natural selection [48]. Previous studies demonstrated that Myanmar was one of the possible migration sources of *S. furcifera* to Yunnan Province, China [13,20]. In the present study, the total number of migrants (N_em) from the YGN, NPT, MDY, MG, SG, and SS populations to Yunnan Province, China, was 7.4, 18.0, 13.1, 79.0, 88.1, and 133.2, respectively. Furthermore, the SG of Myanmar and the CY of Yunnan Province, China had the highest N_em value, while the SS of Myanmar and the JP of Yunnan Province, China had the highest N_em value. All these results suggested that north Myanmar (Sagaing Division and Shan State) might be the major immigration source of the populations in southwestern (e.g., MS and CY) and southern (JP) Yunnan Province, China. The results also confirmed the previous findings on the migration simulations of *S. furcifera* in the Greater Mekong Subregion (GMS) countries [7].

Temperature stress has adverse effects on the survival and reproduction of the WBPH, explaining their adoption of long-distance migration strategies [49]. Under low-temperature conditions (5 °C) after 48 h of treatment, the mortality rate significantly increased (>70%), much higher than the mortality rate under optimal temperature conditions (25 °C) (<5%) [50]. This indicates that the WBPH lacks cold tolerance. Additionally, wind is considered a vector for many migrating organisms, playing a crucial role in their migration process [51]. Every spring, driven by the southwest monsoon, the WBPH migrates northward and can even reach northern China, Japan, and Korea [52]. Analysis based on the mtCOI dataset suggests that the *S. furcifera* population began to expand around 10,000 years ago, which aligns with the results of population expansion after the last glacial period. With the warming climate after the last glacial period, the number of suitable habitats gradually increased. Our study findings are consistent with research conclusions regarding classical expansion patterns after the last glacial period. Similar patterns have been observed in other insects, such as *Stomoxys calcitrans* [53], *Locusta migratoria* [54], *Megacopta cribraria* [55], and *Nesidiocoris tenuis* [56].

Understanding the migration patterns of significant agricultural pests is crucial for developing sustainable pest management strategies. Failure to promptly implement scientifically effective control measures can result in incalculable damage to crops during large-scale migrations, leading to severe economic problems [57–59]. Moreover, such massive pest migrations can trigger serious ecological issues. During their migration, pests may attack various plants, not just limited to crops, but may also pose a threat to wild vegetation, disrupting the balance of natural ecosystems. Therefore, understanding the migration pattern of pests and adopting scientific and reasonable pest management strategies are of vital significance for maintaining ecological balance, guaranteeing the safe production of crops and realizing the sustainable development of agriculture.

5. Conclusions

This study revealed that north and northeast Myanmar might be the major WBPH immigration source of the populations in southwestern and southern Yunnan Province, China. These findings might be useful in designing regional control programs for WBPH populations in these regions.

Author Contributions: Investigation, methodology, software, and writing—original draft, Y.L. (Yue Liu) and K.N.C.; investigation and methodology, X.L. and X.Z.; methodology and writing—review and editing, D.C.; formal analysis, supervision, and validation, A.C.; conceptualization, funding acquisition, supervision, validation, and writing—review and editing, Y.L. (Ying Liu) and Y.Y. All authors have read and agreed to the published version of the manuscript.

Funding: This work was supported by the National Natural Science Foundation of China (31860504), the Special Funds of the Major Science and Technology Project in Yunnan Province (202102AE090003), Yunnan Provincial Science and Technology Talent and Platform Plan (202005AC160031), the Key Laboratory of Prevention and control of invasive alien organisms of the Ministry of Agriculture and Rural Affairs (Kunming), and the Provincial Innovation team of Collaborative Green Control of Agricultural Cross-border Pests of Yunnan Academy of Agricultural Sciences and Talented Young Scientist Program supported by China Science and Technology Exchange Center.

Data Availability Statement: Data are contained within the article.

Conflicts of Interest: The authors declare no conflict of interest.

References

1. Rubia-Sanchez, E.; Suzuki, Y.; Arimura, K.; Miyamoto, K.; Matsumura, M.; Watanabe, T. Comparing *Nilaparvata lugens* (Stal) and *Sogatella furcifera* (Horvath) (Homoptera: Delphacidae) feeding effects on rice plant growth processes at the vegetative stage. *Crop Prot.* **2003**, *22*, 967–974. [CrossRef]
2. Xiao, T.-G.; Tang, J.-X. Effects of the susceptibility of rice varieties to *Sogatella furcifera* on nymphal development and reproduction of *Microvelia horvathi* through a food chain. *Insect Sci.* **2007**, *14*, 317–321. [CrossRef]
3. Dyck, V.A.; Thomas, B. The brown planthopper problem. In *Brown Planthopper: Threat to Rice Production in Asia*; International Rice Research Institute: Los Baños, Philippines, 1979; pp. 3–17.
4. Dingle, H.; Drake, V.A. What Is Migration? *BioScience* **2007**, *57*, 113–121. [CrossRef]
5. Alerstam, T.; Chapman, J.W.; Bäckman, J.; Smith, A.D.; Karlsson, H.; Nilsson, C.; Reynolds, D.R.; Klaassen, R.H.G.; Hill, J.K. Convergent patterns of long-distance nocturnal migration in noctuid moths and passerine birds. *Proc. R. Soc. B Biol. Sci.* **2011**, *278*, 3074–3080. [CrossRef] [PubMed]
6. Hu, G.; Lu, F.; Zhai, B.; Lu, M.; Liu, W.; Zhu, F.; Wu, X.-W.; Chen, G.; Zhang, X. Outbreaks of the Brown Planthopper *Nilaparvata lugens* (Stål) in the Yangtze River Delta: Immigration or Local Reproduction? *PLoS ONE* **2014**, *9*, e88973. [CrossRef] [PubMed]
7. Otuka, A. Migration of rice planthoppers and their vectored re-emerging and novel rice viruses in East Asia. *Front. Microbiol.* **2013**, *4*, 309. [CrossRef] [PubMed]
8. Zhou, G.; Wen, J.; Cai, D.; Li, P.; Xu, D.; Zhang, S. Southern rice black-streaked dwarf virus: A new proposed *Fijivirus* species in the family *Reoviridae*. *Chin. Sci. Bull.* **2008**, *53*, 3677–3685. [CrossRef]
9. Zhou, G.; Xu, D.; Xu, D.; Zhang, M. Southern rice black-streaked dwarf virus: A white-backed planthopper-transmitted fijivirus threatening rice production in Asia. *Front. Microbiol.* **2013**, *4*, 270. [CrossRef] [PubMed]
10. Zhao, X.; Shen, H.; Yin, Y.; Li, X.; Lyu, J.; Chen, A. The occurrence characteristics and population dynamics of *Sogatella furcifera* (Horváth) in Yunnan Province. *Chin. J. Appl. Entomol.* **2013**, *51*, 516–524.
11. Win, S.S.; Muhamad, R.; Ahmad, Z.A.M.; Adam, N.A. Population fluctuations of Brown Plant Hopper *Nilaparvata lugens Stal.* and White Backed Plant Hopper *Sogatella furcifera* Horvath on rice. *J. Entomol.* **2011**, *8*, 183–190. [CrossRef]
12. Hoang, A.T.; Zhang, H.-M.; Yang, J.; Chen, J.-P.; Hébrard, E.; Zhou, G.; Vinh, V.N.; Cheng, J. Identification, Characterization, and Distribution of Southern rice black-streaked dwarf virus in Vietnam. *Plant Dis.* **2011**, *95*, 1063–1069. [CrossRef] [PubMed]
13. Yin, Y.; Li, X.; Chu, D.; Zhao, X.; Sathya, K.; Douangboupha, B.; Kyaw, M.M.; Kongchuensin, M.; Somrith, A.; Ngo, V.V.; et al. Extensive gene flow of white-backed planthopper in the Greater Mekong Subregion as revealed by microsatellite markers. *Sci. Rep.* **2017**, *7*, 15905. [CrossRef]
14. Huang, S.-H.; Cheng, C.-H.; Chen, C.-N.; Wu, W.-J.; Otuka, A. Estimating the immigration source of rice planthoppers, *Nilaparvata lugens* (Stål) and *Sogatella furcifera* (Horváth) (Homoptera: Delphacidae), in Taiwan. *Appl. Entomol. Zool.* **2010**, *45*, 521–531. [CrossRef]
15. Hu, G.; Lu, M.H.; Tuan, H.A.; Liu, W.C.; Xie, M.C.; McInerney, C.E.; Zhai, B.P. Population dynamics of rice planthoppers, *Nilaparvata lugens* and *Sogatella furcifera* (Hemiptera, Delphacidae) in Central Vietnam and its effects on their spring migration to China. *Bull. Entomol. Res.* **2016**, *107*, 369–381. [CrossRef] [PubMed]
16. Hu, C.; Fu, X.; Wu, K. Seasonal migration of white-backed planthopper *Sogatella furcifera* Horváth (Hemiptera: Delphacidae) over the Bohai Sea in northern China. *J. Asia-Pac. Entomol.* **2017**, *20*, 1358–1363. [CrossRef]
17. Shen, H.M.; Lyu, J.P.; Zhou, J.Y.; Cheng, X.N.; Zhang, X.X.; Zhai, B.P. Source areas and landing mechanism of early immigration of white-backed planthoppers *Sogatella furcifera* (Horváth) in Yunnan, 2009. *Acta Ecol. Sin.* **2011**, *31*, 4350–4364.
18. Liu, J.N.; Gui, F.R.; Li, Z.Y. Genetic diversity of the planthopper, *Sogatella furcifera* in the Greater Mekong Subregion detected by inter-simple sequence repeats (ISSR) markers. *J. Insect Sci.* **2010**, *10*, 52. [CrossRef]
19. Matsumoto, Y.; Matsumura, M.; Sanada-Morimura, S.; Hirai, Y.; Sato, Y.; Noda, H. Mitochondrial cox sequences of *Nilaparvata lugens* and *Sogatella furcifera* (Hemiptera, Delphacidae): Low specificity among Asian planthopper populations. *Bull. Entomol. Res.* **2013**, *103*, 382–392. [CrossRef]
20. Li, X.Y.; Chu, D.; Yin, Y.Q.; Zhao, X.Q.; Chen, A.D.; Khay, S.; Douangboupha, B.; Kyaw, M.M.; Kongchuensin, M.; Ngo, V.V.; et al. Possible Source Populations of the White-backed Planthopper in the Greater Mekong Subregion Revealed by Mitochondrial DNA Analysis. *Sci. Rep.* **2016**, *6*, 39167. [CrossRef]
21. Sun, S.; Bao, Y.; Wu, Y.; Lu, M.; Tuan, H. Analysis of the Huge Immigration of *Sogatella furcifera* (Hemiptera: Delphacidae) to Southern China in the Spring of 2012. *Environ. Entomol.* **2018**, *47*, 8–18. [CrossRef]
22. Hu, S.; Fu, D.; Han, Z.; Ye, H. Density, Demography, and Influential Environmental Factors on Overwintering Populations of *Sogatella furcifera* (Hemiptera: Delphacidae) in Southern Yunnan, China. *J. Insect Sci.* **2015**, *15*, 58. [CrossRef]
23. Kranthi, S.; Ghodke, A.B.; Puttuswamy, R.K.; Mandle, M.; Nandanwar, R.; Satija, U.; Pareek, R.K.; Desai, H.; Udikeri, S.S.; Balakrishna, D.J.; et al. Mitochondria COI-based genetic diversity of the cotton leafhopper *Amrasca biguttula biguttula* (Ishida) populations from India. *Mitochondrial DNA Part A* **2018**, *29*, 228–235. [CrossRef]

24. Dumidae, A.; Janthu, P.; Subkrasae, C.; Pumidonming, W.; Dekumyoy, P.; Thanwisai, A.; Vitta, A. Genetic analysis of *Cryptozona siamensis* (Stylommatophora, Ariophantidae) populations in Thailand using the mitochondrial 16S rRNA and COI sequences. *PLoS ONE* **2020**, *15*, e0239264. [CrossRef] [PubMed]
25. An, H.S.; Jee, Y.J.; Min, K.S.; Kim, B.L.; Han, S.J. Phylogenetic analysis of six species of pacific abalone (Haliotidae) based on DNA sequences of 16s rRNA and cytochrome c oxidase subunit I mitochondrial genes. *Mar. Biotechnol.* **2005**, *7*, 373–380. [CrossRef] [PubMed]
26. Wang, Z.-L.; Wang, T.-Z.; Zhu, H.-F.; Pan, H.-B.; Yu, X.-P. Diversity and dynamics of microbial communities in brown planthopper at different developmental stages revealed by high-throughput amplicon sequencing. *Insect Sci.* **2020**, *27*, 883–894. [CrossRef] [PubMed]
27. Simon, C.; Frati, F.; Beckenbach, A.; Crespi, B.; Liu, H.; Flook, P. Evolution, weighting, and phylogenetic utility of mitochondrial gene sequences and a compilation of conserved polymerase chain reaction primers. *Ann. Entomol. Soc. Am.* **1994**, *87*, 651–701. [CrossRef]
28. Othman, S.N.; Putri, E.T.; Messenger, K.R.; Bae, Y.; Yang, Y.; Bova, T.; Reed, T.; Amin, H.; Chuang, M.-F.; Jang, Y.; et al. Impact of the Miocene orogenesis on *Kaloula* spp. radiation and implication of local refugia on genetic diversification. *Integr. Zool.* **2022**, *17*, 261–284. [CrossRef]
29. Bandelt, H.J.; Forster, P.; Röhl, A. Median-joining networks for inferring intraspecific phylogenies. *Mol. Biol. Evol.* **1999**, *16*, 37–48. [CrossRef]
30. Watterson, G.A. On the number of segregating sites in genetical models without recombination. *Theor. Popul. Biol.* **1975**, *7*, 256–276. [CrossRef]
31. Tajima, F. Evolutionary relationship of DNA sequences in finite populations. *Genetics* **1983**, *105*, 437–460. [CrossRef]
32. Nei, M. *Molecular Evolutionary Genetics*; Columbia University Press: New York, NY, USA, 1987.
33. Lynch, M.; Crease, T.J. The analysis of population survey data on DNA sequence variation. *Mol. Biol. Evol.* **1990**, *7*, 377–394. [PubMed]
34. Tajima, F. Statistical method for testing the neutral mutation hypothesis by DNA polymorphism. *Genetics* **1989**, *123*, 585–595. [CrossRef] [PubMed]
35. Fu, Y. Statistical Tests of Neutrality of Mutations Against Population Growth, Hitchhiking and Background Selection. *Genetics* **1997**, *147*, 915–925. [CrossRef] [PubMed]
36. Beerli, P.; Felsenstein, J. Maximum likelihood estimation of a migration matrix and effective population sizes in *n* subpopulations by using a coalescent approach. *Proc. Natl. Acad. Sci. USA* **2001**, *98*, 4563–4568. [CrossRef]
37. Excoffier, L.; Lischer, H.E.L. Arlequin suite ver 3.5: A new series of programs to perform population genetics analyses under Linux and Windows. *Mol. Ecol. Resour.* **2010**, *10*, 564–567. [CrossRef] [PubMed]
38. Wen, H.; Luo, T.; Wang, Y.; Wang, S.; Liu, T.; Xiao, N.; Zhou, J. Molecular phylogeny and historical biogeography of the cave fish genus *Sinocyclocheilus* (Cypriniformes: Cyprinidae) in southwest China. *Integr. Zool.* **2022**, *17*, 311–325. [CrossRef]
39. Stamatakis, A. RAxML version 8: A tool for phylogenetic analysis and post-analysis of large phylogenies. *Bioinformatics* **2014**, *30*, 1312–1313. [CrossRef] [PubMed]
40. Cheng, L.; Connor, T.R.; Sirén, J.; Aanensen, D.M.; Corander, J. Hierarchical and Spatially Explicit Clustering of DNA Sequences with BAPS Software. *Mol. Biol. Evol.* **2013**, *30*, 1224–1228. [CrossRef]
41. Bandelt, H.J.; Macaulay, V.; Richards, M. Median Networks: Speedy Construction and Greedy Reduction, One Simulation, and Two Case Studies from Human mtDNA. *Mol. Phylogenet. Evol.* **2000**, *16*, 8–28. [CrossRef]
42. Drummond, A.J.; Rambaut, A. BEAST: Bayesian evolutionary analysis by sampling trees. *BMC Evol. Biol.* **2007**, *7*, 214. [CrossRef]
43. Papadopoulou, A.; Anastasiou, I.; Vogler, A.P. Revisiting the Insect Mitochondrial Molecular Clock: The Mid-Aegean Trench Calibration. *Mol. Biol. Evol.* **2010**, *27*, 1659–1672. [CrossRef] [PubMed]
44. Polzin, T.; Daneshmand, S.V. On Steiner trees and minimum spanning trees in hypergraphs. *Oper. Res. Lett.* **2003**, *31*, 12–20. [CrossRef]
45. Qi, J.; Shi, F.; Zhang, B.; Chen, X.; Xing, J.; Furumitsu, K.; Corush, J.B.; Yamaguchi, A.; Zhang, J. Insights into genetic variation and demographic history of sharpnose rays: Examinations of three species of *Telatrygon* (Elasmobranchii, Dasyatidae) from the Indo-West Pacific. *Integr. Zool.* **2022**, *17*, 1063–1077. [CrossRef] [PubMed]
46. Sato, M.; Sato, K. Maternal inheritance of mitochondrial DNA by diverse mechanisms to eliminate paternal mitochondrial DNA. *Biochim. Biophys. Acta (BBA)-Mol. Cell Res.* **2013**, *1833*, 1979–1984. [CrossRef] [PubMed]
47. Moore, W.S. Inferring phylogenies from mtDNA variation: Mitochondrial-gene trees versus nuclear-gene trees. *Evolution* **1995**, *49*, 718–726. [CrossRef] [PubMed]
48. Hamrick, J. Allozyme diversity in plant species. In *Plant Population Genetics Breeding & Genetic Resources*; Sinauer Associates Inc.: Sunderland, MA, USA, 1989.
49. Ma, M.; Peng, Z.; He, Y. Effects of Temperature on Functional Response of *Anagrus nilaparvatae* Pang et Wang (Hymenoptera: Mymaridae) on the Eggs of Whitebacked Planthopper, *Sogatella furcifera* Horváth and Brown Planthopper, *Nilaparvata lugens* Stål. *J. Integr. Agric.* **2012**, *11*, 1313–1320. [CrossRef]
50. Xu, D.; Zhong, T.; Feng, W.; Zhou, G. Tolerance and responsive gene expression of *Sogatella furcifera* under extreme temperature stresses are altered by its vectored plant virus. *Sci. Rep.* **2016**, *6*, 31521. [CrossRef]

51. Wu, Y.; Sun, S.; Jiang, Z.; Chen, A.; Ma, M.; Zhang, G.; Zhai, B.; Li, C. Immigration Pathways of White-Backed Planthopper in the Confluence Area of the Two Monsoon Systems. *J. Econ. Entomol.* **2022**, *115*, 1480–1489. [CrossRef]
52. Otuka, A.; Watanabe, T.; Suzuki, Y.; Matsumura, M.; Furuno, A.; Chino, M. Real-time prediction system for migration of rice planthoppers *Sogatella furcifera* (Horváth) and *Nilaparvata lugens* (Stål) (Homoptera: Delphacidae). *Appl. Entomol. Zool.* **2005**, *40*, 221–229. [CrossRef]
53. Dsouli Aymes, N.; Michaux, J.; De Stordeur, E.; Couloux, A.; Veuille, M.; Duvallet, G. Global population structure of the stable fly (*Stomoxys calcitrans*) inferred by mitochondrial and nuclear sequence data. *Infect. Genet. Evol.* **2011**, *11*, 334–342. [CrossRef]
54. Ma, C.; Yang, P.; Jiang, F.; Chapuis, M.-P.; Shali, Y.; Sword, G.A.; Kang, L.E. Mitochondrial genomes reveal the global phylogeography and dispersal routes of the migratory locust. *Mol. Ecol.* **2012**, *21*, 4344–4358. [CrossRef] [PubMed]
55. Li, K.; Lin, C.; Liang, A. Comparative phylogeography of two hemipteran species (*Geisha distinctissima* and *Megacopta cribraria*) in the Zhoushan Archipelago of China reveals contrasting genetic structures despite concordant historical demographies. *Heredity* **2020**, *124*, 207–222. [CrossRef] [PubMed]
56. Xun, H.; Li, H.; Li, S.; Wei, S.; Zhang, L.; Song, F.; Jiang, P.; Yang, H.; Han, F.; Cai, W. Population genetic structure and post-LGM expansion of the plant bug *Nesidiocoris tenuis* (Hemiptera: Miridae) in China. *Sci. Rep.* **2016**, *6*, 26755. [CrossRef] [PubMed]
57. Chapman, J.W.; Bell, J.R.; Burgin, L.E.; Reynolds, D.R.; Pettersson, L.B.; Hill, J.K.; Bonsall, M.B.; Thomas, J.A. Seasonal migration to high latitudes results in major reproductive benefits in an insect. *Proc. Natl. Acad. Sci. USA* **2012**, *109*, 14924–14929. [CrossRef]
58. Chapman, J.W.; Reynolds, D.R.; Wilson, K. Long-range seasonal migration in insects: Mechanisms, evolutionary drivers and ecological consequences. *Ecol. Lett.* **2015**, *18*, 287–302. [CrossRef]
59. Bebber, D.P.; Holmes, T.; Gurr, S.J. The global spread of crop pests and pathogens. *Glob. Ecol. Biogeogr.* **2014**, *23*, 1398–1407. [CrossRef]

genes

MDPI

Article

Comparative Transcriptomic Assessment of Chemosensory Genes in Adult and Larval Olfactory Organs of *Cnaphalocrocis medinalis*

Hai-Tao Du [1], Jia-Qi Lu [1], Kun Ji [1], Chu-Chu Wang [1], Zhi-Chao Yao [1], Fang Liu [1,2,3,*] and Yao Li [1,*]

1 College of Plant Protection, Yangzhou University, Yangzhou 225009, China; dx120180086@stu.yzu.edu.cn (H.-T.D.); mz120221473@stu.yzu.edu.cn (J.-Q.L.); dx120230149@stu.yzu.edu.cn (K.J.); wangchuchu@jscxjszyxy1.wecom.work (C.-C.W.); 008412@yzu.edu.cn (Z.-C.Y.)

2 Jiangsu Co-Innovation Center for Modern Production Technology of Grain Crops, Yangzhou University, Yangzhou 225009, China

3 Joint International Research Laboratory of Agriculture & Agri-Product Safety, Yangzhou University, Yangzhou 225009, China

* Correspondence: liufang@yzu.edu.cn (F.L.); liyao@yzu.edu.cn (Y.L.)

Abstract: The rice leaf folder, *Cnaphalocrocis medinalis* (Lepidoptera: Pyralidae), is a notorious pest of rice in Asia. The larvae and adults of *C. medinalis* utilize specialized chemosensory systems to adapt to different environmental odors and physiological behaviors. However, the differences in chemosensory genes between the olfactory organs of these two different developmental stages remain unclear. Here, we conducted a transcriptome analysis of larvae heads, male antennae, and female antennae in *C. medinalis* and identified 131 putative chemosensory genes, including 32 *OBPs* (8 novel *OBPs*), 23 *CSPs* (2 novel *CSPs*), 55 *ORs* (17 novel *ORs*), 19 *IRs* (5 novel *IRs*) and 2 *SNMPs*. Comparisons between larvae and adults of *C. medinalis* by transcriptome and RT-qPCR analysis revealed that the number and expression of chemosensory genes in larval heads were less than that of adult antennae. Only 17 chemosensory genes (7 *OBPs* and 10 *CSPs*) were specifically or preferentially expressed in the larval heads, while a total of 101 chemosensory genes (21 *OBPs*, 9 *CSPs*, 51 *ORs*, 18 *IRs*, and 2 *SNMPs*) were specifically or preferentially expressed in adult antennae. Our study found differences in chemosensory gene expression between larvae and adults, suggesting their specialized functions at different developmental stages of *C. medinalis*. These results provide a theoretical basis for screening chemosensory genes as potential molecular targets and developing novel management strategies to control *C. medinalis*.

Keywords: *Cnaphalocrocis medinalis*; chemosensory genes; larva; adult; transcriptome analysis; expression pattern

Citation: Du, H.-T.; Lu, J.-Q.; Ji, K.; Wang, C.-C.; Yao, Z.-C.; Liu, F.; Li, Y. Comparative Transcriptomic Assessment of Chemosensory Genes in Adult and Larval Olfactory Organs of *Cnaphalocrocis medinalis*. *Genes* **2023**, *14*, 2165. https://doi.org/10.3390/genes14122165

Academic Editor: Zhiteng Chen

Received: 6 November 2023
Revised: 27 November 2023
Accepted: 28 November 2023
Published: 30 November 2023

1. Introduction

The chemosensory system serves insects to cope with highly intricate and perpetually fluctuating chemical environments [1]. The insect chemosensory system relies on a diversity of proteins expressed in the chemosensory sensilla that are located on sensory organs, such as antennae and mouthparts [2]. These chemosensory proteins group several multi-gene families involved in odor reception, including two classes of ligand-binding proteins: odorant-binding proteins (OBPs) and chemosensory proteins (CSPs), and three classes of membrane receptors: odorant receptors (ORs), ionotropic receptors (IRs), and sensory neuron membrane proteins (SNMPs) [3,4]. The OBP and CSP families, two types of soluble proteins secreted in the sensory lymph, possess six or four conserved cysteines, respectively [5,6]. Two families of chemosensory receptors, namely ORs and IRs, are divergent transmembrane proteins (ORs with a seven-transmembrane domain and IRs with a four-transmembrane domain) that function as heteromeric ligand-gated ion channels with conserved co-receptors [7,8]. Additionally, SNMPs are a smaller receptor family consisting of two members, SNMP1 and SNMP2, which are characterized by two transmembrane domains [9]. These receptors have specific interactions with odorant chemicals [10]. After entering the chemosensory sensilla, external odorant chemicals are captured by ligand-binding proteins, OBPs or CSPs, and delivered to the chemosensory receptors in the dendritic membrane of olfactory sensory neurons (OSNs) [4]. The chemical signals received by the OSNs are transformed into electrical impulses, which are subsequently conveyed to the insect brain, thereby triggering behavioral responses [11].

The larvae and adults of Holometabola, with different morphologies and lifestyles, possess differentiated olfactory organs and specialized chemosensory systems for detecting environmental odors [12]. The larvae primarily forage, feed, and grow for metamorphosis, whereas the adults feed on different substrates to facilitate mate-seeking and reproduction during this developmental stage [13]. Accordingly, larvae and adults exhibit differential sensitivity to chemosensory cues and develop distinct chemosensory systems [14,15]. The larvae of *Drosophila suzukii* and *Aedes aegypti* expressed 34 *ORs* and 24 *ORs*, respectively, which is significantly lower than the number of adult *ORs*: 55 *ORs* and 84 *ORs* [16,17]. These genetic differences in chemosensory genes between larvae and adults have been verified in a few Lepidoptera species, including *Sesamia nonagrioides*, *Cydia pomonella*, *Loxostege sticticalis*, *Spodoptera littoralis*, and *Bombyx mori* [18–22]. Generally, the number and expression level of chemosensory genes in the olfactory organs of larvae are lower than those in adults [18–22]. For instance, *C. pomonella* expressed 16 *ORs* in larval heads, while a significantly higher 57 *ORs* were expressed in the adult antennae [19]. So far, only a limited number of studies have examined the differences in one or several types of chemosensory genes between larvae and adults. However, there has been a lack of systematic investigation into the divergences of the chemosensory systems between larvae and adults.

The rice leaf folder, *C. medinalis* (Guenée), is a destructive migratory pest that causes significant damage to rice cultivation across Asia, Oceania, and Africa [23]. The larvae inflict damage on rice by folding the leaves and scraping the green leaf tissues within the fold, resulting in yield loss by reducing photosynthetic activity [24]. The transcriptome sequencing of antennae from *C. medinalis* adults has successfully identified 102 chemosensory genes, including 30 *OBPs*, 26 *CSPs*, 29 *ORs*, 15 *IRs*, and 2 *SNMPs* [25]. Another transcriptome analysis, reported in 2017, discovered more chemosensory genes, among which 37 newly discovered chemosensory genes (3 *OBPs*, 29 *ORs*, and 5 *IRs*) were identified [26]. There is a lack of systematic studies on the differences in chemosensory genes between *C. medinalis* larval and adult olfactory organs. In this study, we performed a transcriptome analysis of larval heads and male/female adult antennae from *C. medinalis*. Through the transcriptomic data analysis, we identified chemosensory genes in larval heads and adult antennae, investigated phylogenetic relationships and motifs, and compared the expression patterns of chemosensory genes between larval and adult olfactory organs.

2. Materials and Methods

2.1. Insect Specimen Preparation

The *C. medinalis* originated from a research area at Yangzhou University, Yangzhou, China. Insects were reared in a controlled laboratory environment (27 ± 1 °C, 75 ± 5% RH, and 16:8 h L:D) using rice seedlings. For analysis, samples were collected by dissecting male and female antennae from 1 to 5 days (each day-of-age moth weighing approximately 20 mg) and larvae heads of 1st to 5th instar stages (each instar weighing approximately 20 mg). For every sample, biological triplicate duplicates were carried out. The prepared samples were stored at −80 °C.

2.2. Illumina Sequencing, Sequence Assembly, Function Annotation, and Expression Abundance Analysis

RNA isolation, cDNA library creation, and Illumina sequencing of the specimens were conducted at Novogene Co., Ltd. (Beijing, China). The results of Illumina sequencing filter the raw data to obtain clean reads and compare it with our *C. medinalis* genome (unpublished). The new transcript was assembled by StringTie and annotated by the Kyoto

Encyclopedia of Genes and Genomes (KEGG), Gene Ontology (GO), Pfam, SUPERFAMILY, and other databases [27–31]. Gene expression levels were estimated using FPKM (fragments per kb of transcript sequence per million bp sequenced) values [32]. FPKM values equal to or greater than one were considered thresholds to determine if a gene was expressed. We defined FPKM > 100 as high expression, 100 > FPKM > 10 as moderate expression, and FPKM < 10 as low expression. A fold-change threshold of three was set, to identify genes with significant differential expression. Expression of chemosensory genes was revealed by heat mapping using Heml, and the expression profiles of genes were calculated based on $\log_e$FPKM.

2.3. Identification and Phylogenetic Analyses of Chemosensory Genes

The identification of unigenes associated with chemosensory genes was carried out through keyword searches of blastx annotations. Open reading frames (ORF) finder (https://www.ncbi.nlm.nih.gov/orffinder/, accessed on 7 October 2023) and blast were employed to validate the predicted protein sequences further. Signal peptides of OBPs and CSPs were predicted by SignalP-5.0 (https://services.healthtech.dtu.dk/service.php?SignalP-5.0, accessed on 19 October 2023), then the MEME-5.5.4 online server (https://meme-suite.org/meme/tools/meme, accessed on 19 October 2023) was used to discover the motifs of OBP and CSP proteins. TMHMM-2.0 (https://services.healthtech.dtu.dk/services/TMHMM-2.0/, accessed on 20 October 2023) was utilized to predict putative transmembrane domains (TMDs) in ORs, IRs, and SNMPs. The alignment of amino acid sequences encoding OBPs, CSPs, ORs, and IRs was performed using Clustal W (Supplementary Materials SA–SE). For constructing phylogenetic trees, the neighbor joining method with Poisson correction of distances was implemented in MEGA X (https://www.megasoftware.net/, accessed on 22 October 2023). Additionally, node support was evaluated through a bootstrap procedure, consisting of 1000 replicates.

2.4. Quantitative Real-Time Polymerase Chain Reaction Validation

To evaluate the expression patterns of chemosensory genes across various organs (same as those used in transcriptome sequencing), we employed the technique of real-time quantitative PCR (RT-qPCR). Total RNA was extracted by the Trizol method and then reverse transcribed to first-strand cDNA, using the methods provided with the One Step SYBR PrimeScript RT-PCR kit (Takara, Dalian, China). The SYBR Premix Ex Taq (Takara, Dalian, China) and the CFX96™ Real-Time PCR Detection System (Bio-Rad, Hercules, CA, USA) were used for RT-qPCR. *β-actin* was selected as the reference gene and the primers utilized for RT-qPCR were created using an online software program (https://sg.idtdna.com/Scitools/Applications/RealTimePCR/, accessed on 24 October 2023) (Table S1). The RT-qPCR was carried out using the following program: an initial denaturation step at 95 °C for 3 min, followed by 40 cycles of 10 s at 95 °C and 30 s at 60 °C [33]. The relative expression levels of specimens were calculated using the $\Delta\Delta$Ct method [34]. A one-way analysis of variance was conducted in SPSS Statistics 17.0 (SPSS Inc., Chicago, IL, USA), to compare relative expression levels across various organs. A significance level of $p < 0.05$ means a statistically significant difference.

3. Results

3.1. Identification and Comparison of Odorant-Binding Proteins in C. medinalis Larval and Adult Olfactory Organs

A total of 32 different single genes encoding hypothetical OBPs in *C. medinalis* were obtained (8 newly identified and 24 known OBPs), of which 31 *OBP* sequences contained complete ORFs and 28 OBPs possessed signal peptides (Table S2). Sequence alignment showed that most of the hypothetical OBPs shared the classical six-cysteine motifs, except for five OBPs (CmedOBP18/20/33/34/35) belonging to the Minus-C group and one (CmedOBP11) to the Plus-C group (Figure 1). To further research the characteristic regions of the OBP proteins in *C. medinalis*, we uploaded 32 OBPs to MEME and conducted a motif search.

Search results showed seven motifs in the OBPs of *C. medinalis*, and all CmedOBPs had two–four motifs containing four–eight conserved cysteine residues (Figure 2). As expected, the phylogenetic tree showed that four PBPs (CmedPBP1/2/4/5) and two GOBPs (CmedGOBP1/2) clustered into PBP and GOBP clades, respectively. However, CmedPBP3 and CmedGOBP2.1/3 did not cluster into the clades corresponding to their names. In addition, all Minus-C OBPs (CmedOBP18/20/33/34/35) and Plus-C OBPs (CmedOBP11) clustered in the "Plus-C" and "Minus-C" OBP sub-families, in correlation with their cysteine number (Figure 3).

Figure 1. Sequences alignment of candidate CmedOBPs. The conserved cysteine residues were marked with a green shade. All these OBPs were assigned to Classical OBPs, Plus-C OBPs, and Minus-C OBPs.

To reveal the difference of *CmedOBP*s expression between larvae and adults, the expression profiles of 32 *OBP*s in larval heads and adult antennae were compared via heatmap analysis. RNA-Seq results revealed the quantity and transcription level of *CmedOBP*s in different olfactory organs. There were 18, 25, and 26 *OBP*s identified in larval heads and male and female antennae, respectively. Six *OBP*s (*CmedOBP*11/17/19/21/29/30) showed larval head-specific expression. At the same time, 14 *OBP*s expressed exclusively in adult antennae, of which *CmedOBP34* was detected only in female antennae (Figures 4A and 5A). Among the *OBP*s expressed in both the larval heads and adult antennae, *CmedOBP20* displayed larval head-biased expression, while seven *OBP*s (*CmedGOBP2*, *CmedOBP6/13/15/26/27/31*) exhibited preferential expression in adult antennae. In addition, three *OBP*s (*CmedGOBP1*, *CmedPBP2/5*) showed gender differential expression (Figure 4A).

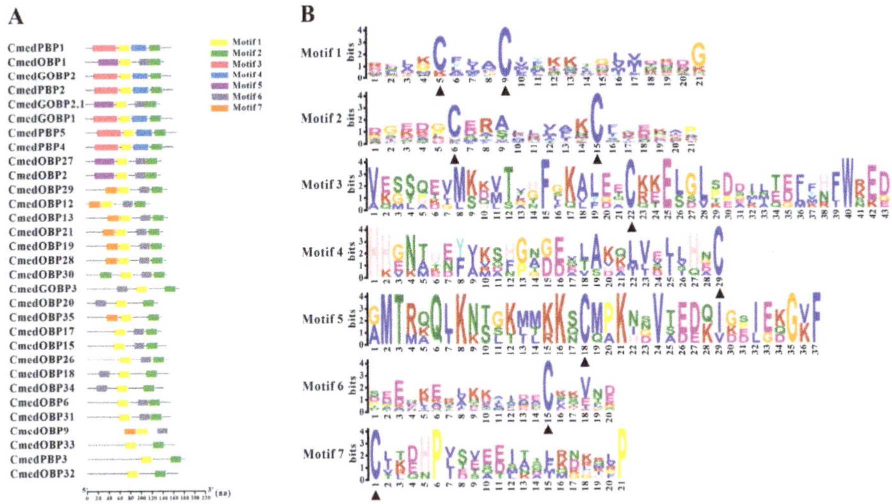

Figure 2. Motif analysis of candidate CmedOBPs. (**A**) Distributions of motifs in candidate CmedOBPs. The x-axis indicates the length of OBP proteins. (**B**) The SeqLogo of the motifs present in candidate CmedOBPs. The black triangle marked the conserved cysteine residues.

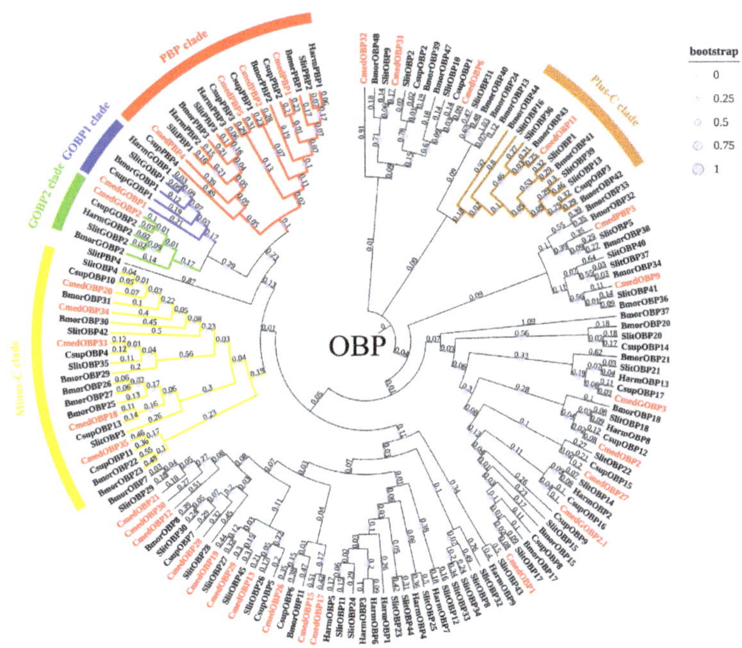

Figure 3. Phylogenetic tree of candidate CmedOBPs with known lepidopteran OBPs. The clade in red indicates the PBP clade; the clade in blue indicates the GOBP1 clade; the clade in green indicates the GOBP2 clade; the clade in yellow indicates the Minus-C OBPs clade; and the clade in brown indicates the Plus-C OBPs clade. The value next to the branch represents the branch length, and the size of the circle on the branch means the bootstrap value. Cmed: *C. medinalis*, Bmor: *B. mori*, Harm: *Helicoverpa armigera*, Csup: *Chilo suppressalis*, Slit: *S. littoralis*.

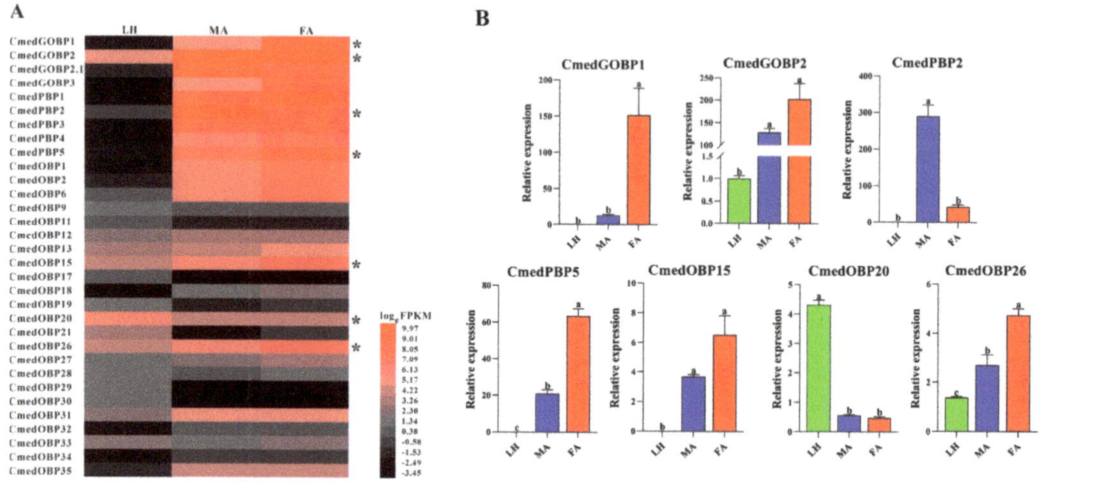

Figure 4. Comparison of candidate *CmedOBPs* expression in different olfactory organs of *C. medinalis*. (**A**) Heatmap based on FPKM values. Red indicates relatively higher expression while black indicates relatively lower expression. Asterisk "*" indicates expression has been validated by RT-qPCR assay. (**B**) Relative expression pattern based on RT-qPCR. The relative expression of each gene normalized to *β-actin* and *CmedGOBP2* transcript levels in larval heads were used as a baseline reference. The significant difference in different olfactory organs was marked on the bars with lowercase letters, $p < 0.05$. LH: larval heads, MA: male antennae, FA: female antennae.

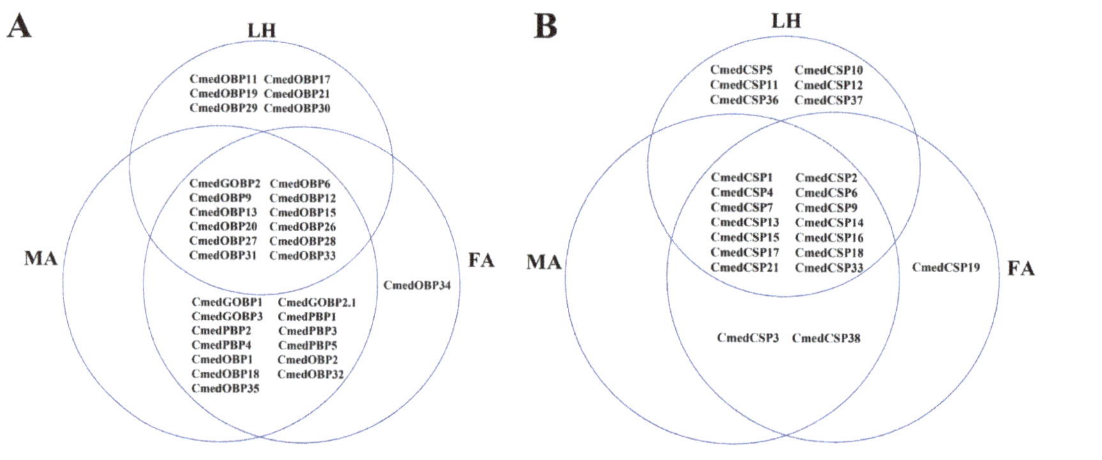

Figure 5. Distribution of *CmedOBPs* and *CmedCSPs* in different olfactory organs of *C. medinalis*. (**A**) *CmedOBPs*. (**B**) *CmedCSPs*. LH: larval heads, MA: male antennae, FA: female antennae.

We investigated the expression patterns of four *OBPs* (*CmedGOBP2*, *CmedOBP15/20/26*) that exhibited high mRNA levels in larval heads and three *OBPs* (*CmedGOBP1*, *CmedPBP2/5*) that displayed gender differential expression by RT-qPCR analysis to confirm the data of transcriptomes. RT-qPCR results showed that the expression levels of *CmedGOBP1* and *CmedPBP5* in female antennae were 1038.4% and 203.2% higher than those in male antennae, respectively. Meanwhile, the expression level of *CmedPBP2* in male antennae was 454.5% higher than in female antennae. In addition, the expression of *CmedOBP20* in larval heads was 798.2% and 678.6% higher than that in female and male antennae, respectively, indicating that *CmedOBP20* expressed preferentially in larval heads. Unexpectedly, *CmedOBP15* was not detected in larval heads by RT qPCR (Figure 4B).

3.2. Identification and Comparison of Chemosensory Proteins in *C. medinalis* Larval and Adult Olfactory Organs

According to transcripts annotation, we identified 23 *CSPs* from *C. medinalis* larval and adult transcriptomes, including 2 novel *CSPs* (*CmedCSP37/38*) and 21 known *CSPs*. Except for *CmedCSP9*, 22 *CSPs* had full-length ORFs (Table S3). The pattern of four conserved cysteine residues was retained among the 22 candidate CSPs, except the remaining CmedCSP9 sequence was too short (Figure 6). Motif analysis using MEME revealed five motifs in 23 putative CSPs, of which two highly conserved motifs (Motif1/2) contained four conserved cysteine residues (Figure 7). No subfamilies were distinguished in the phylogenetic tree, but most CmedCSPs clustered with other Lepidoptera CSPs (Figure 8).

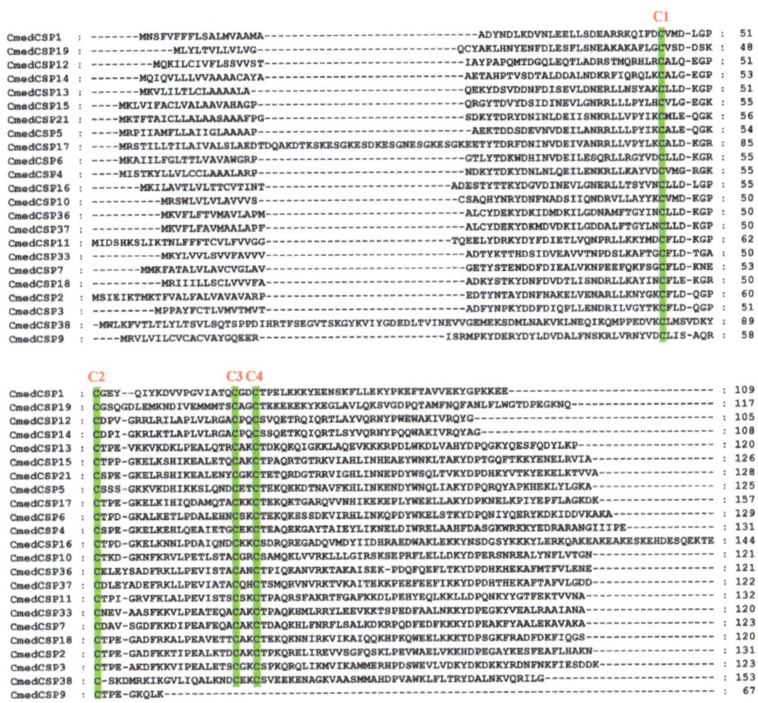

Figure 6. Sequences alignment of candidate CmedCSPs. The conserved cysteine residues were marked with a green shade.

The number of *CSPs* expressed in larval heads and male and female antennae were 20, 16, and 17, respectively. Transcriptome results showed that six *CSPs* (*CmedCSP5/10/11/12/36/37*) specifically expressed in larval heads, while three *CSPs* (*CmedCSP3/19/38*) showed exclusive expression in adult antennae. In addition, *CmedCSP19* exhibited specifical expression in female antennae (Figures 5B and 9A). The transcriptome results also revealed that four *CSPs* (*CmedCSP7/15/18/21*) displayed larval head-biased expression. In comparison, six *CSPs* (*CmedCSP1/6/13/14/16/17*) were expressed at higher levels in adult antennae. No sex-differentially expressed *CSPs* were found (Figure 9A). The transcript expression level of nine highly expressed *CSPs* (*CmedCSP2/4/7/11/15/18/21/33/37*) in larval heads were validated by RT-qPCR. Qualitative data revealed that four *CSPs* (*CmedCSP7/11/15/37*) were detected only in larval heads, and there was no significant difference in the expression levels of three *CSPs* (*CmedCSP2/4/33*) in larval heads compared to adult antennae. Moreover, the expression levels of two *CSPs* (*CmedCSP18/21*) in larval heads were 235.4%/2469.7% higher than male antennae and 544.8%/2304.4% higher than female antennae, respectively (Figure 9B).

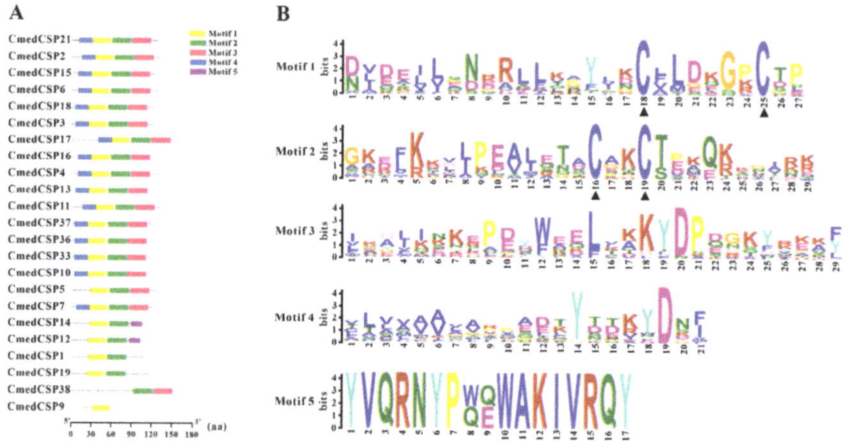

Figure 7. Motif analysis of candidate CmedCSPs. (**A**) Distributions of motifs in candidate CmedCSPs. The x-axis indicates the length of CSP proteins. (**B**) The SeqLogo of the motifs present in candidate CmedCSPs. The black triangle marked the conserved cysteine residues.

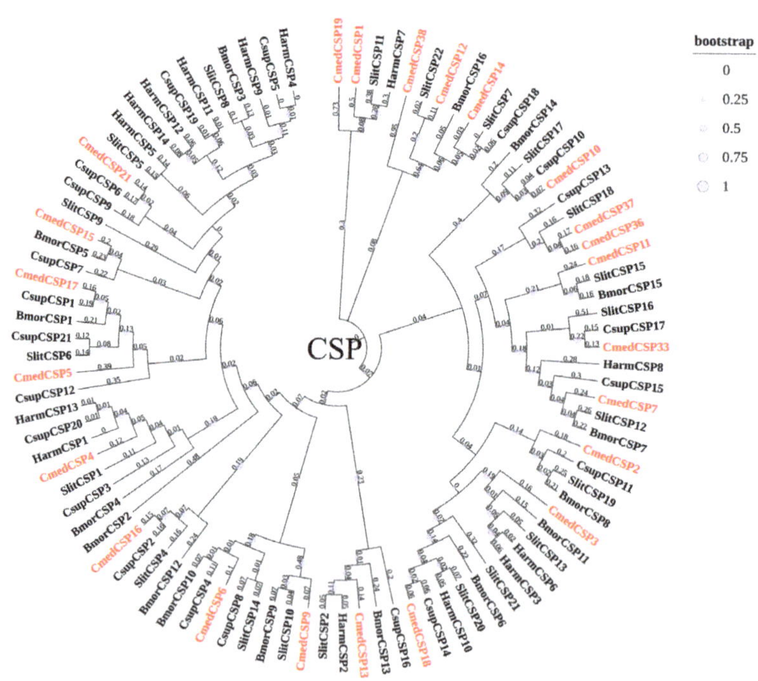

Figure 8. Phylogenetic tree of candidate CmedCSPs with known lepidopteran CSPs. The value next to the branch represents the branch length, and the size of the circle on the branch means the bootstrap value. Cmed: *C. medinalis*, Bmor: *B. mori*, Harm: *H. armigera*, Csup: *C. suppressalis*, Slit: *S. littoralis*.

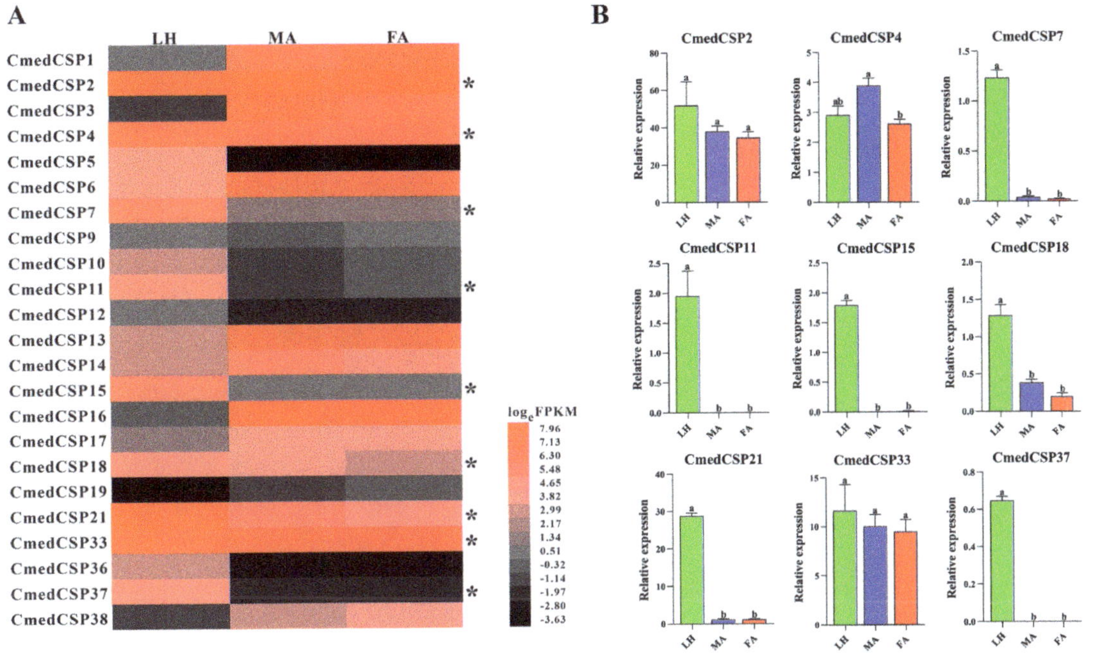

Figure 9. Comparison of candidate *CmedCSPs* expression in different olfactory organs of *C. medinalis*. (**A**) Heatmap based on FPKM values. Red indicates relatively higher expression while black indicates relatively lower expression. Asterisk "*" indicates expression has been validated by RT-qPCR assay. (**B**) Relative expression pattern based on RT-qPCR. The relative expression of each gene normalized to *β-actin* and *CmedGOBP2* transcript levels in larval heads were used as a baseline reference. The significant difference in different olfactory organs was marked on the bars with lowercase letters, $p < 0.05$. LH: larval heads, MA: male antennae, FA: female antennae.

3.3. Identification and Comparison of Odorant Receptors in C. medinalis Larval and Adult Olfactory Organs

Transcriptome analysis of *C. medinalis* olfactory organs revealed 55 candidate *ORs*, of which 17, newly identified, were named *CmedOR54–CmedOR70* (Table S4). The complete ORFs of 48 *ORs* were predicted, and the ORFs of 7 *ORs* (*CmedOR32/45/47/57/61/62/69*) were missing either the 5′ terminal or 3′ terminal. Eleven ORs were expected to contain the seven TMDs, while the other ORs had zero–six TMDs. The phylogenetic tree revealed the relationship between CmedORs and other ORs from Lepidoptera species (*B. mori/H. armigera/C. suppressalis/S. littoralis*) (Figure 10). A highly conserved Orco clade contained CmedOrco and Orcos from other Lepidoptera species. In addition, CmedPR1, CmedOR27, and CmedOR67 clustered with the moths' PR subfamily.

The transcriptome-guided read-mapping showed that 7, 51, and 54 *ORs* were expressed in larval heads and male and female antennae, respectively. There were no specific expression *ORs* in larval heads, while 48 *ORs* exhibited exclusive expression in adult antennae. *CmedPR1* exhibited specifical expression in male antennae, and four *ORs* (*CmedOR54/57/60/63*) showed female antennae-specifical expression (Figures 11A and 12A). In addition, three *ORs* (*CmedOR59/61/67*) had lower expression levels in larval heads than adult antennae (Figure 11A). Two moderately expressed *ORs* (*CmedOR10/39*) and three gender differential expression *ORs* (*CmedPR1/CmedOR27/40*) were validated through RT-qPCR analysis. The results revealed that *CmedPR1* and *CmedOR27* were specifically expressed in male antennae, and the expression level of *CmedOR40* in female antennae was 290.0% higher than that in male antennae. According to RT-qPCR results, the expression level of *CmedOR39* in larval heads was significantly higher than that in adult antennae (Figure 11B).

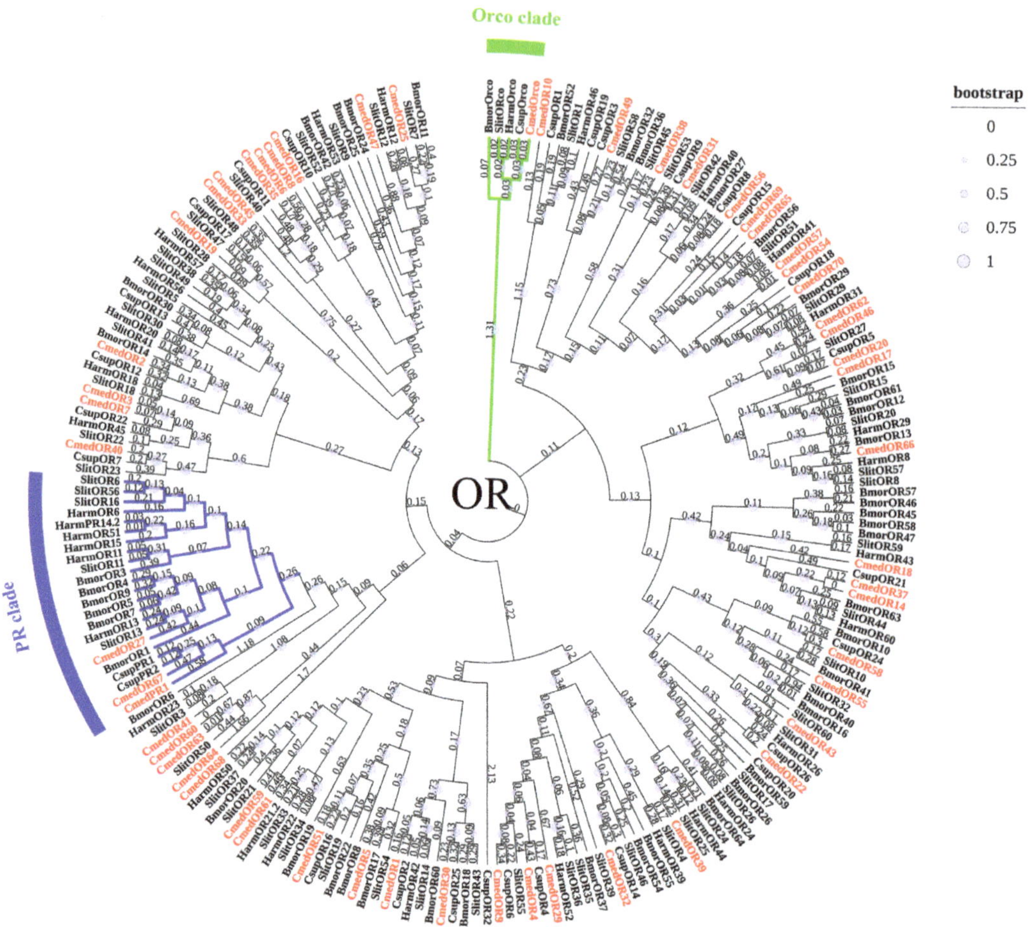

Figure 10. Phylogenetic tree of candidate CmedORs with known lepidopteran ORs. The clade in green indicates the Orco clade; the clade in blue indicates the PR clade. The value next to the branch represents the branch length, and the size of the circle on the branch means the bootstrap value. Cmed: *C. medinalis*, Bmor: *B. mori*, Harm: *H. armigera*, Csup: *C. suppressalis*, Slit: *S. littoralis*.

3.4. Identification and Comparison of Ionotropic Receptors and Sensory Neuron Membrane Proteins in C. medinalis Larval and Adult Olfactory Organs

Nineteen *IR*s were identified in *C. medinalis* and were named based on their blastx best hits to insect IRs and their positions in the phylogenetic tree, including 5 new *IR*s and 14 known *IR*s (Table S5). Of these, 18 *IR*s contained full-length ORFs, except *CmedIR40a*, which lacked a complete 5′ terminal. All hypothetical IRs in *C. medinalis* were confirmed to have one–four TMDs. A phylogenetic tree, indicating evolutionary relationships between *C. medinalis* IRs and those from *Drosophila melanogaster* and other Lepidoptera (*B. mori*/*H. armigera*/*C. suppressalis*/*S. litura*), is shown (Figure 13). This tree indicates that most CmedIRs were separated from each other and clustered in a branch with their homologous genes. A highly conserved IR co-receptors clade and the two large sub-families of IR7d and IR75 clades were present here. *D. melanogaster* divergent IRs clustered phylogenetically into separate clades. According to their phylogenetic position, five new sequences were named "CmedIR1", "CmedIR1.2", "CmedIR7d.1", "CmedIR7d.2.1", and "CmedIR7d.2.2". Transcripts encoding 4, 18, and 19 IRs were identified in larval heads and male and female antennas, respectively. The RNA-Seq results also revealed that 15 *IRs* expressed specifically in adult antennae, in which *CmedIR7d.2.2* was specifically expressed in female antennae. In addition, three *IRs* (*CmedIR25a/64a/75q.1*) showed lower expression levels in larval heads than adult antennae (Figures 12B and 14).

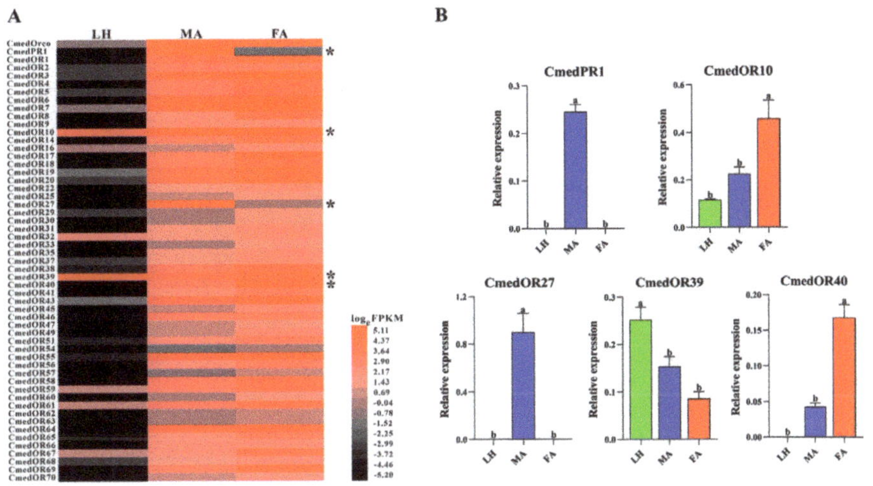

Figure 11. Comparison of candidate *CmedORs* expression in different olfactory organs of *C. medinalis*. (**A**) Heatmap based on FPKM values. Red indicates relatively higher expression while black indicates relatively lower expression. Asterisk "*" indicates expression has been validated by RT-qPCR assay. (**B**) Relative expression pattern based on RT-qPCR. The relative expression of each gene normalized to *β-actin* and *CmedGOBP2* transcript levels in larval heads were used as a baseline reference. The significant difference in different olfactory organs was marked on the bars with lowercase letters, $p < 0.05$. LH: larval heads, MA: male antennae, FA: female antennae.

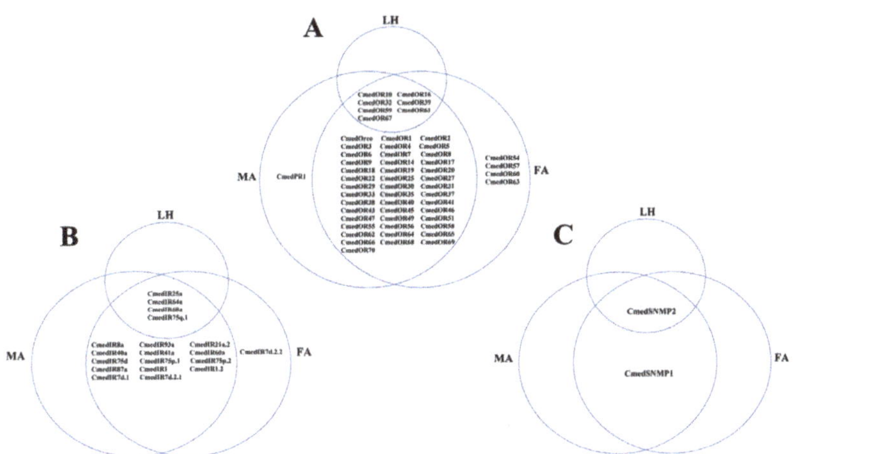

Figure 12. Distribution of *CmedSNMPs*, *CmedORs*, and *CmedIRs* in different olfactory organs of *C. medinalis*. (**A**) *CmedORs*, (**B**) *CmedIRs*, (**C**) *CmedSNMPs*. LH: larval heads, MA: male antennae, FA: female antennae.

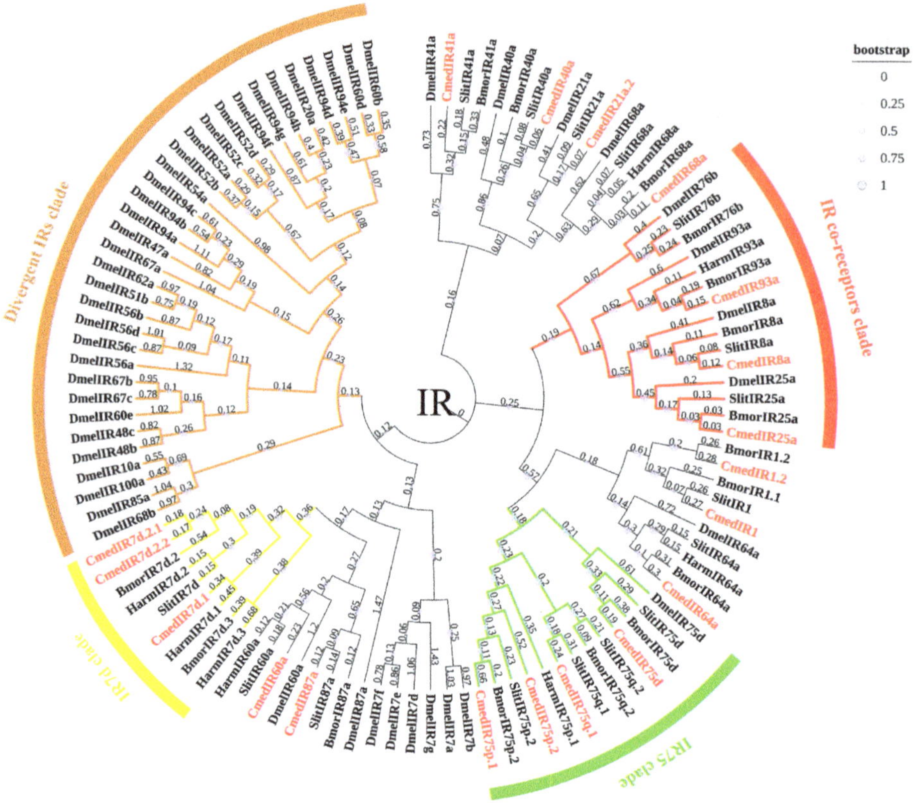

Figure 13. Phylogenetic tree of candidate CmedIRs with known insect's IRs. The clade in red indicates the IR co-receptors clade; the clade in green indicates the IR75 clade; the clade in yellow indicates the IR7d clade; the clade in brown indicates the divergent IRs clade. The value next to the branch represents the branch length, and the size of the circle on the branch means the bootstrap value. Cmed: *C. medinalis*, Bmor: *B. mori*, Harm: *H. armigera*, Slit: *S. littoralis*, Dmel: *D. melanogaster*.

Two previously reported *CmedSNMPs*, *CmedSNMP*1 and *CmedSNMP*2, were identified in our *C. medinalis* transcriptome (Table S6). These two *SNMPs* have complete ORFs and display 100 and 98.85 percent identity to the previously reported sequences at the protein level. Only *CmedSNMP*2 with low mRNA levels (FPKM = 4.58) was detected in larval heads, and the expression level was significantly lower than that in male antennae (FPKM = 1401.96) and female antennae (FPKM = 1314.23) (Figure 12C).

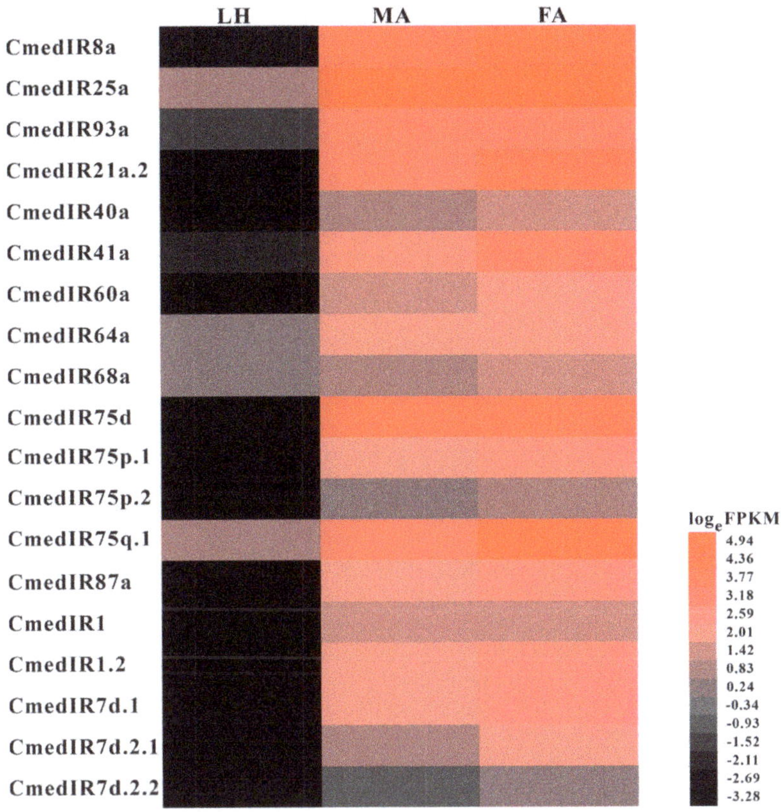

Figure 14. Comparison of candidate *CmedIRs* expression based on FPKM values in different olfactory organs of *C. medinalis*. Red indicates relatively higher expression while black indicates relatively lower expression. LH: larval heads, MA: male antennae, FA: female antennae.

4. Discussion

Using transcriptomic analysis, we identified 131 chemosensory genes from the olfactory organs of *C. medinalis*, including 32 *OBPs*, 23 *CSPs*, 55 *ORs*, 19 *IRs*, and 2 *SNMPs*. The number of chemosensory genes exhibits a comparable magnitude in other lepidopteran species, such as *S. littoralis* (122 chemosensory genes), *C. suppressalis* (116 chemosensory genes), *H. armigera* (139 chemosensory genes), *Helicoverpa assulta* (131 chemosensory genes), and *L. sticticalis* (112 chemosensory genes) [20,35–37]. In previous studies, Zeng and Liu discovered 102 and 90 chemosensory genes in *C. medinalis* adults, respectively [25,26]. More chemosensory genes were identified in our study, including eight novel *OBPs*, two novel *CSPs*, seventeen novel *ORs*, and five novel *IRs*. This may be attributed to transcriptome sequencing analysis referring to our *C. medinalis* genome (Unpublished) and the addition of larval head samples. Further validation is required for the sequence and expression of chemosensory genes in *C. medinalis*.

The studies in most Lepidoptera species demonstrated that the number and expression level of *OBPs* in larval olfactory organs is comparatively lower than those found in adult antennae [18,38,39]. The larvae of *H. armigera* expressed a total of 26 *OBPs* in antennae and mouthparts, while the adults exhibited the expression of 34 *OBPs* in their antennae, of which 6 and 10 *OBPs* showed specific or biased expression in larvae and adults, respectively [38]. An even more striking phenomenon was observed in *S. nonagrioides*, where only *SnonOBP3*, among the identified 12 *OBPs*, showed preferential expression in the antennae and palps of larvae [18]. Two studies in *Spodoptera exigua* produced opposite results in the comparison of specifically or biasedly expressed *OBPs* in larvae and adults [39,40]. The truth in *S. exigua* needs further research. In our study, 18 *CmedOBPs* were identified in *C. medinalis* larval heads, and 26 *CmedOBPs* were identified in the antennae of adults. Seven *CmedOBPs* showed higher expression levels in larval heads compared to those in adult antennae, whereas twenty-one *CmedOBPs* exhibited opposite expression preference. Our results, similar to findings in other Lepidoptera species [18,38], demonstrated that larvae only need a small amount of OBPs to execute the corresponding behaviors and life activities during this development stage.

Studies conducted in *B. mori*, *S. littoralis*, and *S. nonagrioides* found that larvae and adults expressed a similar number of *CSPs* in their olfactory organs [18,21,41]. However, the expression patterns of *CSPs* in adults and larvae exhibited variations among different Lepidoptera species. For instance, the larval heads of *B. mori* expressed eight *CSPs* (*BmorCSP10* showed specific expression in larvae) versus eleven *CSPs* in the adult antennae (four *CSPs* were expressed specifically in adults) [41]. Seventeen *CSPs* of *S. littoralis* presented an overlapping expression between larval and adult chemosensory organs, with only *SlitCSP21* being specifically expressed in larval antennae and palps [21]. In *S. nonagrioides*, thirteen *CSPs* were expressed in larval antennae/palps and eleven *CSPs* were detected in adult antennae, with six and five of them displaying larval-specific/biased or adult-specific/biased expression, respectively [18]. Similarly, larval heads and adult antennae of *C. medinalis* have a comparable magnitude of *CSPs* (larvae expressed 20 *CSPs* versus 17 in adults). Moreover, ten *CSPs* were specifically or preferentially expressed in larval heads, and nine *CSPs* have higher expression levels in adult antennae. Our studies and other reports on various insects indicate that *CSPs* are expressed throughout the insect body and across development stages, implying that they may participate in other physiological processes beyond chemosensory processes [42]. The differences in *CSP* expression patterns suggest their specialized functions at different developmental stages.

The larval *OR* repertoires are much smaller than adult ones. In Lepidoptera, *C. pomonella* larval heads expressed 16 *ORs* versus 57 *ORs* in adult antennae [19], and *S. exigua* larvae expressed 1 ORs versus 7 in adults [39]. In Diptera, the 4 *ORs* detected in *D. suzukii* larval heads were far lower than the 50 *ORs* present in adult antennae (FPKM values greater than one were considered thresholds to determine if a gene was expressed) [16]. Our results showed that the number of *ORs* detected in larval heads of *C. medinalis* was limited to only 7, which is significantly fewer than 55 *ORs* expressed in adult antennae. Like *S. littoralis*, *S. exigua*, and *S. nonagrioides*, no specific expressed *ORs* were detected in larval heads of *C. medinalis*, based on transcriptome results [18,21,40]. All these studies reveal that Lepidoptera and Diptera larvae possess a simpler olfactory system, with a lower number of *ORs* expressed, and have some overlap with adult *ORs*. This may be related to the fact that adults need to undertake more life activities than larvae, which requires more experimental support.

The abundance of other membrane receptors, *IRs* and *SNMPs*, in insect larvae is also significantly lower than in adults. For instance, nine of the ten *IRs* of *S. nonagrioides* were female antennal specifically/biasedly expressed [18]. Compared with 1 *IR* identified from the *D. suzukii* larvae heads, 28 *IRs* were identified from adult antennae (FPKM values greater than one were considered thresholds to determine if a gene was expressed) [16]. Our transcriptome analysis also found that eighteen of the nineteen *CmedIRs* exhibited specific or preferential expression in *C. medinalis* adult antennae, while only four *IRs* showed low expression levels in larval heads. Interestingly, an IR co-receptor, *CmedIR25a*, was detected in low expression in *C. medinalis* larval heads. This mirrors the observed pattern of *IRs* in *S. exigua* and *S. nonagrioides*, suggesting that IR co-receptors play a vital role in the larval heads [18,39]. In addition, *SNMP1* and *SNMP2* were specifically and preferentially expressed in *C. medinalis* adult antennae, respectively, like *SNMPs* in *S. exigua* and *S. nonagrioides* [18,39]. Considering that there are few studies on the functions of IRs and SNMPs in Lepidoptera larvae, we speculate that IRs and SNMPs may perform more functions in the chemosensory processes of adults than those of larvae. Further exploration of IR and SNMP functions is warranted.

5. Conclusions

This study presented a comprehensive investigation of 131 chemosensory genes in the olfactory organs of *C. medinalis*. The comparison of the expression profiles of 131 chemosensory genes revealed differences between larval and adult chemosensory systems, which can be attributed to their different physiological and biochemical characteristics during the development stages. Our findings identified a series of larva or adult-specific chemosensory genes that may be important for developmental stage-specific chemosensory behaviors. These chemosensory genes could serve as potential molecular targets for *C. medinalis* control in different developmental stages.

Supplementary Materials: The following supporting information can be downloaded at: https://www.mdpi.com/article/10.3390/genes14122165/s1, Table S1: oligonucleotide primers used for expression validation analysis; Table S2: list of candidate OBPs in *C. medinalis*; Table S3: list of candidate CSPs in *C. medinalis*; Table S4: list of candidate ORs in *C. medinalis*; Table S5: list of candidate IRs in *C. medinalis*; Table S6: list of candidate SNMPs in *C. medinalis*. Supplementary Materials SA: amino acid sequences of OBPs, CSPs, ORs, IRs, and SNMPs in *C. medinalis*. Supplementary Materials SB: amino acid sequences of OBPs used in phylogenetic analyses. Supplementary Materials SC: amino acid sequences of CSPs used in phylogenetic analyses. Supplementary Materials SD: amino acid sequences of ORs used in phylogenetic analyses. Supplementary Materials SE: amino acid sequences of IRs used in phylogenetic analyses.

Author Contributions: Conceptualization, Y.L. and F.L.; methodology, H.-T.D. and Y.L.; software, H.-T.D. and Y.L.; validation, H.-T.D.; formal analysis, H.-T.D.; investigation, J.-Q.L. and H.-T.D.; resources, K.J. and C.-C.W.; data curation, H.-T.D. and J.-Q.L.; writing—original draft preparation, H.-T.D.; writing—review and editing, H.-T.D., Y.L., Z.-C.Y. and F.L.; visualization, H.-T.D.; supervision, Y.L. and F.L.; project administration, Y.L.; funding acquisition, F.L. All authors have read and agreed to the published version of the manuscript.

Funding: This research was funded by the Jiangsu Province Scientific and Technological Project (Grant No. BE2022425), the Natural Science Foundation of Jiangsu Province (Grant No. BK20220571), and the National Natural Science Foundation of China (Grant No. 32302436).

Institutional Review Board Statement: Not applicable.

Informed Consent Statement: Not applicable.

Data Availability Statement: Transcriptome data in this study have been uploaded to NCBI Sequence Read Archive (SRA) under accession number PRJNA1039023; all sequences (accession on Supplementary Materials SA–SE) have been uploaded.

Conflicts of Interest: The authors declare no conflict of interest.

References

1. Renou, M.; Anton, S. Insect olfactory communication in a complex and changing world. *Curr. Opin. Insect Sci.* **2020**, *42*, 1–7. [CrossRef] [PubMed]
2. Keil, T.A. Morphology and development of the peripheral olfactory organs. In *Insect Olfaction*; Springer: Berlin, Germany, 1999; pp. 5–47.
3. Pelosi, P.; Zhou, J.J.; Ban, L.; Calvello, M. Soluble proteins in insect chemical communication. *Cell. Mol. Life Sci.* **2006**, *63*, 1658–1676. [CrossRef] [PubMed]
4. Leal, W.S. Odorant reception in insects: Roles of receptors, binding proteins, and degrading enzymes. *Annu. Rev. Entomol.* **2013**, *58*, 373–391. [CrossRef]
5. Scaloni, A.; Monti, M.; Angeli, S.; Pelosi, P. Structural analysis and disulfide-bridge pairing of two odorant-binding proteins from *Bombyx mori*. *Biochem. Biophys. Res. Commun.* **1999**, *266*, 386–391. [CrossRef] [PubMed]
6. Pelosi, P.; Iovinella, I.; Felicioli, A.; Dani, F.R. Soluble proteins of chemical communication: An overview across arthropods. *Front. Physiol.* **2014**, *5*, 320. [CrossRef]
7. Mombaerts, P. Seven-transmembrane proteins as odorant and chemosensory receptors. *Science* **1999**, *286*, 707–711. [CrossRef]
8. Rytz, R.; Croset, V.; Benton, R. Ionotropic receptors (IRs): Chemosensory ionotropic glutamate receptors in *Drosophila* and beyond. *Insect Biochem. Mol. Biol.* **2013**, *43*, 888–897. [CrossRef]

9. Rogers, M.E.; Krieger, J.; Vogt, R.G. Antennal SNMPs (sensory neuron membrane proteins) of Lepidoptera define a unique family of invertebrate CD36-like proteins. *J. Neurobiol.* **2001**, *49*, 47–61. [CrossRef]
10. Nakagawa, T.; Vosshall, L.B. Controversy and consensus: Noncanonical signaling mechanisms in the insect olfactory system. *Curr. Opin. Neurobiol.* **2009**, *19*, 284–292. [CrossRef]
11. Skiri, H.T.; Rø, H.; Berg, B.G.; Mustaparta, H. Consistent organization of glomeruli in the antennal lobes of related species of heliothine moths. *J. Comp. Neurol.* **2005**, *491*, 367–380. [CrossRef]
12. Hansson, B.S.; Stensmyr, M.C. Evolution of insect olfaction. *Neuron* **2011**, *72*, 698–711. [CrossRef] [PubMed]
13. Knolhoff, L.M.; Heckel, D.G. Behavioral assays for studies of host plant choice and adaptation in herbivorous insects. *Annu. Rev. Entomol.* **2014**, *59*, 263–278. [CrossRef] [PubMed]
14. Dethier, V.; Schoonhoven, L. Olfactory coding by lepidopterous larvae. *Entomol. Exp. Appl.* **1969**, *12*, 535–543. [CrossRef]
15. Ljungberg, H.; Anderson, P.; Hansson, B. Physiology and morphology of pheromone-specific sensilla on the antennae of male and female *Spodoptera littoralis* (Lepidoptera: Noctuidae). *J. Insect Physiol.* **1993**, *39*, 253–260. [CrossRef]
16. Walker III, W.B.; Mori, B.A.; Cattaneo, A.M.; Gonzalez, F.; Witzgall, P.; Becher, P.G. Comparative transcriptomic assessment of the chemosensory receptor repertoire of *Drosophila suzukii* adult and larval olfactory organs. *Comp. Biochem. Physiol. Part D Genom. Proteom.* **2023**, *45*, 101049. [CrossRef] [PubMed]
17. Bohbot, J.; Pitts, R.; Kwon, H.W.; Rützler, M.; Robertson, H.; Zwiebel, L. Molecular characterization of the *Aedes aegypti* odorant receptor gene family. *Insect Mol. Biol.* **2007**, *16*, 525–537. [CrossRef] [PubMed]
18. Glaser, N.; Gallot, A.; Legeai, F.; Montagné, N.; Poivet, E.; Harry, M.; Calatayud, P.A.; Jacquin-Joly, E. Candidate chemosensory genes in the stemborer *Sesamia nonagrioides*. *Int. J. Biol. Sci* **2013**, *9*, 481–495. [CrossRef]
19. Walker, W.B.; Gonzalez, F.; Garczynski, S.F.; Witzgall, P. The chemosensory receptors of codling moth *Cydia pomonella*–expression in larvae and adults. *Sci. Rep.* **2016**, *6*, 23518. [CrossRef]
20. Wei, H.S.; Li, K.B.; Zhang, S.; Cao, Y.Z.; Yin, J. Identification of candidate chemosensory genes by transcriptome analysis in *Loxostege sticticalis* Linnaeus. *PLoS ONE* **2017**, *12*, e0174036. [CrossRef]
21. Poivet, E.; Gallot, A.; Montagné, N.; Glaser, N.; Legeai, F.; Jacquin-Joly, E. A comparison of the olfactory gene repertoires of adults and larvae in the noctuid moth *Spodoptera littoralis*. *PLoS ONE* **2013**, *8*, e60263. [CrossRef]
22. Tanaka, K.; Uda, Y.; Ono, Y.; Nakagawa, T.; Suwa, M.; Yamaoka, R.; Touhara, K. Highly Selective Tuning of a Silkworm Olfactory Receptor to a Key Mulberry Leaf Volatile. *Curr. Biol.* **2009**, *19*, 881–890. [CrossRef] [PubMed]
23. Khan, Z.; Barrion, A.; Litsinger, J.; Castilla, N.; Joshi, R. A bibliography of rice leaffolders (Lepidoptera: Pyralidae). *Int. J. Trop. Insect Sc.* **1988**, *9*, 129–174. [CrossRef]
24. Padmavathi, C.; Katti, G.; Padmakumari, A.; Voleti, S.; Subba Rao, L. The effect of leaffolder *Cnaphalocrocis medinalis* (Guenee) [Lepidoptera: Pyralidae] injury on the plant physiology and yield loss in rice. *J. Appl. Entomol.* **2013**, *137*, 249–256. [CrossRef]
25. Zeng, F.F.; Zhao, Z.F.; Yan, M.J.; Zhou, W.; Zhang, Z.; Zhang, A.; Lu, Z.X.; Wang, M.Q. Identification and comparative expression profiles of chemoreception genes revealed from major chemoreception organs of the rice leaf folder, *Cnaphalocrocis medinalis* (Lepidoptera: Pyralidae). *PLoS ONE* **2015**, *10*, e0144267. [CrossRef]
26. Liu, S.; Wang, W.L.; Zhang, Y.X.; Zhang, B.X.; Rao, X.J.; Liu, X.M.; Wang, D.M.; Li, S.G. Transcriptome sequencing reveals abundant olfactory genes in the antennae of the rice leaffolder, *Cnaphalocrocis medinalis* (Lepidoptera: Pyralidae). *Entomol. Sci.* **2017**, *20*, 177–188. [CrossRef]
27. Pertea, M.; Pertea, G.M.; Antonescu, C.M.; Chang, T.C.; Mendell, J.T.; Salzberg, S.L. StringTie enables improved reconstruction of a transcriptome from RNA-seq reads. *Nat. Biotechnol.* **2015**, *33*, 290–295. [CrossRef]
28. Kanehisa, M.; Goto, S. KEGG: Kyoto encyclopedia of genes and genomes. *Nucleic Acids Res.* **2000**, *28*, 27–30. [CrossRef]
29. Consortium, G.O. The Gene Ontology (GO) database and informatics resource. *Nucleic Acids Res.* **2004**, *32*, D258–D261. [CrossRef]
30. Bateman, A.; Coin, L.; Durbin, R.; Finn, R.D.; Hollich, V.; Griffiths-Jones, S.; Khanna, A.; Marshall, M.; Moxon, S.; Sonnhammer, E.L. The Pfam protein families database. *Nucleic Acids Res.* **2004**, *32*, D138–D141. [CrossRef]
31. Wilson, D.; Madera, M.; Vogel, C.; Chothia, C.; Gough, J. The SUPERFAMILY database in 2007: Families and functions. *Nucleic Acids Res.* **2007**, *35*, D308–D313. [CrossRef]
32. Roberts, A.; Pimentel, H.; Trapnell, C.; Pachter, L. Identification of novel transcripts in annotated genomes using RNA-Seq. *Bioinformatics* **2011**, *27*, 2325–2329. [CrossRef] [PubMed]
33. Li, Y.; Hu, J.; Xiang, Y.; Zhang, Y.; Chen, D.; Liu, F. Identification and comparative expression profiles of chemosensory genes in major chemoreception organs of a notorious pests, *Laodelphax striatellus*. *Comp. Biochem. Physiol. Part D Genom. Proteom.* **2020**, *33*, 100646. [CrossRef] [PubMed]
34. Livak, K.J.; Schmittgen, T.D. Analysis of relative gene expression data using real-time quantitative PCR and the $2^{-\Delta\Delta CT}$ method. *Methods* **2001**, *25*, 402–408. [CrossRef] [PubMed]
35. Walker, W.B.; Roy, A.; Anderson, P.; Schlyter, F.; Hansson, B.S.; Larsson, M.C. Transcriptome Analysis of Gene Families Involved in Chemosensory Function in *Spodoptera littoralis* (Lepidoptera: Noctuidae). *BMC Genom.* **2019**, *20*, 428. [CrossRef] [PubMed]
36. Cao, D.P.; Liu, Y.; Wei, J.J.; Liao, X.Y.; Walker, W.B.; Li, J.H.; Wang, G.R. Identification of candidate olfactory genes in *Chilo suppressalis* by antennal transcriptome analysis. *Int. J. Biol. Sci.* **2014**, *10*, 846–860. [CrossRef]
37. Zhang, J.; Wang, B.; Dong, S.L.; Cao, D.P.; Dong, J.F.; Walker, W.B.; Liu, Y.; Wang, G.R. Antennal transcriptome analysis and comparison of chemosensory gene families in two closely related noctuidae moths, *Helicoverpa armigera* and *H. assulta*. *PLoS ONE* **2015**, *10*, e0117054. [CrossRef]

38. Chang, H.; Ai, D.; Zhang, J.; Dong, S.; Wang, G. Candidate odorant binding proteins and chemosensory proteins in the larval chemosensory tissues of two closely related noctuidae moths, *Helicoverpa armigera* and *H. assulta*. *PLoS ONE* **2017**, *12*, e0179243. [CrossRef]
39. Liu, N.Y.; Zhang, T.; Ye, Z.F.; Li, F.; Dong, S.L. Identification and Characterization of Candidate Chemosensory Gene Families from *Spodoptera exigua* Developmental Transcriptomes. *Int. J. Biol. Sci.* **2015**, *11*, 1036–1048. [CrossRef]
40. Llopis-Giménez, A.; Carrasco-Oltra, T.; Jacquin-Joly, E.; Herrero, S.; Crava, C.M. Coupling transcriptomics and behaviour to unveil the olfactory system of *Spodoptera exigua* larvae. *J. Chem. Ecol.* **2020**, *46*, 1017–1031. [CrossRef]
41. Gong, D.P.; Zhang, H.j.; Zhao, P.; Lin, Y.; Xia, Q.Y.; Xiang, Z.H. Identification and expression pattern of the chemosensory protein gene family in the silkworm, *Bombyx mori*. *Insect Biochem. Mol. Biol.* **2007**, *37*, 266–277. [CrossRef]
42. Maleszka, J.; Forêt, S.; Saint, R.; Maleszka, R. RNAi-induced phenotypes suggest a novel role for a chemosensory protein CSP5 in the development of embryonic integument in the honeybee (*Apis mellifera*). *Dev. Genes Evol.* **2007**, *217*, 189–196. [CrossRef] [PubMed]

genes

MDPI

Article

The First Three Mitochondrial Genomes for the Characterization of the Genus *Egeirotrioza* (Hemiptera: Triozidae) and Phylogenetic Implications

Zhulidezi Aishan [1,†], Ze-Lu Mu [1,†], Zi-Cong Li [1], Xin-Yu Luo [2] and Ning Huangfu [3,*]

[1] Xinjiang Key Laboratory of Biological Resources and Genetic Engineering, College of Life Science and Technology, Xinjiang University, Urumqi 830017, China; zhulidezi.a@xju.edu.cn (Z.A.); 107552301054@stu.xju.edu.cn (Z.-L.M.); 107552100946@stu.xju.edu.cn (Z.-C.L.)
[2] Department of Entomology, China Agricultural University, Beijing 100193, China; ltquail@126.com
[3] National Natural History Museum of China, Beijing 100050, China
* Correspondence: huangfuning@nnhm.org.cn
† These authors contributed equally to this work.

Abstract: (1) Background: Mitochondrial genomes are important markers for the study of phylogenetics and systematics. Triozidae includes some primary pests of *Populus euphratica*. The phylogenetic relationships of this group remain controversial due to the lack of molecular data. (2) Methods: Mitochondria of *Egeirotrioza* Boselli were sequenced and assembled. We analyzed the sequence length, nucleotide composition, and evolutionary rate of Triozidae, combined with the 13 published mitochondrial genomes. (3) Results: The evolutionary rate of protein-coding genes was as follows: *ATP8* > *ND6* > *ND5* > *ND2* > *ND4* > *ND4L* > *ND1* > *ND3* > *APT6* > *CYTB* > *COX3* > *COX2* > *COX1*. We reconstructed the phylogenetic relationships of Triozidae based on 16 triozid mitochondrial genomes (thirteen ingroups and three outgroups) using the maximum likelihood (ML) and Bayesian inference (BI) approaches. The phylogenetic analysis of the 16 Triozidae mitochondrial genomes showed that *Egeirotrioza* was closely related to *Leptynoptera*. (4) Conclusions: We have identified 13 PCGs, 22 tRNAs, 2 rRNAs, and 1 control region (CR) of all newly sequenced mitochondrial genomes, which were the mitochondrial gene type in animals. The results of this study provide valuable genomic information for the study of psyllid species.

Keywords: Psylloidea; mitochondrial genome; *Populus euphratica*; phylogenomics

Citation: Aishan, Z.; Mu, Z.-L.; Li, Z.-C.; Luo, X.-Y.; Huangfu, N. The First Three Mitochondrial Genomes for the Characterization of the Genus *Egeirotrioza* (Hemiptera: Triozidae) and Phylogenetic Implications. *Genes* **2024**, *15*, 842. https://doi.org/10.3390/genes15070842

Academic Editors: Alfred M. Handler and Zhiteng Chen

Received: 10 May 2024
Revised: 17 June 2024
Accepted: 21 June 2024
Published: 26 June 2024

1. Introduction

Mitochondrial genomes, as one of the important molecular markers, are usually used for studies on speciation, phylogeography, evolutionary history, and phylogeny in insect groups [1–3], benefiting from easy availability, maternal inheritance, and high substitution rates [3,4]. Meanwhile, insect mitochondrial genomes' structural characteristics are conserved, and they can also offer additional evidence of morphological classification. The insect mitochondrial genome ranges mostly from 14,000 to 20,000 bp in length [3], containing 2 ribosomal RNAs (rRNAs), 13 protein-coding genes (PCGs), 22 transfer RNA (tRNAs), and 1 non-coding control region (CR) [4]. Mitochondrial genomes are often referred to as the secondary genetic information system due to their distinctive genetic characteristics [5].

Psyllids (Psylloidea), or jumping plant lice, are an important group within the suborder Sternorrhyncha (Hemiptera), containing more than 4000 described species [6]. All species of Sternorrhyncha are primarily pests, such as whiteflies, aphids, and scales, and they directly inflict damage on host plants through feeding or gall development [7]. During the immature stages, species of psyllids exhibit highly host-specific behavior [8]. These pests not only cause direct damage by siphoning plant sap but also act as vectors for the transmission of plant viruses, inflicting substantial economic losses on agriculture and forestry crops.

Triozidae is a family within Psylloidea. This family comprises about 1000 described species and 70 genera [9]. There are many synonymous genera and species of psyllids, and most phylogenetic studies of Psyllioidea have been based on morphological research, but there has been no evidence of molecular correlation. As Egeirotrioza has never been reported as a molecular marker, the genus was not included in previous phylogenetic studies based on molecular morphology. Some species of this group are the main pests of *P. euphratica*. Some psyllid species, in order to create a good, stable growth and development environment, can make galls, but psyllid galls are mainly deposited on the leaves of the host plant. A gall is the abnormal tissue formed by the plant when stimulated through psyllid feeding or egg-laying; the nymphs live in the galls as adults [10]. *P. euphratica* is an important forest resource in desert areas, playing a crucial role in windbreak and sand fixation, regulating the climate of oases, and stabilizing the ecological balance in desert zones [10,11]. Investigations into the endangered situation of *P. euphratica* have revealed significant damage caused by psyllids in northern and southern Xinjiang [12].

We conducted a field survey of poplar forests in Luntai County and Shaya County, Xinjiang, China, from 2021 to 2023. Through field research, we found that infestation rates on new shoots of saplings often exceed 80% (unpublished), with a high propensity for rapid dissemination, leading to widespread infestations that result in seedling mortality and wilting and thereby causing catastrophic impacts on the growth and development of *P. euphratica*. In Xinjiang, both natural and artificially planted *P. euphratica* forests are affected to varying extents by psyllids, including *Egeirotrioza rufa* and *Egeirotrioza gracilis*, leading to poor growth and abnormal development in young trees. During psyllid outbreaks, when pest density is high, *P. euphratica* leaves are covered with galls on both sides, and severe infestations lead to premature leaf yellowing, resulting in weakened tree vigor and slow growth (Figure 1). These pests not only harm *P. euphratica* but also threaten surrounding species such as tamarisk and Haloxylon, posing a threat to the desert forests of Xinjiang and the ecological environment of western China's desert regions.

Figure 1. *Populus euphratica* leaves covered with galls.

Despite Triozidae being important pests, molecular data, and especially mitochondrial genome resources, on Triozidae are still deficient. To date, only a few mitochondrial genomes in this group have been made available on GenBank and other databases. The lack of molecular data seriously restricts our understanding of biological characteristics' evolution.

In this study, we newly sequenced, assembled, and annotated three species of *Egeirotrioza* within Triozidae and also analyzed the features of their mitochondrial genome structure. Combined with 10 previously published mitochondrial genomes among Triozidae, we compared the substitution, main features, and evolutionary rates. Finally, we reconstructed the phylogenetic relationships of Triozidae based on 16 mitochondrial genomes (13 ingroups and 3 outgroups) via maximum likelihood (ML) and Bayesian inference (BI) approaches.

2. Materials and Methods

2.1. Taxon Sampling and Sequencing

In this study, we analyzed 16 taxa of Triozidae. We newly sequenced three species within *Egeirotrioza* which were collected from Luntai County, Xinjing, China (84.3081° E, 41.5911° N), by sweeping (detailed information shown in Table 1). In addition, the mitochondrial genomes of 13 Triozidae species were downloaded from GenBank for phylogenetic and comparative mitogenomic analyses. We selected three species that are closely related to Triozidae as outgroups. All species were identified by Xin-yu Luo, and the voucher specimen was deposited at the College of Life Science and Technology, Xinjing University, Xinjiang, China. The samples were stored in 95% ethanol at −20 °C until morphological examination and DNA extraction.

Table 1. Collected information of newly sequenced species in this study.

Species	Location	Longitude and Latitude	Elevation (m)	Data	Collector
Egeirotrioza gracilis	Luntai County, Xinjiang, China	84.3081° E, 41.5911° N	931.4	27.V.2023	Chen-Hong Wang
Egeirotrioza rufa	Luntai County, Xinjiang, China	84.2050° E, 41.2537° N	928.1	26.V.2023	Jin-Ling Wang
Egeirotrioza xingi	Luntai County, Xinjiang, China	84.2050° E, 41.2537° N	931.4	27.V.2023	Chen-Hong Wang

Legs were used to extract the whole genomic DNA using a Qiagen DNeasy Blood & Tissue Kit (Qiagen, Dusseldorf, Germany) according to the manufacturer's protocol. The Qubit® DNA Assay Kit in a Qubit® 2.0 Flurometer (ThermoFisher, Waltham, MA, USA) was applied to measure the concentration of the DNA (Berry Genomics, Beijing, China). The Illumina NovaSeq 6000 (Illumina, San Diego, CA, USA) platform was used to generate sequencing libraries of 150 bp paired-end reads with an insert size of 350 bp. Trimmomatic v0.32 [13] was applied to remove short reads, adapters, and low-quality reads of raw data.

2.2. Assembly, Annotation, and Composition Analyses

Two methods were used for de novo assembly: (1) NOVOPlasty v3.8.3 (Brussels, Belgium) [14] was utilized to assemble Illumina reads and k-mer sizes of 23–39 bp; (2) IDBA-UD v1.1.3 (Boston, MA, USA) [15] was implemented for mitochondrial genome assembly ("–mink 40 –maxk 120"). Mitochondrial genome sequences were compared, which were obtained by the aforementioned methods, using Geneious 2020.2.1 [16] and merged into a single sequence. Then, tRNAscan SE 2.0 [17] and MITOS WebServer were utilized to annotate and analyze the secondary structure of tRNAs. Clustal Omega in Geneious was applied to annotate rRNAs and PCGs based on *Bactericera cockerelli*. The boundaries of rRNAs and PCGs were manually proofread using MEGA X software [18]. SeqKit v0.16.0 (Chongqing, China) [19] was used to examine the bias of the nucleotide composition and the nucleotide composition of each gene. AT-skew and GC-skew were calculated by two formulas: AT-skew = $(A - T)/(A + T)$; GC-skew = $(G - C)/(G + C)$. The base composition, AT- and GC-skew, and relative synonymous codon usage (RSCU) of 10 species of Nitidulidae were calculated using MEGA X. The nucleotide diversity (Pi) of 13 PCGs and 2 rRNAs of three *Egeirotrioza* species was calculated using DnaSP v6.0 with a sliding window and a step size window of 200 bp and 20 bp, respectively.

DnaSP 6.0 [20] was utilized to calculate the non-synonymous substitution rate (Ka)/synonymous substitution rate (Ks) for each PCG. CGview (https://cgview.ca/, accessed on 28 February 2024), an online server, was used to generate the visual sequence features of the mitochondrial genomes. Finally, all newly sequenced mitochondrial genomes were submitted to GenBank (for accession number).

2.3. Phylogenetic Analyses

To reconstruct the phylogenetic relationship of Triozidae, we sampled 13 Triozidae taxa, including species of *Trioza*, *Aacanthocnema*, *Pariaconus*, *Leptynoptera*, *Bactericera*, and *Paratrioza*. Based on the published phylogenies of Psylloidea [6], we selected two *Diphorina* species and one *Arytainilla* species as outgroups. The phylogenetic analyses of Triozidae were conducted using 2 rRNAs and 13 PCGs from 16 mitochondrial genomes. MAFFT v7.450 (Osaka, Japan) [21] was applied to align nucleotide and protein sequences with the L-INS-I method. Trimal v1.4.1 (Barcelona, Spain) [22] was utilized for sequence trimming with the "-automated1" strategy. FASconCAT-G v1.04 (Santa Cruz, CA, USA) [23] was used to concatenate the matrices finally for phylogeny analysis: (1) the cds_faa matrix, including all PCG amino acid reads; (2) the cds_fna matrix, containing all PCG nucleotide reads; (3) the cds_rrna matrix, including all PCG and two rRNA nucleotide reads; (4) the cds12_fna matrix, containing all PCG nucleotide reads except the third codon positions; (5) the cds12_rrna matrix, including all PCG nucleotide reads with the third codon positions removed and two rRNA genes. Heterogeneity among the matrices was calculated via AliGROOVE v1.06 (Bonn, Germany) [24].

Here, we used the ML and BI methods to reconstruct the phylogenetic relationship within Triozidae. In the ML analysis, ModelFinder [25] in IQ-TREE 2 (Canberra, ACT, Australia) [26] was used to select befitting substitution models. To minimize long-branch attraction artifacts, we also used the posterior mean site frequency (PMSF) [27] model ('−m − mtART + C60 + FO + R') in IQ-TREE for the cds_faa matrix. A BI tree was generated via PhyloBayes-MPI (Montréal, QC, Canada) [28] with the site-heterogeneous mixture model (−m CAT + GTR). An available online website, iTOL, was utilized to fine-tune the final phylogenetic tree (https://itol.embl.de/upload.cgi, accessed on 15 February 2024).

3. Results and Discussion

3.1. Mitogenomic Organization

Here, we sequenced approximately 6 GB of raw data for each species. A total of three mitochondrial genomes of Triozidae were obtained, of which two were complete mitochondrial genomes (*E. xingi* and *E. rufa*), while that of *E. gracilis* was a linear mitochondrial genome; all of them have been deposited in GenBank with the accession numbers PP471964-PP471966. The whole lengths of the newly sequenced genomes are as follows: *E. rufa*, 15,830 bp; *E. gracilis*, 15,355 bp; *E. xingi*, 15,301 bp (Table 2). The unstable size of the CR was the primary reason for the whole length, as previously observed from other triozid species [6]. We identified 13 PCGs, 22 tRNAs, 2 rRNAs, and 1 CR, which were the typeset mitochondrial genes in animals (Figure 2), and the three mitochondrial genomes exhibit a high degree of conservation. Our newly assembled mitochondrial genomes are similar to those of previously published species of Hemiptera and other insects in length and gene order [2,3,6,29–31]. The mitochondrial features of the represented species are depicted in Figure 2. The trnS1 secondary structures of the three mitochondrial genome sequences of *Egeirotrioza* species lack stem–loop structures with dihydrouracil, and the other 21 tRNA secondary structures are typical clover structures. The stem–loop structure with dihydrouracil deletion of trnS1 is a typical feature in insects' mitochondrial genomes [6].

Table 2. Nucleotide compositions of three newly sequenced species in this study.

Species	Whole Genome					PCG					tRNA					rRNA					CR				
	Length (bp)	AT%	AT-Skew	GC%	GC-Skew	Length (bp)	AT%	AT-Skew	GC%	GC-Skew	Length (bp)	AT%	AT-Skew	GC%	GC-Skew	Length (bp)	AT%	AT-Skew	GC%	GC-Skew	Length (bp)	AT%	AT-Skew	GC%	GC-Skew
Egeirotrioza rufa	15,830	76.920	0.010	23.06	−0.154	10,825	78.343	−0.134	21.64513	−0.080	1377	79.450	0.013	20.55	0.145	1892	79.080	−0.019	20.92	0.314	1764	71.090	−0.065	28.74	0.014
Egeirotrioza gracilis	15,355	74.460	0.033	25.53	−0.23854	10,822	73.553	−0.135	26.44744	−0.110	1361	78.170	0.024	21.82	0.145	1891	76.975	−0.016	23.03	0.331	1303	82.880	0.006	16.96	−0.177
Egeirotrioza xingi	15,301	76.4	0.031	23.63	−0.243	10,818	75.8	−0.131	24.19436	−0.094	1367	79.6	0.035	20.41	0.154	1898	78.6	−0.035	21.4	0.357	1249	84.7	−0.009	15.14	−0.100

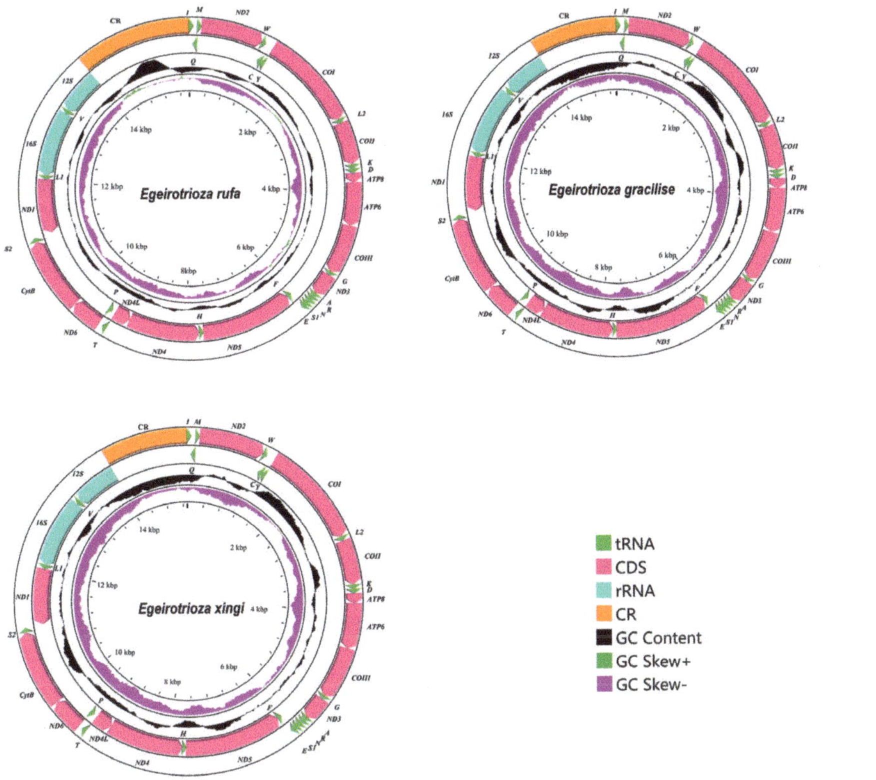

Figure 2. Mitochondrial genome map showing the mitochondrial genome characteristics of representative species within the genus *Egeirotrioza*. The arrows indicate the direction of gene transcription. Normative abbreviations are used to represent PCGs and rRNAs, and single letter abbreviations are used to represent tRNAs. Red, green, blue, and orange represent PCGs, tRNA, rRNA, and CR, respectively. The GC content of the complete mitochondrial genome is shown in the second circle. The GC-skew of the complete mitochondrial genome is shown in the third circle. The innermost circle shows the length of the complete mitochondrial genome.

The newly sequenced mitochondrial genomes were found to have similar nucleotide compositions (Table 2), revealing the characteristic AT-biased composition in Triozidae and other insects [3,31,32]. The AT contents (%) of the three newly reported genomes are 76.92 (*E. rufa*), 74.46 (*E. gracilis*), and 76.40 (*E. xingi*) (Table 2). All the newly reported mitochondrial genomes have a positive AT-skew, while the GC-skew is negative. In most hemipteran species, there are fewer T than A and fewer G than C bases. The AT-skew and GC-skew values of the three newly reported genomes are 0.010 and −0.154 for *E. rufa*, −0.239 and 0.033 for *E. gracilis*, and 0.031 and −0.243 for *E. xingi*, respectively (Table 2), suggesting that the chain asymmetry was reversed [33].

3.2. Protein-Coding Genes, Composition, and Evolutionary Rates

There are no remarkable differences in the sizes of tRNAs, PCGs, and rRNAs among each species when compared. The PCG lengths of the three newly obtained species are 10,825 (*E. rufa*), 10,822 (*E. gracilis*), and 10,818 bp (*E. xingi*). All newly obtained mitochondrial genomes exhibit a negative GC-skew and AT-skew; the GC-skew values are −0.08 (*E. rufa*), −0.11 (*E. gracilis*), and −0.09 (*E. xingi*), while the AT-skew values are −0.13 (*E. rufa*), −0.14 (*E. gracilis*), and 0.13 (*E. xingi*) (Table 2). The AT contents (%) are 78.34 (*E. rufa*), 73.55 (*E. gracilis*), and 75.80 (*E. xingi*), and the GC contents (%) are 21.65 (*E. rufa*), 26.45 (*E. gracilis*), and 24.19 (*E. xingi*) (Table 2). Compared with published data, we found that the AT content of the third codon positions was higher than that of the first and second positions in PCGs (Figure 3).

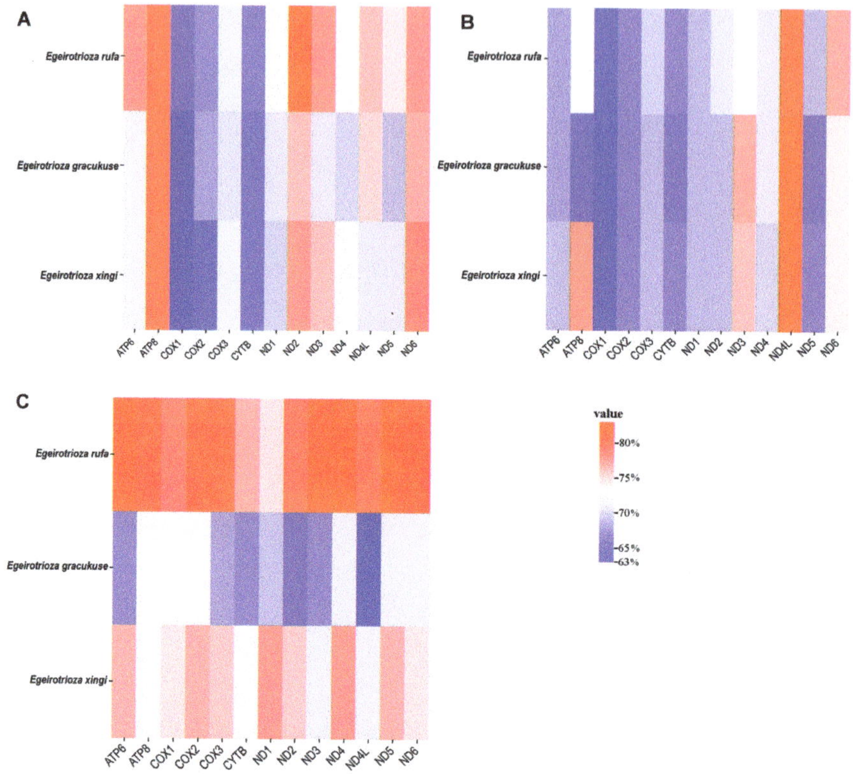

Figure 3. Difference in AT contents of protein-coding genes of *Egeirotrioza* mitochondrial genomes. (**A**) First codon positions; (**B**) second codon positions; (**C**) third codon positions.

Similar to insect mitochondria, the three newly obtained mitochondrial genomes' PCGs started with ATN [2,3,31]. However, diverse start codons were discovered: COI used ATG as the start codon in three species; the *ND3* gene used ATT in two species and ATA in one species; *ND4L* used TTG in three species; *ND5* used ATC in one species and ATT in two species; *ND2* used ATG in one species and ATA in two species, etc. The codon sizes of the three newly sequenced species were 1764 (*E. rufa*), 1303 (*E. gracilis*), and 1290 (*E. xingi*). Most PCGs in this group used TAA or TAG as the stop codon. However, *ND5*, *ND1*, and *COX2* in *Egeirotrioza* have an incomplete termination codon of TA or T; for example, that for *ND1* in *E. xingi* was TA and in *E. rufa* was T, and that for *COX2* in *E. xingi* and *E. rufa* was T. Incomplete termination codons of PCGs are frequently observed in insects and are typically completed through polyadenylation following the excision of the downstream tRNA gene [34–36].

The Ka/Ks value (ω) is commonly employed to gauge the rate of sequence evolution driven by natural selection [37]. The Ka/Ks analysis results showed that all 13 PCGs have a value of less than one, ranging from 0.07 (*COX1*) to 0.55 (*ATP8*) (Figure 4), and the evolution rate is as follows: *ATP8* > *ND6* > *ND5* > *ND2* > *ND4* > *ND4L* > *ND1* > *ND3* > *APT6* > *CYTB* > *COX3* > *COX2* > *COX1*. This indicates that each gene has undergone purifying selection, and some genes such as *ATP8* and *ND6* experienced relatively relaxed selection pressure. The DNA barcoding gene *COX1* underwent the strongest purifying selection, which is consistent with previous studies of psyllids [6].

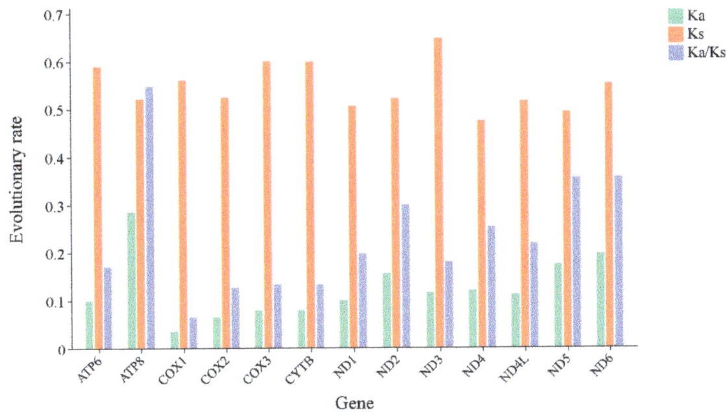

Figure 4. Evolution rates of 13 PCGs of *Egeirotrioza*. Ka refers to non-synonymous nucleotide substitutions, Ks refers to synonymous nucleotide substitutions, and Ka/Ks refers to the selection pressure of each PCG. The abscissa represents the 13 PCGs, and the ordinate represents the Ka/Ks values.

The 22 tRNAs ranged from 58 to 70 bp in length. The AT contents (%) of the newly obtained mitochondrial genomes are as follows: 79.45 (*E. rufa*), 78.17 (*E. gracilis*), and 79.60 (*E. xingi*). All the newly assembled mitochondrial genomes exhibit positive AT-skew and GC-skew; the AT-skew values are 0.013 (*E. rufa*), 0.024 (*E. gracilis*), and 0.035 (*E. xingi*), and the GC-skew values are 0.145 (*E. rufa*), 0.145 (*E. gracilis*), and 0.154 (*E. xingi*) (Table 2). The rRNA lengths are as follows: 1892 (*E. rufa*), 1891 (*E. gracilis*), and 1898 (*E. xingi*). The AT contents (%) are 79.08 (*E. rufa*), 76.95 (*E. gracilis*), and 78.60 (*E. xingi*). The AT-skew values of all mitochondrial genomes are negative (−0.035 to −0.016), while the GC-skew values are positive (0.0314 to 0.0357) (see Table 2).

3.3. Nucleotide Diversity and Codon Usage

The sliding window analysis showed that the nucleotide diversity (Pi) of the 13 PCGs in *Egeirotrioza* is highly variable, with the highest Pi obtained for *ATP8* (0.390) followed by *ND1* (0.367) and *ND4* (0.280), and the lowest Pi for *COX1* (0.100) (Figure 5). The relative synonymous codon usage (RSCU) patterns exhibited by the three mitochondrial genomes are largely similar, as depicted in Figure 6, which includes the RSCU values for all possible synonymous codons corresponding to the 22 amino acids and 62 available codons used in the 13 PCGs of *Egeirotrioza* species. Among the *Egeirotrioza* species, UUA is the preferred codon, while Leu2, Ile, Phe, and Mte are the most frequently utilized amino acids. When considering the three species, *E. gracilis* and *E. xingi* favor the codons AUU, UUA, UUU, and AUA, whereas *E. rufa* prefers the codons UUA, AUU, UUU, and AUA. Prior to this study, the genus *Egeirotrioza* never had any molecular markers published in GenBank and BOLD systemv4. Our results on nucleotide diversity show that the *Egeirotrioza COX1* gene can be used as an effective molecular marker for classification. This study has generated barcode reference data for *Egeirotrioza* species in order to use DNA barcoding as a rapid tool for accurate identification of the psyllid to aid phytosanitary measures. Meanwhile, the published mitochondrial genome of *Egeirotrioza* also facilitates the application of multi-marker DNA meta-barcoding technology for rapid pest monitoring. Codons ending in A/T are used more frequently, and the AT content at the third codon position is higher than that at the first and second codons, indicating that the third codon is more vulnerable to AT alterations [38].

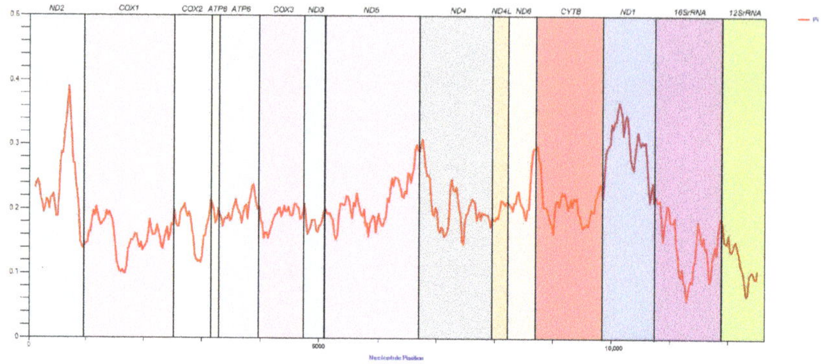

Figure 5. Sliding window analysis of 13 PCGs and 2 rRNAs in the mitochondrial genomes of 3 species of Triozidae. The red line represents the nucleotide diversity (Pi) value (window size = 200 bp, step size = 20 bp).

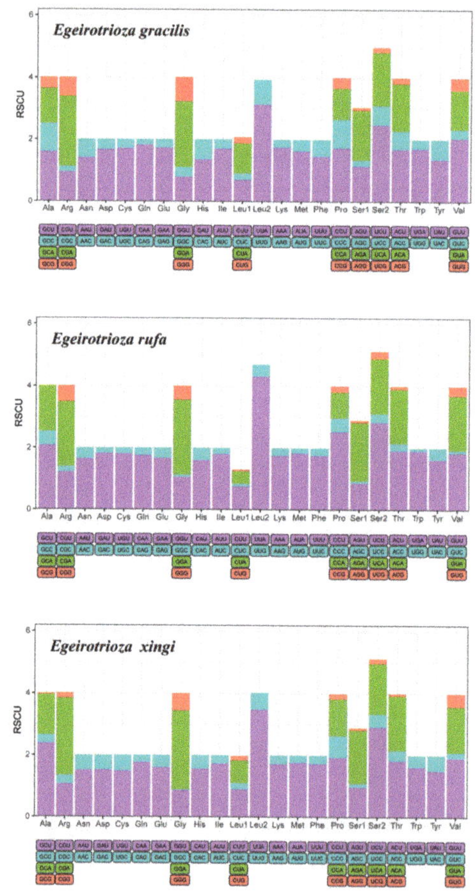

Figure 6. Relative synonymous codon usage (RSCU) of mitochondrial protein-coding genes of 3 Triozidae species. The *X*-axis shows different amino acids, and the *Y*-axis shows the RSCU value (the number of times a certain synonymous codon is used/the average number of times that all codons encoding the amino acid are used).

3.4. Phylogenetic Relationships

The heterogeneous divergence analysis using AliGROOVE found that the cds_faa matrix exhibited lower heterogeneity than the cds_fna, cds12_fna, cds12_rrna, and cds_rrna matrices (Figure 7). We used a supermatrix including cds_faa (5375 sites), cds_fna (7148 sites), cds_rrna (12,580 sites), cds12_fna (7148 sites), and cds12_rrna (9006 sites) to reconstruct the phylogenetic relationship of Triozidae based on BI and ML methods. The two approaches using these five matrices generated five BI and six ML trees, respectively (Figure 8 and Figures S1–S10).

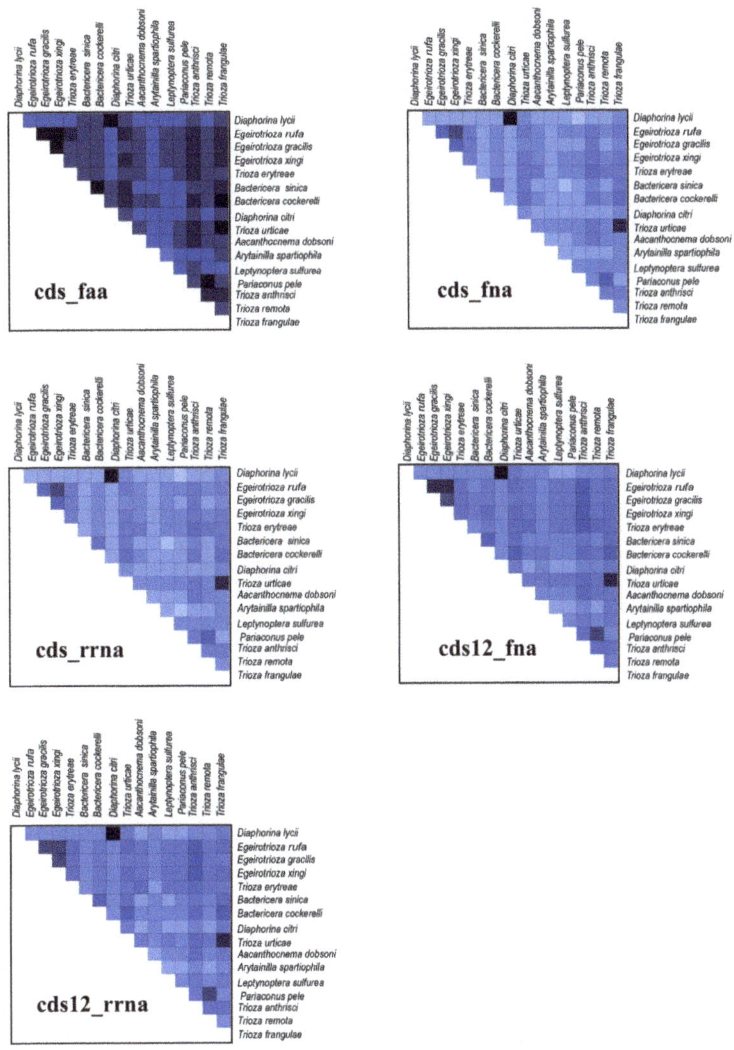

Figure 7. Heterogeneity analysis for different matrices. Colored squares represent pairwise Aliscore values. Score values range from −1 (indicating fully random similarity, dark blue) to +1 (indicating non-random similarity).

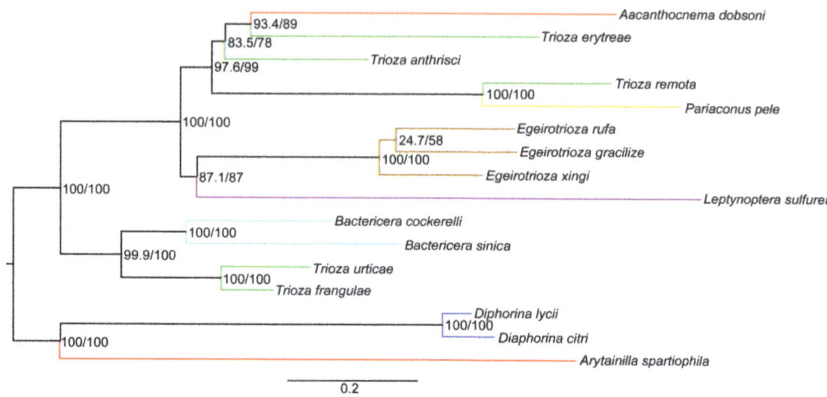

Figure 8. Phylogenetic trees of *Egeirotrioza* based on the cds_faa matrix with the PMSF model in IQ-TREE. Support values on nodes indicate r SH-aLRT/UFBoot2, respectively.

The phylogenetic analyses using all the matrices produced highly consistent topologies, thus supporting the monophyly of *Egeirotrioza*. The phylogenetic relationships show that *Egeirotrioza* is closely related to *Leptynoptera* (Figure 8). The genus *Evegeirotrioza* was established by Li [10]. Subsequently, Burckhardt and Ouvrard [11] proposed the synonymy of *Evegeirotrioza* with *Egeirotrioza* based on the morphological character of the genital plate, which is conical when viewed from the side. *Evegeirotrioza* was morphologically considered to be synonymous with *Egeirotrioza*, but there was no evidence of molecular correlation. In our phylogenetic studies, we formed the structure (*Egeirotrioza xingi* + (*Evegeirotrioza rufa* + *Evegeirotrioza gracilis*)) (Figures 8 and S1–S10). *Evegeirotrioza* has been well restored to the taxa of *Egeirotrioza*, so we reject *Evegeirotrioza* at both morphological and molecular levels and restore the monophyly of *Egeirotrioza* [39]. As *Egeirotrioza* has never been reported as a molecular marker, the genus was not included in previous phylogenetic studies based on molecular morphology.

Trioza is considered by most authors as an artificial receptacle for species not showing any particular morphological modifications. In our results, the position of *Trioza* (five species) was located in four groups, and the relationships of *Leptynoptera* and *Bactericera* are highly consistent with previous studies [6]. In Triozidae, a species-rich, probably monophyletic family, most of the genera are ill-defined and artificial, and the phylogenetic relationships between genera remain largely unknown. Here, we have provided the mitochondrial DNA of *Egeirotrioza*, the first molecular analysis of its phylogenetic position, and also new fundamental data on the phylogenetic relationships of Psyllidae.

Due to the limited samples, the position of the genus *Trioza* was unclear and it also did not support the monophyly of *Trioza*, which was located in four groups (Figure 8). Additional taxon sampling, data, and analyses are necessary to resolve this ambiguity in future studies. For instance, low-coverage sequences are often utilized to extract more molecular markers to reconstruct stable phylogenetic relationships, as successfully demonstrated in other groups [40–42].

4. Conclusions

Three mitochondrial genomes of *Egeirotrioza* were obtained, including two complete mitochondrial genomes and one linear mitochondrial genome. The newly sequenced mitochondrial genomes exhibit similar structural features and nucleotide compositions to the previously published data of Triozidae. The nucleotide diversity (Pi) analysis showed that the diversity of *ATP8* is the highest, and that of *COX1* is the lowest. The three mitochondrial genomes display similar relative synonymous codon usage (RSCU) patterns. In adding published data, we could also reconstruct the phylogenetic relationships among

Triozidae. In our results, the monophyly of *Egeirotrioza* was well supported, while the position of the genus *Trioza* was unclear due to the limited samples.

Supplementary Materials: The following supporting information can be downloaded at: https://www.mdpi.com/article/10.3390/genes15070842/s1, Figure S1: Bayesian inference phylogenetic tree of *Egeirotrioza* based on the analysis of cds_faa with a GTR + CAT model in PhyloBayes. Support values on nodes indicate Bayesian posterior probabilities; Figure S2: Maximum likelihood phylogenetic tree of Egeirotrioza based on the analysis of cds_faa with the Partition model in IQ-TREE. Support values on nodes indicate SH-aLRT/UFBoot2, respectively; Figure S3: Bayesian infer-ence phylogenetic tree of Egeirotrioza based on the analysis of cds_fna with a GTR + CAT model in PhyloBayes. Support values on nodes indicate Bayesian posterior probabilities; Figure S4: Maximum likelihood phylogenetic tree of Egeirotrioza based on the analysis of cds_fna with the Partition model in IQ-TREE. Support values on nodes indicate SH-aLRT/UFBoot2, respectively; Figure S5: Bayesian inference phylogenetic tree of Egeirotrioza based on the analysis of cds_rrna with a GTR + CAT model in PhyloBayes. Support values on nodes indicate Bayesian posterior probabilities; Figure S6: Maxi-mum likelihood phylogenetic tree of Egeirotrioza based on the analysis of cds_rrna with the Partition model in IQ-TREE. Support values on nodes indicate SH-aLRT/UFBoot2, respectively; Figure S7: Bayesian inference phylogenetic tree of Egeirotrioza based on the analysis of cds12_fna with a GTR + CAT model in PhyloBayes. Support values on nodes indicate Bayesian posterior probabilities; Figure S8: Maximum likelihood phylogenetic tree of Egeirotrioza based on the analysis of cds12_fna with the Partition model in IQ-TREE. Support values on nodes indicate SH-aLRT/UFBoot2, respectively; Figure S9: Bayesian inference phylogenetic tree of Egeirotrioza based on the analysis of cds12_rrna with a GTR + CAT model in PhyloBayes. Support values on nodes indicate Bayesian posterior probabilities; Figure S10: Maximum likelihood phylogenetic tree of Egeirotrioza based on the analysis of cds12_rrna with the Partition model in IQ-TREE. Support values on nodes indicate SH-aLRT/UFBoot2, respectively.

Author Contributions: Software, Z.-C.L.; formal analysis, Z.A. and Z.-L.M.; resources, Z.A. and Z.-L.M.; data curation, Z.A. and Z.-L.M.; writing—original draft preparation, Z.A. and Z.-L.M.; writing—review and editing, X.-Y.L. and N.H.; visualization, Z.-C.L.; supervision, N.H. All authors have read and agreed to the published version of the manuscript.

Funding: This research was funded by the Natural Science Foundation of Xinjiang Uygur Autonomous Region (grant number 2022D01C403), the Beijing Natural Science Foundation (grant number 5244031), and the Beijing Government. The APC was funded by the Beijing Government.

Institutional Review Board Statement: Not applicable.

Informed Consent Statement: Not applicable.

Data Availability Statement: Regarding the availability of DNA sequences, the new mitochondrial genomes are deposited in GenBank (NCBI), and the accession numbers are PP471964-PP471966.

Acknowledgments: We sincerely thank Chen-Hong Wang and Jin-Lin Wang for collecting the materials.

Conflicts of Interest: The authors declare no conflicts of interest.

References

1. Cameron, S.L. Insect Mitochondrial Genomics: Implications for Evolution and Phylogeny. *Annu. Rev. Entomol.* **2014**, *59*, 95–117. [CrossRef]
2. Ge, X.; Peng, L.; Vogler, A.P.; Morse, J.C.; Yang, L.; Sun, C.; Wang, B. Massive gene rearrangements of mitochondrial genomes and implications for the phylogeny of Trichoptera (Insecta). *Syst. Entomol.* **2023**, *48*, 278–295. [CrossRef]
3. Lin, X.-L.; Liu, Z.; Yan, L.-P.; Duan, X.; Bu, W.-J.; Wang, X.-H.; Zheng, C.-G. Mitogenomes provide new insights of evolutionary history of Boreheptagyiini and Diamesini (Diptera: Chironomidae: Diamesinae). *Ecol. Evol.* **2022**, *51*, 119–132. [CrossRef]
4. Boore, L. Animal mitochondrial genomes. *Nucleic Acids Res.* **1999**, *27*, 1767–1780. [CrossRef]
5. Kim, K.Y.; Lee, S.Y.; Bang, I.C.; Nam, Y.K. Complete mitogenome sequence of an endangered freshwater fish, Iksookimia choii (Teleostei; Cypriniformes; Cobitidae). *Mitochondrial DNA* **2008**, *19*, 438–445. [CrossRef]
6. Percy, D.M.; Crampton-Platt, A.; Sveinsson, S.; Lemmon, A.R.; Lemmon, E.M.; Ouvrard, D.; Burckhardt, D. Resolving the psyllid tree of life: Phylogenomic analyses of the superfamily Psylloidea (Hemiptera). *Syst. Entomol.* **2018**, *43*, 762–776. [CrossRef]

7. Perilla-Henao, L.M.; Casteel, C.L. Vector-Borne Bacterial Plant Pathogens: Interactions with Hemipteran Insects and Plants. *Front. Plant Sci.* **2016**, *64*, 325–338. [CrossRef]
8. Hodkinson, I.D. The biology of the Psylloidea (Homoptera): A review. *Bull. Entomol. Res.* **1974**, *64*, 325–338. [CrossRef]
9. Ouvrard, D. Psyl'list-The World Psylloidea Database. Available online: http://flow.hemiptera-databases.org/flow/ (accessed on 19 February 2015).
10. Li, F.S. *Psyllydomorpha of China*; Science Press: Beijing, China, 2011.
11. Ren, L.L.; Li, Z.Y.; Li, Y.C.; Guo, Y.M. Revision of Scientific Nosmes for the main Insect Species in the Monograph "Forest Insects of China (2rd Edition, 1992)". *Sci. Silvae Sin.* **2016**, *52*, 110–115. [CrossRef]
12. Zhang, B.K.; Cui, X.P.; Wang, P.L.; Sun, H.Y.; Zhou, L.; Pang, H.L.; Tong, L. Morphological Identification and Damage Characteristics of Three Populus Euphratica Psyllid in Northern Xinjiang. *Xinjiang Agric. Sci.* **2012**, *49*, 1887–1890. [CrossRef]
13. Bolger, A.M.; Marc, L.; Bjoern, U. Trimmomatic: A flexible trimmer for Illumina sequence data. *Bioinformatics* **2014**, *30*, 2114–2120. [CrossRef]
14. Nicolas, D.; Patrick, M.; Guillaume, S. NOVOPlasty: De novo assembly of organelle genomes from whole genome data. *Nucleic Acids Res.* **2017**, *45*, 18. [CrossRef]
15. Peng, Y.; Leung, H.C.M.; Yiu, S.M.; Chin, F.Y.L. IDBA-UD: A de novo assembler for single-cell and metagenomic sequencing data with highly uneven depth. *Bioinformatics* **2012**, *28*, 1420–1428. [CrossRef]
16. Kearse, M.; Moir, R.; Wilson, A.; Stones-Havas, S.; Cheung, M.; Sturrock, S.; Buxton, S.; Cooper, A.; Markowitz, S.; Duran, C.; et al. Geneious Basic: An integrated and extendable desktop software platform for the organization and analysis of sequence data. *Bioinformatics* **2012**, *28*, 1647–1649. [CrossRef]
17. Chan, P.P.; Lowe, T.M. tRNAscan-SE: Searching for tRNA genes in genomic sequences. In *Gene Prediction;* Springer: Berlin/Heidelberg, Germany, 2019; pp. 1–14. [CrossRef]
18. Kumar, S.; Stecher, G.; Li, M.; Knyaz, C.; Tamura, K. MEGA X: Molecular Evolutionary Genetics Analysis across computing platforms. *Mol. Biol. Evol.* **2018**, *35*, 1547–1549. [CrossRef]
19. Shen, W.; Le, S.; Li, Y.; Hu, F.Q. SeqKit: A Cross-Platform and Ultrafast Toolkit for FASTA/Q File Manipulation. *PLoS ONE* **2016**, *11*, 10. [CrossRef]
20. Rozas, J.; Ferrer-Mata, A.; Sánchez-DelBarrio, J.C.; Guirao-Rico, S.; Librado, P.; Ramos-Onsins, S.E.; Sánchez-Gracia, A. DnaSP 6: DNA Sequence Polymorphism Analysis of Large Data Sets. *Mol. Biol. Evol.* **2017**, *34*, 3299–3302. [CrossRef]
21. Standley, D. MAFFT multiple sequence alignment software version 7: Improvements in performance and usability. *Mol. Biol. Evol.* **2013**, *30*, 772–780. [CrossRef]
22. Capella-Gutierrez, S.; Silla-Martinez, J.M.; Gabaldon, T. trimAl: A tool for automated alignment trimming in large-scale phylogenetic analyses. *Bioinformatics* **2009**, *25*, 1972–1973. [CrossRef]
23. Longo, P.K.G. FASconCAT-G: Extensive functions for multiple sequence alignment preparations concerning phylogenetic studies. *Front. Zool.* **2014**, *11*, 812. [CrossRef]
24. Kück, P.; Meid, S.A.; Groß, C.; Wägele, J.W.; Misof, B. AliGROOVE—Visualization of heterogeneous sequence divergence within multiple sequence alignments and detection of inflated branch support. *BMC Bioinform.* **2014**, *15*, 294. [CrossRef]
25. Kalyaanamoorthy, S.; Minh, B.Q.; Wong, T.K.; Von Haeseler, A.; Jermiin, L.S. ModelFinder: Fast Model Selection for Accurate Phylogenetic Estimates. *Nat. Methods* **2017**, *14*, 587–589. [CrossRef]
26. Minh, B.Q.; Schmidt, H.A.; Chernomor, O.; Schrempf, D.; Woodhams, M.D.; von Haeseler, A.; Lanfear, R. IQ-TREE 2: New Models and Efficient Methods for Phylogenetic Inference in the Genomic Era. *Mol. Biol. Evol.* **2020**, *37*, 1530–1534, Erratum in *Mol. Biol. Evol.* **2020**, *37*, 2461.
27. Wang, H.-C.; Minh, B.Q.; Susko, E.; Roger, A.J. Modeling Site Heterogeneity with Posterior Mean Site Frequency Profiles Accelerates Accurate Phylogenomic Estimation. *Syst. Biol.* **2018**, *67*, 216–235. [CrossRef]
28. Nicolas Lartillot, N.R.; Stubbs, D.; Richer, J. PhyloBayes MPI: Phylogenetic reconstruction with infinite mixtures of profiles in a parallel environment. *Syst. Biol.* **2013**, *62*, 611–615. [CrossRef]
29. Cui, Y.; Xie, Q.; Hua, J.; Dang, K.; Zhou, J.; Liu, X.; Wang, G.; Yu, X.; Bu, W. Phylogenomics of Hemiptera (Insecta: Paraneoptera) based on mitochondrial genomes. *Syst. Entomol.* **2013**, *38*, 233–245. [CrossRef]
30. Wang, S.; Zhu, R.; Xue, H.; Li, Y.; Bu, W. Mitogenomics of Chinch Bugs from China and Implications for Its Coevolutionary Relationship with Grasses. *Insects* **2022**, *13*, 643. [CrossRef]
31. Zhang, D.; He, F.-X.; Li, X.-B.; Aishan, Z.; Lin, X.-L. New Mitogenomes of the Polypedilum Generic Complex (Diptera: Chironomidae): Characterization and Phylogenetic Implications. *Insects* **2023**, *14*, 238. [CrossRef]
32. Ge, X.; Zang, H.; Ye, X.; Peng, L.; Wang, B.; Lian, G.; Sun, C. Comparative Mitogenomic Analyses of Hydropsychidae Revealing the Novel Rearrangement of Protein-Coding Gene and tRNA (Trichoptera: Annulipalpia). *Insects* **2022**, *13*, 759. [CrossRef]
33. Yu, P.; Zhou, L.; Zhou, X.-Y.; Yang, W.-T.; Zhang, J.; Zhang, X.-J.; Wang, Y.; Gui, J.-F. Unusual AT-skew of Sinorhodeus microlepis mitogenome provides new insights into mitogenome features and phylogenetic implications of bitterling fishes. *Int. J. Biol. Macromol.* **2019**, *129*, 339–350. [CrossRef]
34. Bratic, A.; Clemente, P.; Calvogarrido, J.; Maffezzini, C.; Felser, A.; Wibom, R.; Wedell, A.; Freyer, C.; Wredenberg, A. Mitochondrial Polyadenylation Is a One-Step Process Required for mRNA Integrity and tRNA Maturation. *PLoS Genet.* **2016**, *12*, e1006028. [CrossRef] [PubMed]

35. Ji, H.; Xu, X.; Jin, X.; Yin, H.; Luo, J.; Liu, G.; Zhao, Q.; Chen, Z.; Bu, W.; Gao, S. Using high-resolution annotation of insect mitochondrial DNA to decipher tandem repeats in the control region. *RNA Biol.* **2019**, *16*, 830–837. [CrossRef] [PubMed]
36. Toompuu, M.; Tuomela, T.; Laine, P.; Paulin, L.; Dufour, E.; Jacobs, H.T. Polyadenylation and degradation of structurally abnormal mitochondrial tRNAs in human cells. *Nucleic Acids Res.* **2018**, *46*, 5209–5226. [CrossRef] [PubMed]
37. Yang, Z.H.; Bielawski, J.P. Statistical methods for detecting molecular adaptation. *Trends Ecol. Evol.* **2000**, *15*, 496–503. [CrossRef] [PubMed]
38. Eyre-Walker, A. Differentiating between selection and mutation bias. *Genetic* **1997**, *147*, 1983–1987. [CrossRef] [PubMed]
39. Ouvrard, D.B.D. A revised classification of the jumping plant-lice (Hemiptera: Psylloidea). *ZOOTAXA* **2012**, *3509*, 1–34. [CrossRef]
40. Sun, X.; Ding, Y.; Orr, M.C.; Zhang, F. Streamlining universal single-copy orthologue and ultraconserved element design: A case study in Collembola. *Mol. Ecol. Resour.* **2020**, *20*, 706–717. [CrossRef] [PubMed]
41. Zhang, D.; Niu, Z.-Q.; Luo, A.-R.; Orr, M.C.; Ferrari, R.R.; Jin, J.-F.; Wu, Q.-T.; Zhang, F.; Zhu, C.-D. Testing the systematic status of Homalictus and Rostrohalictus with weakened cross-vein groups within Halictini (Hymenoptera: Halictidae) using low-coverage whole-genome sequencing. *Insect Sci.* **2022**, *29*, 1819–1833. [CrossRef]
42. Zhang, F.; Ding, Y.; Zhu, C.D.; Zhou, X.; Orr, M.C.; Scheu, S.; Luan, Y.X. Phylogenomics from low-coverage whole-genome sequencing. *Methods Ecol. Evol.* **2019**, *10*, 507–517. [CrossRef]

genes

MDPI

Article

Sequencing and Description of the Mitochondrial Genome of *Orthopodomyia fascipes* (Diptera: Culicidae)

Fábio Silva da Silva [1,2], Bruna Laís Sena do Nascimento [2], Ana Cecília Ribeiro Cruz [1,2], Sandro Patroca da Silva [2], Carine Fortes Aragão [2], Daniel Damous Dias [1,2], Lucas Henrique da Silva e Silva [1,2], Lúcia Aline Moura Reis [1,2], Hanna Carolina Farias Reis [1,2], Liliane Leal das Chagas [2], José Wilson Rosa Jr. [2], Durval Bertram Rodrigues Vieira [2], Roberto Carlos Feitosa Brandão [2], Daniele Barbosa de Almeida Medeiros [1,2] and Joaquim Pinto Nunes Neto [1,2,*]

1 Graduate Program in Parasitary Biology in the Amazon Region, Center of Biological and Health Sciences, State University of Pará, Belém 66095-663, Brazil; fabiodasilva@iec.gov.br (F.S.d.S.); anacecilia@iec.gov.br (A.C.R.C.); damous1994@gmail.com (D.D.D.); biohenriquesilva@hotmail.com (L.H.d.S.e.S.); luciaalinereis@gmail.com (L.A.M.R.); hanna_carolina@hotmail.com (H.C.F.R.); danielemedeiros@iec.gov.br (D.B.d.A.M.)

2 Evandro Chagas Institute—IEC/MS/SVSA, Department of Arbovirology and Hemorragic Fevers, Ananindeua 67030-000, Brazil; brunanascimento@iec.gov.br (B.L.S.d.N.); sandrosilva@iec.gov.br (S.P.d.S.); carinearagao@iec.gov.br (C.F.A.); lilianechagas@iec.gov.br (L.L.d.C.); josejr@iec.gov.br (J.W.R.J.); durvalvieira@iec.gov.br (D.B.R.V.); robertobrandao@iec.gov.br (R.C.F.B.)

* Correspondence: joaquimneto@iec.gov.br

Abstract: The genus *Orthopodomyia* Theobald, 1904 (Diptera: Culicidae) comprises 36 wild mosquito species, with distribution largely restricted to tropical and temperate areas, most of which are not recognized as vectors of epidemiological importance due to the lack of information related to their bionomy and involvement in the cycle transmission of infectious agents. Furthermore, their evolutionary relationships are not completely understood, reflecting the scarcity of genetic information about the genus. Therefore, in this study, we report the first complete description of the mitochondrial genome of a Neotropical species representing the genus, *Orthopodomyia fascipes* Coquillet, 1906, collected in the Brazilian Amazon region. Using High Throughput Sequencing, we obtained a mitochondrial sequence of 15,598 bp, with an average coverage of 418.5×, comprising 37 functional subunits and a final portion rich in A + T, corresponding to the control region. The phylogenetic analysis, using Maximum Likelihood and Bayesian Inference based on the 13 protein-coding genes, corroborated the monophyly of Culicidae and its two subfamilies, supporting the proximity between the tribes Orthopodomyiini and Mansoniini, partially disagreeing with previous studies based on the use of molecular and morphological markers. The information generated in this study contributes to a better understanding of the taxonomy and evolutionary history of the genus and other groups of Culicidae.

Keywords: Culicidae; mitogenome; molecular taxonomy; mosquitoes; phylogenetics

Citation: Silva, F.S.d.; Nascimento, B.L.S.d.; Cruz, A.C.R.; Silva, S.P.d.; Aragão, C.F.; Dias, D.D.; Silva, L.H.d.S.e.; Reis, L.A.M.; Reis, H.C.F.; Chagas, L.L.d.; et al. Sequencing and Description of the Mitochondrial Genome of *Orthopodomyia fascipes* (Diptera: Culicidae). *Genes* **2024**, *15*, 874. https://doi.org/10.3390/genes15070874

Academic Editor: Zhiteng Chen

Received: 3 May 2024
Revised: 3 June 2024
Accepted: 4 June 2024
Published: 3 July 2024

1. Introduction

Mosquitoes (Diptera: Culicomorpha: Culicoidea: Culicidae) [1,2] comprise a monophyletic taxon consisting of around 3700 formally recognized species and are taxonomically organized into two subfamilies: Anophelinae Grassi, 1900, and Culicinae Meigen, 1818 [3,4], with records of occurrence in most continents and bioregions, with the exception only in permanently frozen areas [5], and being particularly recognized for their great importance in transmission cycles of various infectious agents on a global scale, with this factor being the main driver of most studies and investigations carried out on this group of insects [6].

However, despite its great medical-epidemiological relevance, there are still few studies that address the evolutionary relationships between these organisms. In this context, taxonomic knowledge, as well as the survey of hypotheses about the phylogenetic relationships between mosquitoes, constitute fundamental tools in understanding basic patterns associated with aspects related to their ecology, such as specialization in larval habitat [7], vector competence [8,9], and population dynamics [10,11].

Most of the studies and investigations carried out focus on evaluating the epidemiological potential of mosquitoes, especially the genera *Aedes* Meigen, 1818, *Anopheles* Meigen, 1818, and *Culex* Linnaeus, 1758. However, it is important to consider that factors such as the increased deforestation and the unrestrained exploitation of natural resources, directly related to the consequent intensification of urbanization processes and human interference in forest ecosystems, act to significantly influence the dynamics and behavior of wild populations of mosquito species that were previously considered minor concerns, and today are increasingly gaining attention in the field of public health [12–15].

The genus *Orthopodomyia* as the only representative of the tribe Orthopodomyiini, comprises 36 species distributed according to Zavortink (1968) [16] into eight groups, largely or totally restricted, each one to a single zoogeographic region of the planet. Species of the genus can be easily distinguished from adults of other mosquito genera based on the observation of the unique ornamentation in the bands and stripes of white, silver, and gold scales in the thorax region, with some species also presenting distinctive patches of light and dark scales, particularly on the wings. Additionally, they are distinguished by the first tarsomere (T_1) of the fore and middle legs being longer than the sum of the lengths of the four posterior tarsal segments combined ($T_{2\text{–}5}$), with the T_4 segment significantly shorter in relation to T_5 [16–18].

Although aspects of the biology and ecological habits of representatives of the genus are not well known, it is generally observed that the immature stages are found in a variety of habitats, such as tree holes, bamboo stumps and internodes, bromeliad axils, Heliconia flower spathes, and occasionally in artificial containers [16,19]. The larval stages apparently feed by filtering microorganisms and small particles removed from the water, and the pupal stage takes five to eight days for the adult to emerge, which in turn is essentially wild and apparently presents predominantly nocturnal activity [16].

Given the widespread lack of information on the vectorial capacity of *Orthopodomyia*, it is assumed that representatives of the genus play a role in the transmission cycles of avian arboviruses due to the ornithophilic behavior exhibited by females [16,20], with only two species, *Orthopodomyia albipes* Leicester, 1904, and *Orthopodomyia andamanensis* Barraud, 1934, belonging to the oriental *Alpibes* group, which are also known to feed on humans [16]. In general, species of the genus are not recognized as vectors of medical-epidemiological importance, largely due to the difficulty of collection and the lack of records of their involvement in the transmission of infectious agents to humans and domestic animals [16]. However, the species *Orthopodomyia signifera* Coquillett, 1896, in particular, in addition to demonstrating strong ornithophilic behavior, has also been reported as a competent vector for transmission of *Eastern Equine Encephalitis virus* (EEEV) and *Western Equine Encephalitis virus* (WEEV) [21], with records of detection of these viruses from the species in regions of Mexico [22] and the United States [23].

The evolutionary relationships of *Orthopodomyia* in relation to other genera of Culicidae are not completely understood, and are mainly based on morphological aspects of the larval and adult stages [24,25]. These former studies had classified Orthopodomyiini along with Aedeomyiini, Culisetini, Ficalbiini, Toxorhynchitini, Uranotaeniini and Hodgesiini as non-related groups [3,25]. New technologies allowed for some studies to evaluate the evolutionary positioning of *Orthopodomyia* based on few molecular data, which in turn placed *Orthopodomyia* close to *Coquillettidia* Dyar, 1905 genus, and these two related to the Aedini tribe [6]. The increasing use of genomic tools applied to evolutionary studies, particularly of arthropods of medical-epidemiological importance, has significantly contributed to the development of control strategies, mainly based on evaluating properties related to the population variability of vectors and the co-evolutionary association between them and the infectious agents they can harbor and potentially transmit [26–28]. In this context, the use of molecular markers from genomic sets such as the mitochondrial genome, particularly applied in the development of taxonomic and evolutionary studies of metazoans, has been increasingly consolidated, due to factors such as its high rate of accumulation of mutations (particularly synonymous), absence of introns and recombinant processes, maternal uniparental inheritance, and easy amplification due to its large number of copies per cell compared to the nuclear genome (nDNA) [29,30].

The mitochondrial genome (mtDNA) consists of a compact double-stranded circular molecule, with a length ranging from 15 to 20 kb in insects, consisting of 37 functional subunits, including 13 protein-coding genes (PCGs), 22 transport RNAs (tRNAs), and 2 ribossomal RNAs (rRNAs), in addition to a final portion rich in Adenine and Thymine (A + T), associated with replicative processes [31]. Some of its main regions used in genomic investigations are active in protein coding, and belong to complexes I (*NADH* 4 and 5), III (*CytB*), and especially IV (*COI* and *COII*) [32], with some of these, specially *COI*, considering their 5′ terminal portion, being named DNA barcodes, based on the proposal that they can be used as universal identification markers of metazoans at higher levels of genus and species [33]. However, with significant advances in the development of genomic sequencing methodologies and bioinformatics tools, it is now possible for rapid and complete characterization of the mitochondrial genomes, particularly of mosquito species, supporting studies in population genetics and molecular taxonomy on the representatives for both subfamilies, Anophelinae [11,34] and Culicinae [35–43].

To date, studies on the application of molecular markers for identifying species and reconstructing the evolutionary relationships of *Orthopodomyia* have focused mainly on the *Signifera* (Holarctic) and *Albipes* (Oriental) groups, and as for the available mitochondrial data, existing information is limited to partial sequences, mainly from the *COI* gene [44]. Therefore, considering the great potential of the applicability of molecular methods and markers such as the mitochondrial genome in the development of taxonomic and evolutionary investigations, especially of mosquitoes, this study presents the characterization of the first complete mitochondrial genome of a Neotropical species of the genus *Orthopodomyia* found in the Brazilian Amazon region: *Or. fascipes*, using High Throughput Sequencing (HTS), and considers the reconstructing of the Culicidae phylogeny based on the maximum totality of coding regions of the investigated genome.

2. Materials and Methods

2.1. Sample Collection and Total DNA Extraction

The specimens of *Or. fascipes* were collected during ecoepidemiological expeditions conducted in the ecological settlement Expedito Ribeiro (1°12′32.4″ S 48°16′18.2″ W) (Figure 1), located in the municipality of Santa Bárbara do Pará, State of Pará, Brazil, from 17–19 March 2021, under authorization granted by the Biodiversity Authorization and Information System of the Ministry of the Environment (SISBIO/IBAMA), registration no. 56504-6. The collection procedures involved the installation of larvitraps made from the bark of the fruit of the chestnut tree species *Bertholletia excelsa*, locally known as *Castanha-do-Pará*, in shaded, poorly lit areas protected from rain, being left in place for five days until they were retrieved, and transported to the Medical Entomology laboratory of the Arbovirology and Hemorrhagic Fevers Section of the Evandro Chagas Institute (IEC/SVSA/MS) in the city of Ananindeua, State of Pará, Brazil (1°22′32.8″ S 48°23′03.8″ W). The immature forms collected were reared in entomological cages in the laboratory until the emergence of adults, which were subsequently identified with the assistance of the taxonomic keys developed by Lane (1953) [17] and Zavortink (1968) [16], individualized in cryogenic tubes, and stored in a freezer at −70 °C for preservation.

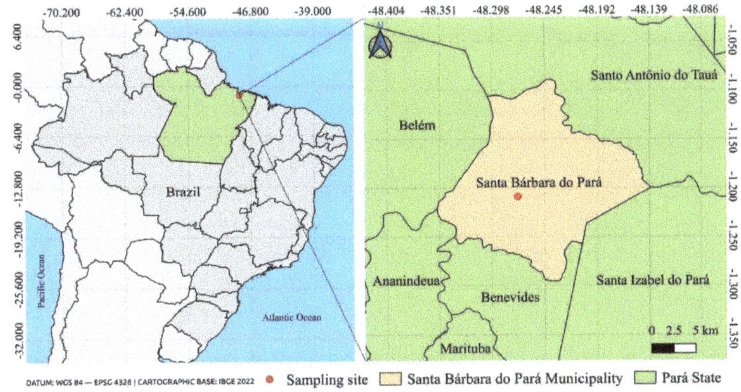

Figure 1. Location map of the municipality of Santa Bárbara do Pará demarcating the collection site of the investigated species. This Figure was created using the software QGIS v.3.10.4 (available at https://qgis.org/en/site/, accessed on 30 March 2024) in conjunction with the IBGE 2022 cartographic database (available at https://www.ibge.gov.br/, accessed on 30 March 2024).

A single female specimen of the species *Or. fascipes* was reserved and macerated together with 360 µL of phosphate saline solution (PBS) using a 3 mm diameter stainless steel sphere in the TissueLyzer II equipment (Qiagen, Hilden, Germany). Then, 180 µL of the supernatant were reserved to carry out the total DNA extraction procedures, using the commercial kit DNeasy Blood & Tissue (Qiagen, Hilden, Germany). Subsequently, the extraction products were quantified using the dsDNA Hs Assay kit (Life Technologies, Waltham, MA, USA) on Qubit 2.0 fluorometer equipment (Life Technologies, Waltham, MA, USA), following the protocol recommendations established by the manufacturer.

2.2. Genomic Library Preparation and Sequencing

Total DNA extraction products were standardized to a concentration of 0.2 ng/µL and then fragmented and labeled with adapter sequences (i7 and i5) for post-sequencing identification using the Nextera XT DNA Library Preparation kit (Illumina, San Diego, CA, USA) according to the manufacturer's recommendations. Subsequently, the genomic libraries obtained were quantified using the Qubit 2.0 fluorometer equipment (Life Technologies, Waltham, MA, USA) and qualitatively evaluated using the High Sensitivity DNA analysis kit (Agilent Technologies, Santa Clara, CA, USA) on the BioAnalyzer 2100 equipment (Agilent Technologies, Santa Clara, CA, USA). Finally, the libraries were sequenced using the NextSeq 500/550 High Output v.2.5 kit (Illumina, San Diego, CA, USA) in 300 cycles (2×150) on the NextSeq 500 platform System (Illumina, San Diego, CA, USA).

2.3. Data Processing and mtDNA Characterization

The sequencing files obtained were initially subjected to quality assessment using Fastp v.0.23.4 [45], configured to remove adapter sequences and reads with *Phred quality* base < 20 and length less than 50 nt. Subsequently, assembly was performed *de novo* using MEGAHIT v.1.2.9 [46] in its standard configuration (*k-mers* of 21, 29, 39, 59, 79, 99, 119, and 141 nt). The contig corresponding to the mtDNA of the investigated species was identified using DIAMOND v.2.1.9.163 [47] (Blastx modality, *e-value* of 10^{-5}) and Krona v.2.8.1 [48], and manually inspected using Geneious v.11.1.5 [49]. The mitochondrial sequence obtained was annotated using the *MITOchondrial genome annotation Server* (MITOS) [50] and circularized by identifying overlaps between the ends using Blastn v.2.15.0 [51]. Coverage metrics by refmap and the circular structural representation of the genome were obtained using Bowtie2 v.2.5.3 [52] and CGview [53], respectively. In turn, the nucleotide composition and relative use of synonymous codons (RSCU) metrics were obtained using MEGA v.10.2.6 [54] and Geneious v.11.1.5, and the general nucleotide composition biases were estimated using the formulations AT skew = (A% − T %)/(A% + T%) and

GC skew = (G% − C%)/(G% + C%) [55]. Additionally, the codon usage bias based on the effective number of codons (ENc), as well as the composition metrics GC12 (average between GC contents of the first and second codon positions), GC3 (GC content of the third codon positions), and GC3s (GC content of the third position in synonymous codons) were obtained using CodonW v.1.4.4 [56]. The evaluation of the evolutionary pressure acting on PCGs was carried out based on the calculation of the ratio between non-synonymous (*dN*) and synonymous (*dS*) substitutions ($\omega = dN/dS$) using the CodeML application included in the PAML v.4.9j [57]. Finally, the graphs illustrating the results obtained were produced using the R v.4.3.3 [58], and the libraries *circlize* v.0.4.16 [59], *ggplot2* v.3.5.0 [60], *pheatmap* v.1.0.12 [61], and *reshape2* v.1.4.4 [62].

2.4. Phylogenetic Analysis

Phylogenetic analyses were performed based on the use of all 13 PCGs from the newly sequenced genome, along with another 53 mtDNAs, including assembled reference sequences and sequencing data (*rawdata*) available in the GenBank and SRA repositories (NCBI) (Table S1). Initially, the sequences were aligned using the MAFFT v.7.520 algorithm [63] and then manually inspected using Aliview v.1.28 [64]. The nucleotide substitution saturation of the set of aligned sequences was evaluated using the test developed by Xia et al. (2003) [65] in DAMBE v.7.3.11 [66]. The graphical representation of transitions and transversions in relation to genetic distance (model K80) was represented using *ape* library v.5.7.1 [67] implemented in R language. Then, using IQ-TREE v.1.6.12 [68], the best nucleotide substitution model (GTR + F + R5) was determined according to the Akaike information criterion (AIC). Subsequently, the phylogenetic signal was evaluated [69], and the phylogeny was reconstructed using the Maximum Likelihood method, with support values (bootstrapping) defined for 1000 replicates. In a parallel analysis, the phylogeny was also reconstructed using the Bayesian Inference method in MrBayes v.3.2.7a [70] in two independent and simultaneous runs, each with four chains (three hot chains and one cold one) and established for 1,000,000 generations, with sampling every 100 generations, and calculation of posterior probabilities after an initial relative burn of 25%. Convergence and estimated sample size (considering *Ess* > 200) were monitored using Tracer v.1.7.2 [71], and the obtained topologies were visualized using FigTree v.1.4.4 [72], with rooting based on the definition of the midpoint of the distances. *Dixella aestivalis* Meigen, 1818 (Diptera: Dixidae), was intentionally included in this analysis to define the anchoring group of the topologies, as it belongs to a group morphologically and phylogenetically related to Culicidae, and shares common ancestry with Chaoboridae and Corethrellidae, all belonging to the superfamily Culicoidea, suborder Culicomorpha [1,2].

3. Results

3.1. Assembly, Genomic Organization and General Composition of the Sequence Obtained

A total of 62.5 million paired reads were generated from the genomic sequencing of the investigated species, with approximately 93.1% being suitable for subsequent analysis after the quality control stage. Following genomic assembly, 623,807 contigs were obtained with lengths ranging from 200 to 1,236,672 bp (N50 = 667 bp). The contig corresponding to the mitochondrial genome of *Or. fascipes* (GenBank ID: PP749023) was 15,598 bp long, with an average coverage of 418.5× (7×~2506×), comprising 61,744 mapped reads (0.10% of the total reads generated), and was structurally organized in a compact circular molecule consisting of 37 functional subunits: 13 PCGs, 22 rRNAs, and two rRNAs, with a clear absence of inversion events and/or translocation of genes along the coding chains, including a final portion corresponding to the replication control region rich in Adenine and Thymine (A + T) (Figure 2A, Table S2). The taxonomic identification of the sequence obtained was confirmed with a comparison based on the 5′ terminal portion of the *COI* gene (639 bp) with other sequences of the genus available in the GenBank repository (NCBI) (Table S1), resulting in nucleotide and amino acid identities of 99.5% and 100%, respectively, for the investigated species (Figure 2B).

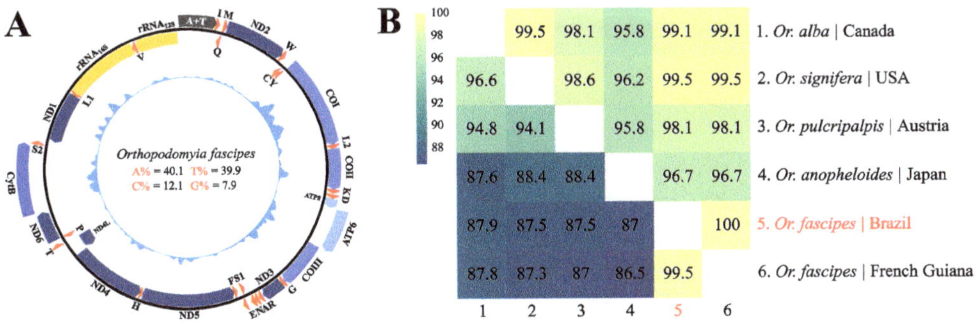

Figure 2. (**A**) Structural representation of *Or. fascipes* mtDNA. Internal values indicate the content of nucleotide bases. The blue inner graph indicates the distribution of genomic coverage by region. The blue, red, yellow, and gray blocks indicate the PCGs, tRNAs, rRNAs, and A + T region, respectively. Genes outside and inside the circle have forward and reverse transcription directions, respectively. (**B**) Heatmap of nucleotide/amino acid identity between taxa of the genus *Orthopodomyia*, based on the barcode region of the *COI* gene. The lower and upper triangles contain the percentages of nucleotide and amino acid identity, respectively. The taxon highlighted in red indicates the obtained sequence.

The sequence obtained presented nucleotide composition metrics and biases following the pattern typically observed for other species of mosquitoes with previously characterized mtDNA (Figure 3A–C, Table S3), exhibiting an AT content of 80% when evaluating the total genome (including the A + T region) and known patterns of asymmetry, with positive AT-skew and negative GC-skew, indicating the existence of higher proportions of Adenine and Cytosine in relation to Thymine and Guanine along the majority chain. Excluding the A + T region, the sequence contains 136 non-coding bp distributed in 21 intergenic regions, ranging from 1 to 21 bp (Table S2).

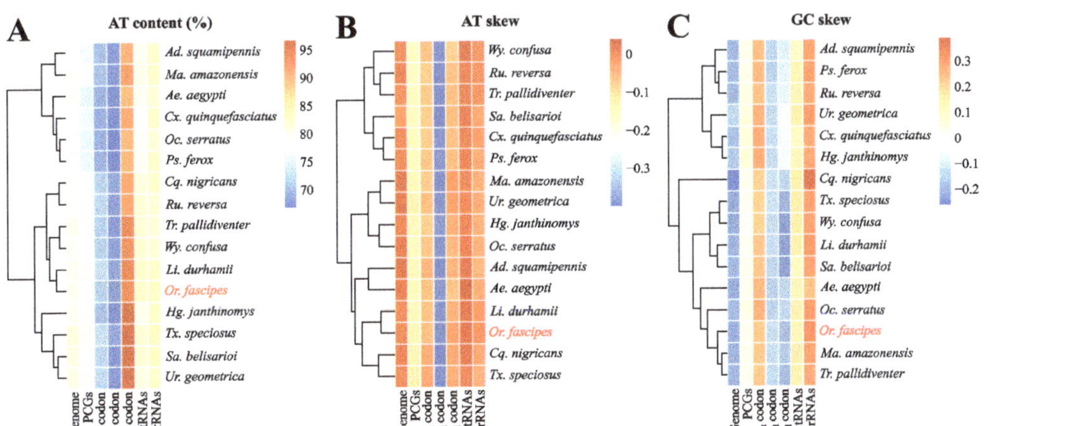

Figure 3. (**A**) AT% composition, (**B**) AT-skew, and (**C**) GC-skew of 16 previously characterized mosquito mtDNAs, including *Or. fascipes* (highlighted in red). Hierarchical groupings (clusters) of species (*y*-axis) are established based on the quantity of each metric per region evaluated (*x*-axis). The scales, particularly AT-skew and GC-skew, consider the real (non-normalized) values obtained in each analysis.

Most of the 22 tRNA subunits identified throughout the obtained sequence could be arranged in the typical "cloverleaf" secondary structure, with the exception of only $tRNA^{Ser1}$, which, as observed in previously described mtDNAs, presented a reduction in the dihydrouridine arm (DHU), being replaced by an equivalent DHU loop (Figure S1). The total length considering the 22 concatenated tRNAs was 1487 bp, with an AT content of 80%, positive AT/GC-skews, and 19 base incompatibilities (18 GU and 01 UU). Additionally, concatenated rRNA subunits totaling 2172 bp exhibited negative and positive AT/GC-skews and structural arrangements akin to those found in other mosquito species (Figure S1).

3.2. Description of Protein-Coding Genes (PCGs)

Maintaining the classic pattern of genomic organization in mosquitoes, nine PCGs presented forward transcriptional sense (*ND2, COI, COII, ATP8, ATP6, COIII, ND3, ND6,* and *CytB*) and four reversed sense (*ND5, ND4, ND4L,* and *ND1*), varying in length (excluding stop codons) from 159 bp (ATP8) to 1734 bp (ND5), and AT content from 70.4% (*COI*) to 86.5% (*ND6*). With the exception of just *COI* (TTG), all other PCGs presented start codons in the ATN pattern, more precisely ATG (*ATP6, COII, COIII, ND3, ND5, ND4, ND4L,* and *CytB*) and ATT (*ND2, ATP8, ND6,* and *ND1*), and the complete TAA-type stop codon terminated the coding chain in 12 PCGs, except in *COII* (TAG) (Table S2).

Disregarding stop codons, the analysis of the relative use of synonymous codons (RSCU) recorded 3732 codons in use by the PCGs of *Or. fascipes*, expressing 58 of the 62 amino acid triplets predicted for the mitochondrial genetic code of invertebrates, with GCG (Alanine), CGC (Arginine), CUG (Leucine), and AGG (Serine) were also absent. When comparing the RSCU metrics of *Or. fascipes* with those of other species of Culicinae, it is observed that there is a pattern in the expression of amino acids, where the majority of codons whose third base is Adenine or Uracil (Thymine) are expressed at a higher frequency (mostly with RSCU > 1) than those ending in Cytosine or Guanine, taking as an example the triplets UUA (RSCUμ = 5.14) and UUG (RSCUμ = 0.19), which although expressing the same amino acid (Leucine), showed significantly different frequency metrics (Figure 4, Table S4).

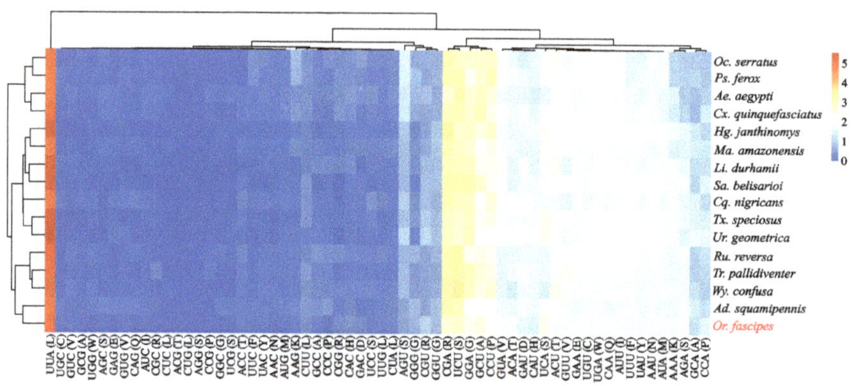

Figure 4. Relative use of synonymous codons of *Or. fascipes* (highlighted in red) in comparison to other representatives of the Culicinae subfamily. Hierarchical groupings (clusters) of species (y-axis) are established based on the RSCU of each codon (x-axis), with indications regarding the scale values.

The analysis of codon usage bias based on the effective number of codons (ENc) and considering all PCGs of taxa representing the subfamily Culicinae, including the sequence recently obtained, resulted in an average of 30.4 (21.8~42.7) (Table S5), reflecting a strong trend in the use of synonymous codons (ENc $\leq$ 35). The evaluation of the influence of the GC content present in the third base of the synonymous codons (GC3s) on the ENc resulted in values mostly close to the base and far from the standard curve (Figure 5A), indicating that not only the mutation pressure, but also factors such as natural selection pressure may probably be involved in the formation of codon bias, particularly of the evaluated species. Additionally, the action of the main evolutionary forces that act on PCG complexes could be better observed from a neutrality analysis based on the correlation between GC12 content (average between GC content of the first and second codon positions) and GC3 (GC content third codon position). The regression analysis between these two metrics resulted in positive correlations for all PCG complexes, being weak and significant in complexes I (*NADH*, $r = 0.2130$, $p < 0.05$) and IV (*COX*, $r = 0.1896$, $p = 0.0334$) and weak and non-significant in complexes III (*CytB*, $r = 0.2617$, $p = 0.0940$) and V (*ATP*, $r = 0.1345$, $p = 0.2223$), still presenting slope coefficients lower than 0.5 (Figure 5B–E), suggesting a greater impact of natural selection pressure when compared to mutation pressure.

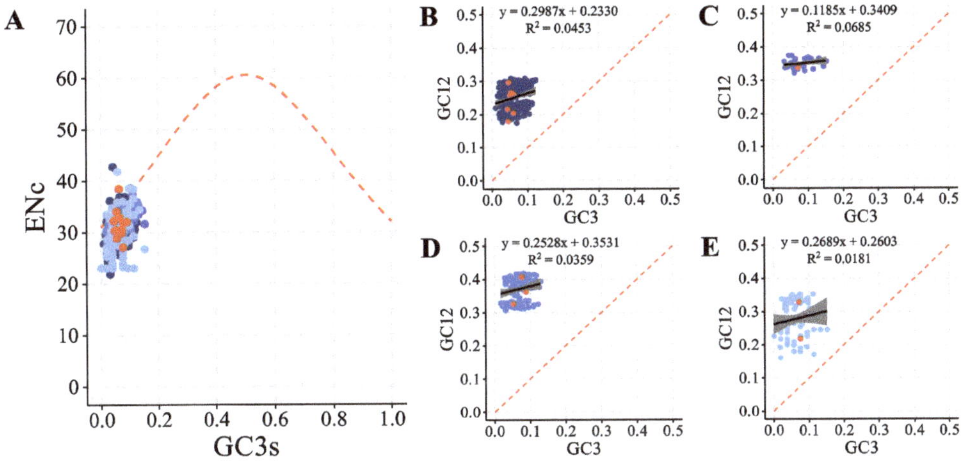

Figure 5. (**A**) ENc-GC3s graph. The red dashed line indicates the curve expected when codon usage bias is affected only by mutation pressure. GC12-GC3 neutrality plots are arranged in the order of mitochondrial complexes: (**B**) I (*NADH*), (**C**) III (*Cytb*), (**D**) IV (*COX*), and (**E**) V (*ATP*). In the graphs, each blue point (varying in tone according to the complex) represents a PCG independent of the Culicinae mtDNAs used as a reference in the analyses. PCGs of *Or. fascipes* are highlighted in red.

The evolutionary pressure that acts on the PCGs of the investigated species and other representatives of Culicinae could also be evaluated based on the estimate of the proportions of non-synonymous to synonymous substitutions (*dN*/*dS*). The results obtained indicate that the different coding regions are evolving globally under the influence of negative (or purifying) pressure ($\omega < 1$), with proportions ranging from 0.0186 ± 0.0847 in *COI* to 0.0941 ± 0.9467 in *ATP8* (Figure 6A). Purifying selection was particularly strong ($\omega_{\mu} < 0.1$) on the PCGs of complexes III and IV and on the *ATP6* subunit belonging to complex V. In turn, *ATP8* and the genes belonging to complex I showed higher *dN*/*dS* ratios ($0.1 < \omega_{\mu} < 0.5$), indicating the presence of less conservative evolutionary constraints in these regions. However, it was also observed that the rates of synonymous substitutions (*dS*) on PCGs were still significantly higher compared to the occurrence of non-synonymous substitutions (*dN*) (Figure 6B).

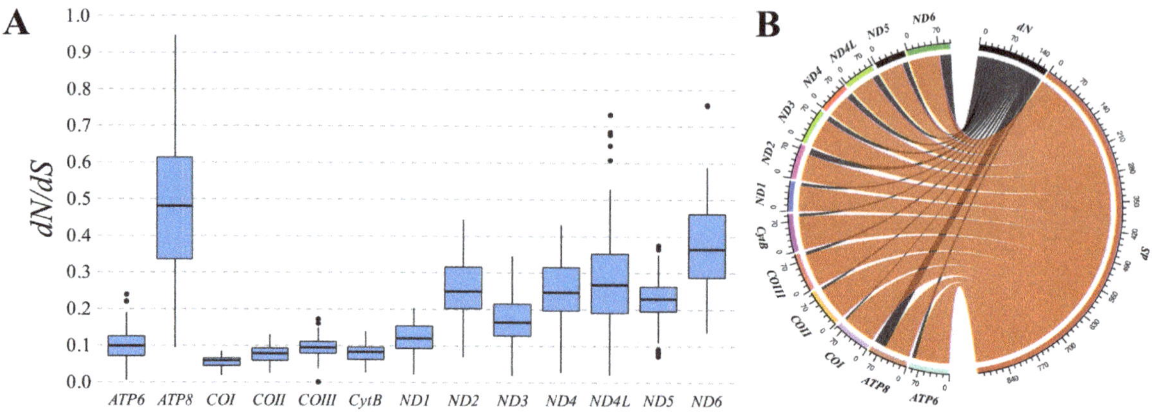

Figure 6. (**A**) dN/dS ratios calculated from pairwise analyses between 17 Culicinae mtDNAs. The ω ratios and PCGs are arranged along the y and x axes, respectively. (**B**) Representation of the proportion of non-synonymous (*dN*) (gray strings) and synonymous (*dS*) (red strings) substitutions that occur in each PCG. This Figure serves as a visual complement to (**A**), indicating that the accumulation of synonymous substitutions is predominantly greater in all Culicinae PCGs, directly influencing the observed negative (or purifying) selection pressure ($\omega < 1$).

3.3. Phylogenetic Analysis

Analysis of nucleotide substitution saturation of the set of sequences evaluated resulted in the transition and transversion rates being linearly associated with genetic distance (Figure 7A). Furthermore, based on the test proposed by Xia et al. (2003) [65], performing multiple randomizations on subsets of 4 to 32 operational taxonomic units (OTUs) resulted in a saturation index (*Iss*) significantly lower ($p = 0.00001$) than the critical value (*Iss.c*), both for symmetric topologies (*Iss.cSym*) and asymmetric (*Iss.cAsym*) (Figure 7B), indicating the unsaturation of the set of sequences evaluated, and qualifying it for phylogenetic reconstruction analysis.

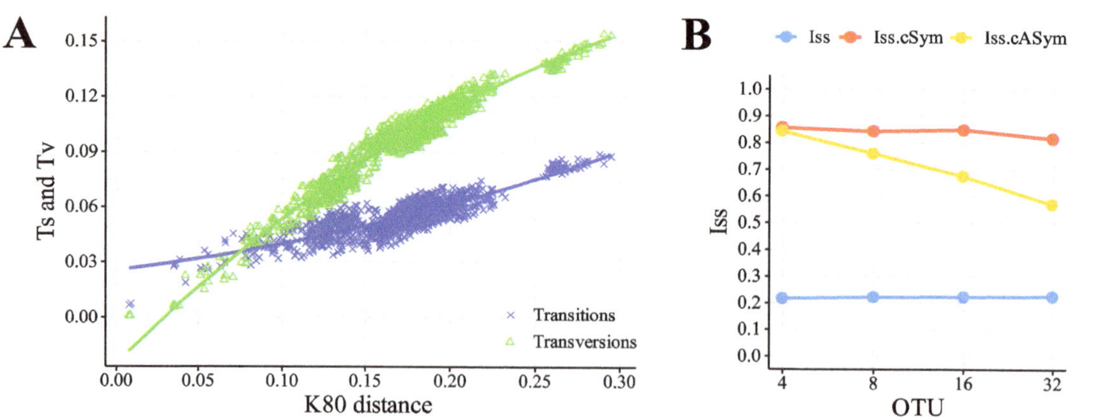

Figure 7. (**A**) Nucleotide substitution saturation graph of the set of evaluated sequences, considering all PCGs, demonstrating the increase and linear association of transition (*Ts*) and transversion (*Tv*) rates (y-axis) in relation to genetic distance (x-axis). (**B**) Graphical representation of the results obtained from the application of the test proposed by Xia et al. (2003) [65], statistically supported (*Iss* < *Iss.c*, $p = 0.00001$), and showing the low occurrence of nucleotide saturation in the set of sequences evaluated.

Reconstruction of the phylogeny using Maximum Likelihood (Figure 8A) and Bayesian Inference (Figure S2) resulted in identical topologies with high internal anchoring values, consisting of a main group, externally anchored by the taxon *D. aestivalis* and containing 53 taxa distributed in two well-structured and monophyletic clades corresponding to the subfamilies Culicinae (with 42 taxa divided into eight tribes) (BS = 100%, BPP = 100%) and Anophelinae (with 11 taxa) (BS = 100%, BPP = 100%). The evaluation of the quality of the phylogenetic signal of the set of sequences used, particularly in Maximum Likelihood phylogeny, resulted in 97.2% of resolved quartets with an approximate distribution at the vertices of the diagram (Figure 8B), demonstrating the reconstruction of well-resolved topologies and high reliability during analysis.

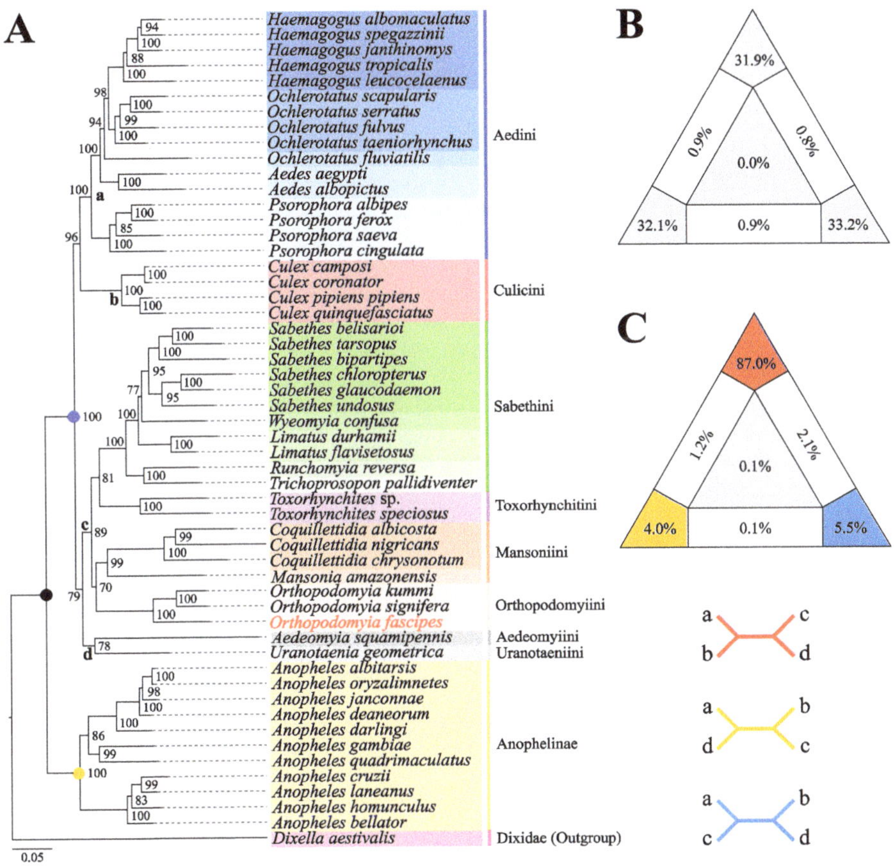

Figure 8. (**A**) Phylogeny reconstructed using the Maximum Likelihood method based on the 13 PCGs of *Or. fascipes* (highlighted in red) and other taxa available in the GenBank repository (NCBI). Bootstrapping support values (BS) are shown on each node. The colored dots indicate the main reconstructed taxonomic groupings: family Culicidae (black) and subfamilies Culicinae (blue), and Anophelinae (yellow). (**B**) Maximum Likelihood mapping diagram demonstrating the quality of the phylogenetic signal resulting from the quartet analysis, with 98.2% resolved quartets. (**C**) Four-cluster likelihood mapping (FcLM) diagram representing four taxon groupings in Culicinae. The analyzed groups are indicated in the topology by the letters "a" (Aedini), "b" (Culicini), "c" (Orthopodomyiini + Mansoniini + Toxorhynchitini + Sabethini), and "d" (Aedeomyiini + Uranotaeniini). The bidirectional arrows indicate different hypotheses regarding the relationships among the four clusters, with their respective colors in the diagram corresponding to the support percentages.

In both analyses, the Aedini + Culicini group was externally anchored by a subclade made up of Sabethini + Toxorhynchitini, with Mansoniini + Orthopodomyiini as its sister group, and was externally anchored by Aedeomyiini + Uranotaeniini. This phylogenetic conformation in Culicinae was strongly supported by FcLM analysis, showing a support value of 87% (Figure 8C). *Or. fascipes*, together with other taxa of its genus, constituted a reasonably well-supported subclade (BS = 70%, BPP = 96%) with representatives of the genus *Coquillettidia* and *Mansonia* Blanchard, 1901, which was corroborated by the smaller nucleotide distance of these taxa in relation to the tribe Mansoniini when compared with other groups such as Aedini, Culicini, and Sabethini (Figure S3, Table S6).

4. Discussion

The present study describes, for the first time, the complete mitochondrial sequence of the species *Or. fascipes*, collected during ecoepidemiological expeditions in the northeast region of the State of Pará, Brazil, and characterized using HTS. When compared with mitochondrial genomes from previously described mosquito species, the sequence obtained exhibited conserved characteristics in several aspects, including the arrangement of its 37 functional subunits in two transcription chains, the length and metrics of nucleotide composition, the arrangement of secondary structures of tRNAs, and the use of codons by PCGs [37–43], with no translocation events observed, especially between subunits of tRNAs [32], as evidenced in mtDNAs from representatives of the Sabethini tribe [36].

The predominance of unevenly distributed nucleotide bases, commonly resulting in a high A + T content, is a phenomenon frequently observed in insect mitochondrial genomes, significantly impacting the constitution of protein-coding regions [35,73]. In this context, in addition to the majority of *Or. fascipes* PCGs having start codons in the ATN pattern (except *COI*, which presented TTG) and complete stop codons of the TAA type (except *COII*, which presented TAG), we also observed a substantial increase in AT content in the third position of amino acid codons. These characteristics corroborate the pattern observed in previous studies characterizing the mtDNA of mosquito species [34–43], and are justified, respectively, by the occurrence of post-transcriptional polyadenylation in the formation of complete stop codons [74] and by the high availability of free ATP, which together with other NTPs, promotes an increase in the use of Adenines, especially in the third codon position, contributing positively to the efficiency of transcription processes [75].

Considering the clear pattern in amino acid expression in relative codon usage (RSCU) analysis among representatives of Culicinae, including *Or. fascipes*, it is suggested that the high AT content in the third codon position may be associated with bias in usage of synonymous codons, thus influencing the reduction in the intensity of purifying selection against deleterious mutations in this same region [35,76]. Furthermore, the effective number of codons (ENc) and neutrality analyses, used, respectively, to detect heterogeneity in the use of codons [77] and determine the main evolutionary forces [78] acting on the PCG complexes of Culicinae, showed a strong tendency in the use of synonymous codons (ENc $\leq$ 35), further indicating that factors such as mutational pressure and natural selection may be influencing the formation of the codon bias observed in Culicinae, with natural selection appearing to have a greater impact, as indicated by the slope coefficients obtained. Given these findings, this study complements the efforts of a previous evaluation carried out by Hao et al. (2017) [35], who obtained similar results when analyzing 50 mtDNAs from mosquitoes, although these were representatives of the Anophelinae.

Additionally, it became evident that PCGs from *Or. fascipes* and other representatives of Culicinae are evolving globally under negative (or purifying) selection pressure ($dN/dS < 1$), resulting in high rates of synonymous substitutions within the evaluated regions [79]. As previously noted in other studies, this purifying selection pressure was particularly stronger in complexes III (*CytB*) and IV (*COX*), while there were fewer conservative constraints and a tendency towards positive (or adaptive) selection pressure, especially in complex I (*NADH*) and *ATP8*. However, it is important to note that different rates of mutation accumulation occur among PCG complexes, depending on the role each plays in metabolic processes and in the reactions of the oxidative phosphorylation chain of cellular respiration [80].

In this study, the joint use of PCGs from 54 mitochondrial genomes, including the newly sequenced species, and two phylogenetic reconstruction approaches (ML and BI) resulted in topologies with high internal anchoring values and significant support for the basal dichotomy between the subfamilies Culicinae and Anophelinae, and also corroborated the monophyly, in Culicinae, of the tribes Aedini [40,43], Culicini [41,81], Sabethini [36,42], and Mansoniini [38,82]. However, our findings suggest that the Aedini and Culicini tribes make up a sister group of the monophyletic group formed by Aedeomyiini + Uranotaeniini + (Mansoniini + Orthopodomyiini + (Sabethini + Toxorhynchitini)), corroborating other studies that used the same genomic set in previous investigations, with the continued absence of representatives of the Orthopodomyiini tribe [37–40,83], and partially disagreeing with the results obtained in other investigations based on the use of mitochondrial markers [41].

The monophyletic grouping Sabethini + Toxorhynchitini, externally anchored by Mansoniini, is consistent with results obtained in recent studies that used mitochondrial and nuclear markers, corroborating the evolutionary positioning of these tribes [39,41,82]. On the other hand, these results contrast with previous analyses based on the observation of morphological characters and nuclear markers in a smaller sample, where Mansoniini was classified as a sister group to Aedini, as well as Sabethini in relation to Culicini, and Toxorhynchitini appeared as a basal taxon [6,25].

Furthermore, outside the clade constituted by the set of tribes initially mentioned, Aedeomyiini and Uranotaeniini were recovered in a single group, diverging in evolutionary positioning in relation to previous studies. Reidenbach et al. (2009) [6], by employing six nuclear markers to reconstruct the phylogeny of Culicidae, demonstrated that Aedeomyiini and Toxorhynchitini, externally anchored by a representative of the genus *Mimomyia* Theobald, 1903 (tribe Ficalbiini), formed a monophyletic group with Sabethini, while Uranotaeniini grouped with Culicini. These observations were later corroborated by Lorenz et al. (2021) [39] in a temporal analysis covering 102 taxa and considering the use of all 13 mitochondrial PCGs.

In contrast, the evolutionary positioning of Uranotaeniini varied according to the configuration of the mitochondrial dataset and the phylogenetic reconstruction methods used by Chen et al. (2023) [41]. In this study, the phylogeny reconstructed with the Maximum Likelihood method recovered the monophyletic grouping Uranotaeniini + Sabethini, while in the topologies obtained with Bayesian Inference, Uranotaeniini was positioned outside the group formed by Mansoniini + (Sabethini + Toxorhynchitini). Additionally, in previous studies based on mitochondrial [38,83], nuclear [82], and morphological data [25], Aedeomyiini and Uranotaeniini were consistently observed as external and basal groups in relation to the monophyletic grouping formed by all other tribes of Culicinae. In this context, although our results contribute to clarifying the evolutionary history of these tribes, we recognize the need to carry out more comprehensive investigations and broader sampling efforts, aiming to establish a solid basis for their molecular taxonomy.

In the present study, in the context of the evolutionary order, particularly of the second subclade of tribes in Culicinae, the taxon *Or. fascipes*, together with other taxa of its genus, constituted a monophyletic and reasonably well-supported grouping (BS = 70%, BPP = 96%), together with representatives of the Mansoniini tribe, which was also evidenced when evaluating the nucleotide distance metrics. The results obtained corroborate molecular studies carried out previously, including representatives of the genus, where their proximity to Mansoniini was also observed, albeit with analyses based on the use of nuclear markers and significantly different sampling of genes [6,82]. On the other hand, studies that evaluated the evolutionary positioning of Orthopodomyiini based on the observation of morphological characters considered the possibility that representatives of the tribe occupied basal positions in relation to the other tribes of Culicinae [25]. This was even supported in a molecular study carried out by Munstermann et al. (1985) [84], who, after evaluating the structure of the well-resolved polytene chromosomes of *Orthopodomyia pulcripalpis* Rondani, 1872, belonging to the *Signifera* group, suggested that the species may have gone through potentially greater evolutionary periods compared to other species of the subfamily. These observations are further supported by the fact that *Orthopodomyia* is a monotypic taxon that does not present subgeneric classifications to date, and is considered somewhat primitive since its current and wide geographic distribution may be the result of its dissemination before geological events associated with the division of Africa and South America in the Cretaceous [6].

Or. fascipes has three taxonomic synonyms: *longipalpis*, *townsendi*, and *bacigalupoi*, and curiously, due to the presence of characteristics similar to representatives of the Mansoniini tribe, in its first description records it was named *Mansonia fascipes* by Coquillett (1905) [85]. The species has a geographic distribution that ranges from Nicaragua to the south of Bolivia, and extends through the Brazilian Amazon region to the states of Goiás and Minas Gerais, being classified according to Zavortink (1968) [16] as one of the two representative species of the group (or section) Neotropical *Thomasina* responsible for most of the known characteristics of the group, mainly in relation to the morphological aspects of the immature stages. In this context, little is known about the initial developmental stages of *Orthopodomyia sampaioi* da Costa Lima, 1935, with the distinction between the two species being precisely evidenced only based on the morphology of the adults, and although they are basically allopatric, their distribution areas apparently overlap in the Brazilian state of Goiás.

The evolutionary affinity relationships of *Orthopodomyia* with other genera of Culicidae are largely misunderstood, and to date have been concluded to be mainly based on the observation of morphological characters [16,24,25], relying on very few studies available to assess its evolutionary position using molecular markers. In this context, the unavailability of genetic information about this group of mosquitoes is possibly due to the lack of bioecological records and their involvement in the transmission and/or maintenance of infectious agents, particularly in wild environments, neglecting their medical-epidemiological and evolutionary importance.

Taxonomy based on morphological aspects continues to be, without a doubt, the main means of classifying and validating species identification processes. However, this method is not free from possible flaws, especially when dealing with a group of organisms as diverse and ancient as the Culicidae family. The advent of molecular tools and computational analysis methods has made it possible to carry out increasingly deeper investigations into the taxonomic relationships between these organisms. In this context, the characterization and application of mitochondrial genomes have contributed greatly to the supply of information at various levels of the system. Despite this, it is evident that mtDNA may not be sufficient to recover deeper phylogenetic relationships. Therefore, integration with other markers may be appropriate, considering that this also represents a future objective in the development of evolutionary studies, especially of insects [73,86].

This study represents the first record of the complete mitochondrial genome recovery of a Neotropical species belonging to the genus *Orthopodomyia* collected in the Brazilian Amazon region. Until this study, there were no records of complete mitochondrial sequences of the genus deposited in public repositories, except for only two records of genomic sequencing data (*rawdata*) of species from the *Signifera* group obtained by Soghigian et al. [82] that are available in the SRA database (NCBI). These data were included in the analysis of this study, and although they did not constitute complete mitochondrial sequences after assembly using the *de novo* method and/or reference mapping, it was possible to recover a large part of the PCGs content.

We consider that with the increasing number of molecular taxonomy studies of Culicidae, based on the acquisition of new mtDNAs, the phylogenetic relationships observed in this study could potentially change. Consequently, it will always be necessary to extensively expand taxon sampling to reconstruct more reliable Culicidae phylogenies, focusing particularly on areas rich in high biodiversity, such as the Amazon region.

Supplementary Materials: The following supporting information can be downloaded at: https://www.mdpi.com/article/10.3390/genes15070874/s1, Figure S1—Secondary structure of tRNAs and rRNAs of *Or. fascipes*; Figure S2—Phylogeny reconstructed by the Bayesian Inference method based on the 13 PCGs of *Or. fascipes* and other taxa available in the GenBank repository (NCBI); Figure S3—Intra/intergroup distances between four tribes of Culicinae and *Or. fascipes*; Table S1—Taxa used in the evolutionary analyzes; Table S2—*Or. fascipes* mtDNA nucleotide composition metrics; Table S3—Nucleotide composition metrics of *Or. fascipes* in comparison with other species of the Culicinae; Table S4—Relative use of synonymous codons (RSCU) of *Or. fascipes* in comparison with other species of the Culicinae; Table S5—ENc, GC12 and GC3s metrics of *Or. fascipes* in comparison with other species of the Culicinae; Table S6—Maximum Likelihood Composition nucleotide distance matrix.

Author Contributions: Conceptualization, F.S.d.S., B.L.S.d.N., and J.P.N.N.; methodology, F.S.d.S., B.L.S.d.N., S.P.d.S., C.F.A., D.D.D., L.H.d.S.e.S., L.A.M.R., H.C.F.R., L.L.d.C., J.W.R.J., D.B.R.V., R.C.F.B., and D.B.d.A.M.; validation, F.S.d.S., B.L.S.d.N., and J.P.N.N.; formal analysis, F.S.d.S.; investigation, F.S.d.S. and B.L.S.d.N.; resources, A.C.R.C. and J.P.N.N.; data curation, F.S.d.S. and S.P.d.S.; writing—original draft preparation, F.S.d.S.; writing—review and editing, F.S.d.S., B.L.S.d.N., and J.P.N.N.; visualization, F.S.d.S., B.L.S.d.N., and J.P.N.N.; supervision, J.P.N.N.; project administration, J.P.N.N.; funding acquisition, J.P.N.N. All authors have read and agreed to the published version of the manuscript.

Funding: This research was funded by the Coordination for the Improvement of Higher Education Personnel (CAPES) (process No. 88887.756782/2022-00), Graduate Program in Parasitary Biology in the Amazon Region (PPGBPA) of State University of Pará (edital No. 013/2021-UEPA), and Evandro Chagas Institute (IEC/SVSA/MS).

Institutional Review Board Statement: Not applicable.

Informed Consent Statement: Not applicable.

Data Availability Statement: All the data obtained during this study are available in the tables and figures included in the text, and in the Supplementary Materials. The mitochondrial sequence of *Or. fascipes* obtained here was deposited in the GenBank database under accession code PP749023, and the raw sequence reads generated are available in the NCBI Sequence Read Archive (SRA) database under BioProject PRJNA1116825, BioSample SAMN41551723, and SRA acession SRR29190358.

Conflicts of Interest: The authors declare no conflicts of interest.

References

1. Amorim, D.d.S.; Yeates, D. Pesky Gnats: Ridding Dipteran Classification of the Nematocera. *Stud. Dipterol.* **2006**, *13*, 1–7.
2. Kutty, S.N.; Wong, W.H.; Meusemann, K.; Meier, R.; Cranston, P.S. A Phylogenomic Analysis of Culicomorpha (Diptera) Resolves the Relationships among the Eight Constituent Families. *Syst. Entomol.* **2018**, *43*, 434–446. [CrossRef]
3. Harbach, R.E. The Culicidae (Diptera): A Review of Taxonomy, Classification and Phylogeny. *Zootaxa* **2007**, *638*, 591–638. [CrossRef]
4. Harbach, R.E. Mosquito Taxonomic Inventory. Available online: https://mosquito-taxonomic-inventory.myspecies.info/ (accessed on 11 April 2024).
5. Forattini, O.P. *Culicidologia Médica: Identificação, Biologia, Epidemiologia*; Edusp: São Paulo, Brazil, 2002; Volume 2.
6. Reidenbach, K.R.; Cook, S.; Bertone, M.A.; Harbach, R.E.; Wiegmann, B.M.; Besansky, N.J. Phylogenetic Analysis and Temporal Diversification of Mosquitoes (Diptera: Culicidae) Based on Nuclear Genes and Morphology. *BMC Evol. Biol.* **2009**, *9*, 298. [CrossRef] [PubMed]
7. Soghigian, J.; Andreadis, T.G.; Livdahl, T.P. From Ground Pools to Treeholes: Convergent Evolution of Habitat and Phenotype in *Aedes* Mosquitoes. *BMC Evol. Biol.* **2017**, *17*, 262. [CrossRef] [PubMed]
8. Lobo, F.P.; Mota, B.E.F.; Pena, S.D.J.; Azevedo, V.; Macedo, A.M.; Tauch, A.; Machado, C.R.; Franco, G.R. Virus-Host Coevolution: Common Patterns of Nucleotide Motif Usage in *Flaviviridae* and Their Hosts. *PLoS ONE* **2009**, *4*, e6282. [CrossRef] [PubMed]
9. Severson, D.W.; Behura, S.K. Genome Investigations of Vector Competence in *Aedes aegypti* to Inform Novel Arbovirus Disease Control Approaches. *Insects* **2016**, *7*, 58. [CrossRef] [PubMed]
10. Behura, S.K. Molecular Marker Systems in Insects: Current Trends and Future Avenues. *Mol. Ecol.* **2006**, *15*, 3087–3113. [CrossRef]
11. Ditter, R.E.; Campos, M.; Pinto, J.; Cornel, A.J.; Rompão, H.; Lanzaro, G.C. Mitogenome Analyses Reveal Limited Introduction of *Anopheles coluzzii* Into the Central African Islands of São Tomé and Príncipe. *Front. Trop. Dis.* **2022**, *3*, 855272. [CrossRef]
12. Patz, J.A.; Graczyk, T.K.; Geller, N.; Vittor, A.Y. Environmental Changes & Parasitic Diseases. *Int. J. Parasitol.* **2000**, *30*, 1395–1405.

13. Vasconcelos, P.F.; Travassos da Rosa, A.P.; Rodrigues, S.G.; Travassos da Rosa, E.S.; Dégallier, N.; Travassos da Rosa, J.F. Inadequate Management of Natural Ecosystem in the Brazilian Amazon Region Results in the Emergence and Reemergence of Arboviruses. *Cad. Saúde Pública/Ministério Saúde Fundação Oswaldo Cruz Esc. Nac. Saúde Pública* **2001**, *17*, 155–164. [CrossRef]
14. Jones, K.E.; Patel, N.G.; Levy, M.A.; Storeygard, A.; Balk, D.; Gittleman, J.L.; Daszak, P. Global Trends in Emerging Infectious Diseases. *Nature* **2008**, *451*, 990–993. [CrossRef] [PubMed]
15. Ortiz, D.; Piche-Ovares, M.; Romero-Vega, L.M.; Wagman, J.; Troyo, A. The Impact of Deforestation, Urbanization, and Changing Land Use Patterns on the Ecology of Mosquito and Tick-Borne Diseases in Central America. *Insects* **2022**, *13*, 20. [CrossRef]
16. Zavortink, T.J. Mosquito Studies (Diptera, Culicidae) VIII. A Prodrome of the Genus *Orthopodomyia*. *Contrib. Am. Entomol. Inst.* **1968**, *3*, 221.
17. Lane, J. *Neotropical Culicidae*; University of São Paulo: São Paulo, Brazil, 1953.
18. Yeo, H.; Tan, C.H.; Chong, C.S.; Lam-Phua, S.G. Identification Key to the Genera of Adult Female Mosquitoes (Diptera: Culicidae) of Singapore. *Zootaxa* **2019**, *4624*, 407–423. [CrossRef] [PubMed]
19. Hanson, S.M.; Novak, R.J.; Lampman, R.L.; Vodkin, M.H. Notes on the Biology of *Orthopodomyia* in Illinois. *J. Am. Mosq. Control Assoc.* **1995**, *11*, 375–376. [PubMed]
20. Becker, N.; Petric, D.; Zgomba, M.; Boase, C.; Madon, M.; Dahl, C.; Kaiser, A. *Mosquitoes and Their Control*, 2nd ed.; Springer: Berlin/Heidelberg, Germany, 2010.
21. Chamberlain, R.W.; Sikes, R.K.; Nelson, D.B.; Sudia, W.D. Studies on the North American Arthropod-Borne Encephalitides. VI. Quantitative Determinations of Virus-Vector Relationships. *Am. J. Trop. Med. Hyg.* **1954**, *60*, 278–285. [CrossRef]
22. Vargas, L. Los Mosquitos de Sonora En Relacion Con El Problem de Encephalitis. *Rev. Mex. Med.* **1960**, *40*, 338–345.
23. Centers for Disease Control and Prevention (CDC). West Nile Virus. Available online: https://www.cdc.gov/westnile/vectorcontrol/index.html (accessed on 17 April 2024).
24. Edwards, F.W. *Genera Insectorum. Diptera, Fam. Culicidae. Fascicle 194*; Desmet-Verteneuil: Brussels, Belgium, 1932.
25. Harbach, R.E.; Kitching, I.J. Phylogeny and Classification of the Culicidae (Diptera). *Syst. Entomol.* **1998**, *23*, 327–370. [CrossRef]
26. Yan, G.; Chadee, D.D.; Severson, D.W. Evidence for Genetic Hitchhiking Effect Associated with Insecticide Resistance in *Aedes aegypti*. *Genetics* **1998**, *148*, 793–800. [CrossRef]
27. Barzon, L.; Lavezzo, E.; Militello, V.; Toppo, S.; Palù, G. Applications of Next-Generation Sequencing Technologies to Diagnostic Virology. *Int. J. Mol. Sci.* **2011**, *12*, 7861–7884. [CrossRef]
28. Tabachnick, W.J. Nature, Nurture and Evolution of Intra-Species Variation in Mosquito Arbovirus Transmission Competence. *Int. J. Environ. Res. Public Health* **2013**, *10*, 249–277. [CrossRef]
29. Brown, W.M.; George, M.; Wilson, A.C. Rapid Evolution of Animal Mitochondrial DNA. *Proc. Natl. Acad. Sci. USA* **1979**, *76*, 1967–1971. [CrossRef]
30. Avise, J.C. Phylogeography: Retrospect and Prospect. *J. Biogeogr.* **2009**, *36*, 3–15. [CrossRef]
31. Boore, J.L. Animal Mitochondrial Genomes. *Nucleic Acids Res.* **1999**, *27*, 1767–1780. [CrossRef]
32. Mandal, S.D.; Chhakchhuak, L.; Gurusubramanian, G.; Kumar, N.S. Mitochondrial Markers for Identification and Phylogenetic Studies in Insects—A Review. *DNA Barcodes* **2014**, *2*, 1–9. [CrossRef]
33. Hebert, P.D.N.; Cywinska, A.; Ball, S.L.; DeWaard, J.R. Biological Identifications through DNA Barcodes. *Proc. R. Soc. B Biol. Sci.* **2003**, *270*, 313–321. [CrossRef]
34. Hao, Y.J.; Zou, Y.L.; Ding, Y.R.; Xu, W.Y.; Yan, Z.T.; Li, X.D.; Fu, W.B.; Li, T.J.; Chen, B. Complete Mitochondrial Genomes of *Anopheles stephensi* and *An. dirus* and Comparative Evolutionary Mitochondriomics of 50 Mosquitoes. *Sci. Rep.* **2017**, *7*, 7666. [CrossRef]
35. Demari-Silva, B.; Foster, P.G.; de Oliveira, T.M.P.; Bergo, E.S.; Sanabani, S.S.; Pessôa, R.; Sallum, M.A.M. Mitochondrial Genomes and Comparative Analyses of *Culex camposi*, *Culex coronator*, *Culex usquatus* and *Culex usquatissimus* (Diptera: Culicidae), Members of the *Coronator* Group. *BMC Genomics* **2015**, *16*, 831. [CrossRef]
36. Lorenz, C.; Alves, J.M.P.; Foster, P.G.; Sallum, M.A.M.; Suesdek, L. First Record of Translocation in Culicidae (Diptera) Mitogenomes: Evidence from the Tribe Sabethini. *BMC Genomics* **2019**, *20*, 721. [CrossRef]
37. Silva, F.S.; Cruz, A.C.R.; Medeiros, D.B.d.A.; Silva, S.P.; Nunes, M.R.T.; Martins, L.C.; Chiang, J.O.; Lemos, P.d.S.; Cunha, G.M.; Araujo, R.F.; et al. Mitochondrial Genome Sequencing and Phylogeny of *Haemagogus albomaculatus*, *Haemagogus leucocelaenus*, *Haemagogus spegazzinii*, and *Haemagogus tropicalis* (Diptera: Culicidae). *Sci. Rep.* **2020**, *10*, 16948. [CrossRef] [PubMed]
38. Nascimento, B.L.S.; Silva, F.S.; Nunes-Neto, J.P.; Medeiros, D.B.d.A.; Cruz, A.C.R.; Silva, S.P.; Silva, L.H.d.S.; Monteiro, H.A.d.O.; Dias, D.D.; Vieira, D.B.R.; et al. First Description of the Mitogenome and Phylogeny of Culicinae Species from the Amazon Region. *Genes* **2021**, *12*, 1983. [CrossRef] [PubMed]
39. Lorenz, C.; Alves, J.M.P.; Foster, P.G.; Suesdek, L.; Sallum, M.A.M. Phylogeny and Temporal Diversification of Mosquitoes (Diptera: Culicidae) with an Emphasis on the Neotropical Fauna. *Syst. Entomol.* **2021**, *46*, 798–811. [CrossRef]
40. Silva, L.H.d.S.; Silva, F.S.; Medeiros, D.B.d.A.; Cruz, A.C.R.; Silva, S.P.; Aragão, A.d.O.; Dias, D.D.; Nascimento, B.L.S.; Rosa, J.W., Jr.; Vieira, D.B.R.; et al. Description of the Mitogenome and Phylogeny of *Aedes* spp. (Diptera: Culicidae) from the Amazon Region. *Acta Trop.* **2022**, *232*, 106500. [CrossRef] [PubMed]
41. Chen, D.H.; He, S.L.; Fu, W.B.; Yan, Z.T.; Hu, Y.J.; Yuan, H.; Wang, M.B.; Chen, B. Mitogenome-Based Phylogeny of Mosquitoes (Diptera: Culicidae). *Insect Sci.* **2024**, *31*, 599–612. [CrossRef] [PubMed]

42. Silva, F.S.; Nascimento, B.L.S.; Cruz, A.C.R.; Silva, S.P.; Aragão, A.d.O.; Dias, D.D.; Silva, L.H.d.S.; Reis, L.A.M.; Rosa, J.W.; Vieira, D.B.R.; et al. Sequencing and Description of the Complete Mitochondrial Genome of *Limatus durhamii* (Diptera: Culicidae). *Acta Trop.* **2023**, *239*, 106805. [CrossRef] [PubMed]
43. Sousa, A.A.d.; Cruz, A.C.R.; Silva, F.S.; Silva, S.P.; Nunes-Neto, J.P.; Barros, M.C.; Fraga, E.d.C.; Sampaio, I. Sequencing and Analysis of the Mitochondrial Genome of *Aedes aegypti* (Diptera: Culicidae) from the Brazilian Amazon Region. *Insects* **2023**, *14*, 938. [CrossRef]
44. National Center for Biotechnology Information (NCBI). Orthopodomyia. Available online: https://www.ncbi.nlm.nih.gov/datasets/taxonomy/139053/ (accessed on 11 April 2024).
45. Chen, S.; Zhou, Y.; Chen, Y.; Gu, J. Fastp: An Ultra-Fast All-in-One FASTQ Preprocessor. *Bioinformatics* **2018**, *34*, i884–i890. [CrossRef]
46. Li, D.; Liu, C.M.; Luo, R.; Sadakane, K.; Lam, T.W. MEGAHIT: An Ultra-Fast Single-Node Solution for Large and Complex Metagenomics Assembly via Succinct de Bruijn Graph. *Bioinformatics* **2015**, *31*, 1674–1676. [CrossRef]
47. Buchfink, B.; Reuter, K.; Drost, H.G. Sensitive Protein Alignments at Tree-of-Life Scale Using DIAMOND. *Nat. Methods* **2021**, *18*, 366–368. [CrossRef]
48. Ondov, B.D.; Bergman, N.H.; Phillippy, A.M. Interactive Metagenomic Visualization in a Web Browser. *BMC Bioinform.* **2011**, *12*, 385. [CrossRef] [PubMed]
49. Kearse, M.; Moir, R.; Wilson, A.; Stones-Havas, S.; Cheung, M.; Sturrock, S.; Buxton, S.; Cooper, A.; Markowitz, S.; Duran, C.; et al. Geneious Basic: An Integrated and Extendable Desktop Software Platform for the Organization and Analysis of Sequence Data. *Bioinformatics* **2012**, *28*, 1647–1649. [CrossRef] [PubMed]
50. Bernt, M.; Donath, A.; Jühling, F.; Externbrink, F.; Florentz, C.; Fritzsch, G.; Pütz, J.; Middendorf, M.; Stadler, P.F. MITOS: Improved de Novo Metazoan Mitochondrial Genome Annotation. *Mol. Phylogenet. Evol.* **2013**, *69*, 313–319. [CrossRef]
51. Camacho, C.; Coulouris, G.; Avagyan, V.; Ma, N.; Papadopoulos, J.; Bealer, K.; Madden, T.L. BLAST+: Architecture and Applications. *BMC Bioinform.* **2009**, *10*, 421. [CrossRef] [PubMed]
52. Langmead, B.; Salzberg, S.L. Fast Gapped-Read Alignment with Bowtie 2. *Nat. Methods* **2012**, *9*, 357–359. [CrossRef] [PubMed]
53. Stothard, P.; Grant, J.R.; Van Domselaar, G. Visualizing and Comparing Circular Genomes Using the CGView Family of Tools. *Brief. Bioinform.* **2018**, *20*, 1576–1582. [CrossRef] [PubMed]
54. Kumar, S.; Stecher, G.; Li, M.; Knyaz, C.; Tamura, K. MEGA X: Molecular Evolutionary Genetics Analysis across Computing Platforms. *Mol. Biol. Evol.* **2018**, *35*, 1547–1549. [CrossRef] [PubMed]
55. Perna, N.T.; Kocher, T.D. Patterns of Nucleotide Composition at Fourfold Degenerate Sites of Animal Mitochondrial Genomes. *J. Mol. Evol.* **1995**, *41*, 353–358. [CrossRef]
56. Peden, J.F. Analysis of Codon Usage. Ph.D. Thesis, University of Nottingham, Nottingham, UK, 1999.
57. Yang, Z. User Guide PAML: Phylogenetic Analysis by Maximum Likelihood. *Mol. Biol. Evol.* **2007**, *4*, 1–70.
58. R Core Team. R: A Language and Environment for Statistical Computing. Available online: https://www.r-project.org/ (accessed on 6 January 2024).
59. Gu, Z.; Gu, L.; Eils, R.; Schlesner, M.; Brors, B. Circlize Implements and Enhances Circular Visualization in R. *Bioinformatics* **2014**, *30*, 2811–2812. [CrossRef]
60. Wickham, H. *Ggplot2 Elegant Graphics for Data Analysis*, 2nd ed.; Gentleman, R., Hornik, K., Parmigiani, G., Eds.; Springer: Houston, TX, USA, 2016.
61. Kolde, R. Package "Pheatmap": Pretty Heatmaps. R Package Version 1.0.12. Available online: https://cran.r-project.org/web/packages/pheatmap/ (accessed on 30 March 2024).
62. Wickham, H. Reshaping Data with the Reshape Package. *J. Stat. Softw.* **2007**, *21*, 1–20. [CrossRef]
63. Katoh, K.; Standley, D.M. MAFFT Multiple Sequence Alignment Software Version 7: Improvements in Performance and Usability. *Mol. Biol. Evol.* **2013**, *30*, 772–780. [CrossRef] [PubMed]
64. Larsson, A. AliView: A Fast and Lightweight Alignment Viewer and Editor for Large Datasets. *Bioinformatics* **2014**, *30*, 3276–3278. [CrossRef] [PubMed]
65. Xia, X.; Xie, Z.; Salemi, M.; Chen, L.; Wang, Y. An Index of Substitution Saturation and Its Application. *Mol. Phylogenet. Evol.* **2003**, *26*, 1–7. [CrossRef] [PubMed]
66. Xia, X. DAMBE7: New and Improved Tools for Data Analysis in Molecular Biology and Evolution. *Mol. Biol. Evol.* **2018**, *35*, 1550–1552. [CrossRef] [PubMed]
67. Paradis, E.; Claude, J.; Strimmer, K. APE: Analyses of Phylogenetics and Evolution in R Language. *Bioinformatics* **2004**, *20*, 289–290. [CrossRef] [PubMed]
68. Nguyen, L.T.; Schmidt, H.A.; Von Haeseler, A.; Minh, B.Q. IQ-TREE: A Fast and Effective Stochastic Algorithm for Estimating Maximum-Likelihood Phylogenies. *Mol. Biol. Evol.* **2015**, *32*, 268–274. [CrossRef]
69. Strimmer, K.; Von Haeseler, A. Likelihood-Mapping: A Simple Method to Visualize Phylogenetic Content of a Sequence Alignment. *Proc. Natl. Acad. Sci. USA* **1997**, *94*, 6815–6819. [CrossRef]
70. Ronquist, F.; Teslenko, M.; Van Der Mark, P.; Ayres, D.L.; Darling, A.; Höhna, S.; Larget, B.; Liu, L.; Suchard, M.A.; Huelsenbeck, J.P. Mrbayes 3.2: Efficient Bayesian Phylogenetic Inference and Model Choice across a Large Model Space. *Syst. Biol.* **2012**, *61*, 539–542. [CrossRef]

71. Rambaut, A.; Drummond, A.J.; Xie, D.; Baele, G.; Suchard, M.A. Posterior Summarization in Bayesian Phylogenetics Using Tracer 1.7. *Syst. Biol.* **2018**, *67*, 901–904. [CrossRef]
72. Rambaut, A. FigTree v.1.4.4. Available online: http://tree.bio.ed.ac.uk/software/figtree/ (accessed on 6 January 2024).
73. Cameron, S.L. Insect Mitochondrial Genomics: Implications for Evolution and Phylogeny. *Annu. Rev. Entomol.* **2014**, *59*, 95–117. [CrossRef] [PubMed]
74. Ojala, D.; Montoya, J.; Attardi, G. TRNA Punctuation Model of RNA Processing in Human Mitochondria. *Nature* **1981**, *290*, 470–474. [CrossRef] [PubMed]
75. Sun, Z.; Wan, D.G.; Murphy, R.W.; Liang, M.; Zhang, X.S.; Huang, D.W. Comparison of Base Composition and Codon Usage in Insect Mitochondrial Genomes. *Genes Genomics* **2009**, *31*, 65–71. [CrossRef]
76. Wang, H.; Meng, T.; Wei, W. Analysis of Synonymous Codon Usage Bias in Helicase Gene from Autographa Californica Multiple Nucleopolyhedrovirus. *Genes Genomics* **2018**, *40*, 767–780. [CrossRef] [PubMed]
77. Wright, F. The "effective Number of Codons" Used in a Gene. *Gene* **1990**, *87*, 23–29. [CrossRef]
78. Sueoka, N. Directional Mutation Pressure and Neutral Molecular Evolution. *Proc. Natl. Acad. Sci. USA* **1988**, *85*, 2653–2657. [CrossRef]
79. Zhangtl, Z.; Yu, J. Evaluation of Six Methods for Estimating Synonymous and Non-Synonymous Substitution Rates Comparative Results. *Genomics Proteomics Bioinform.* **2006**, *4*, 173–181. [CrossRef]
80. Wolstenholme, D.R. Animal Mitochondrial DNA: Structure and Evolution. *Int. Rev. Cytol.* **1992**, *141*, 173–216. [CrossRef] [PubMed]
81. Li, L.-Y.; Deng, Y.-P.; Zhang, Y.; Wu, Y.; Fu, Y.-T.; Liu, G.-H.; Liu, J.-H. Characterization of the Complete Mitochondrial Genome of *Culex vishnui* (Diptera: Culicidae), One of the Major Vectors of Japanese Encephalitis Virus. *Parasitol. Res.* **2023**, *122*, 1403–1414. [CrossRef]
82. Soghigian, J.; Sither, C.; Justi, S.A.; Morinaga, G.; Cassel, B.K.; Vitek, C.J.; Livdahl, T.; Xia, S.; Gloria-Soria, A.; Powell, J.R.; et al. Phylogenomics Reveals the History of Host Use in Mosquitoes. *Nat. Commun.* **2023**, *14*, 6252. [CrossRef]
83. Silva, A.F.; Machado, L.C.; de Paula, M.B.; Vieira, C.J.d.S.P.; Bronzoni, R.V.d.M.; Santos, M.A.V.d.M.; Wallau, G.L. Culicidae Evolutionary History Focusing on the Culicinae Subfamily Based on Mitochondrial Phylogenomics. *Sci. Rep.* **2020**, *10*, 18823. [CrossRef]
84. Munstermann, L.E.; Marchi, A.; Sabatini, A.; Coluzzi, M. Polytene Chromosomes of *Orthopodomyia pulcripalpis* (Diptera, Culicidae). *Parassitologia* **1985**, *27*, 267–277. [PubMed]
85. Coquillett, D.W. New Culicidae from the West Indies and Central America. *Proc. Entomol. Soc. Wash.* **1906**, *7*, 182–186.
86. Kjer, K.M.; Simon, C.; Yavorskaya, M.; Beutel, R.G. Progress, Pitfalls and Parallel Universes: A History of Insect Phylogenetics. *J. R. Soc. Interface* **2016**, *13*, 20160363. [CrossRef] [PubMed]

genes MDPI

Article

Identification and Prediction of Differentially Expressed MicroRNAs Associated with Detoxification Pathways in Larvae of *Spodoptera frugiperda*

Yan-Ping Wang [1], Xing-Yu Chen [2], De-Qiang Pu [1], Chun-Yan Yi [1], Chang-Hua Liu [1], Cui-Cui Zhang [1], Zhen-Zhen Wei [1], Jing-Wei Guo [1], Wen-Juan Yu [1], Song Chen [1] and Hong-Ling Liu [1,*]

[1] Key Laboratory of Integrated Pest Management of Southwest Crops, Institute of Plant Protection, Sichuan Academy of Agricultural Sciences, Chengdu 610066, China; singular2023@outlook.com (Y.-P.W.); pdqpudeqiang@163.com (D.-Q.P.); yichunyan2023@outlook.com (C.-Y.Y.); liuchanghua2023@outlook.com (C.-H.L.); zhangcuicui2023@outlook.com (C.-C.Z.); zhenzhen-wei@hotmail.com (Z.-Z.W.); jingwei.guo66@hotmail.com (J.-W.G.); wjyu0906@163.com (W.-J.Y.); chensong523@126.com (S.C.)

[2] Science and Technology Security Center, Sichuan Academy of Agricultural Sciences, Chengdu 610066, China; chenxingyu2023@outlook.com

* Correspondence: liuhongling@ippsaas.org.cn

Abstract: *Spodoptera frugiperda* poses a severe threat to crops, causing substantial economic losses. The increased use of chemical pesticides has led to resistance in *S. frugiperda* populations. Micro ribonucleic acids (MicroRNAs or miRNAs) are pivotal in insect growth and development. This study aims to identify miRNAs across different developmental stages of *S. frugiperda* to explore differential expression and predict target gene functions. High-throughput sequencing of miRNAs was conducted on eggs, 3rd instar larvae, pupae, and adults. Bioinformatics analyses identified differentially expressed miRNAs specifically in larvae, with candidate miRNAs screened to predict target genes, particularly those involved in detoxification pathways. A total of 184 known miRNAs and 209 novel miRNAs were identified across stages. Comparative analysis revealed 54, 15, and 18 miRNAs differentially expressed in larvae, compared to egg, pupa, and adult stages, respectively. Eight miRNAs showed significant differential expression across stages, validated by quantitative reverse transcription PCR (qRT-PCR). Gene Ontology and Kyoto Encyclopedia of Genes and Genomes enrichment analyses predicted target genes' functions, identifying eight differentially expressed miRNAs targeting 10 gene families associated with detoxification metabolism, including P450s, glutathione S-transferase (GSTs), ATP-binding cassette (ABC) transporters, and sodium channels. These findings elucidate the species-specific miRNA profiles and regulatory mechanisms of detoxification-related genes in *S. frugiperda* larvae, offering insights and strategies for effectively managing this pest.

Keywords: detoxification; larva; microRNAs; resistance; *S. frugiperda*; the fall armyworm

Citation: Wang, Y.-P.; Chen, X.-Y.; Pu, D.-Q.; Yi, C.-Y.; Liu, C.-H.; Zhang, C.-C.; Wei, Z.-Z.; Guo, J.-W.; Yu, W.-J.; Chen, S.; et al. Identification and Prediction of Differentially Expressed MicroRNAs Associated with Detoxification Pathways in Larvae of *Spodoptera frugiperda*. *Genes* **2024**, *15*, 1021. https://doi.org/10.3390/genes15081021

Academic Editors: Wei Zhang and Valentina G. Kuznetsova

Received: 6 July 2024
Revised: 29 July 2024
Accepted: 2 August 2024
Published: 3 August 2024

1. Introduction

MicroRNAs (miRNAs) are endogenous, single-stranded, non-protein-coding small RNAs approximately 19 to 24 nucleotides in length [1]. These miRNAs are ubiquitously present in animals, plants, and microorganisms, playing a crucial role in the post-transcriptional regulation of genes [2–8]. Studies have demonstrated that miRNAs are integral to various biological processes, including tissue growth, germ cell development, hormone action, and the development and function of the central nervous system, primarily through gene regulation [9,10]. It is estimated that miRNAs regulate the expression of more than 50% of protein-coding genes in animals [11].

Research on insect miRNAs has garnered increasing attention, leading to the discovery of numerous miRNAs across different insect species. These miRNAs regulate a wide range of physiological functions throughout insect development, such as molting, metamorphosis, oogenesis, embryogenesis, behavior, and host–pathogen interactions [12–18]. For instance, Dicer-1, a key gene in miRNA processing, when interfered with RNA, was found to partially inhibit the metamorphosis of *Blattella germanica* (L.) (Blattaria: Blattidae) after molting [4]. The miR-2 family influences insect metamorphosis and development via juvenile hormones [19–21], while miR-14 regulates insect molting by participating in the regulation of ecdysone receptors [22,23]. Additionally, miRNAs play vital roles in the growth and development of other insects such as *Bactrocera dorsalis* Hendel (Diptera: Tephritidae), *Nilaparvata lugens* (Hemiptera: Delphacidae), and *Spodoptera exigua* (Lepidoptera: Noctuidae) [24–26].

Given their significant role in post-transcriptional gene regulation, binding to specific sequences of target mRNA, miRNA can degrade target mRNA or inhibit its normal translation, thereby regulating the expression of target genes. miRNAs are also implicated in the molecular mechanisms that enhance drug resistance by regulating detoxification genes in insects [27,28]. Detoxification metabolic resistance is crucial for pest resistance development. For instance, miR-7a and miR-8519 could upregulate the expression of the ryanodine receptor (RyR), enhancing resistance to chlorantraniliprole [29]. miR-998-3p can negatively regulate the expression of ABCC2, contributing to the detoxification of BtCry1Ac toxins in three typical lepidopteran pests: *Helicoverpa armigera* (Lepidoptera: Noctuidae), *S. exigua*, and *Plutella xylostella* (Lepidoptera: Plutellidae) [30]. Furthermore, miRNAs novel-85 and novel-191 can target CYP6ER1 and carboxylesterase 1 (CarE1), significantly altering *N. lugens'* susceptibility to nitenpyram [31]. In addition, miR-4133-3p could regulate the detoxification of gossypol and tannic acid by targeting the cytochrome P450 4CJ1 (CYP4CJ1) in Aphis gossypii Glover (Hemiptera: Aphididae) [32].

The fall armyworm, *S. frugiperda* Smith, 1797 (Lepidoptera: Noctuidae), is a major migratory agricultural pest of global concern [33]. This omnivorous pest has a strong migratory ability, a high reproductive rate, and a short life cycle, damaging 353 species of plants across 76 families, including major crops such as corn, sorghum, sugarcane, barley, rice, pepper, wild oat, and potato [34,35]. The primary method of controlling this pest relies on chemical pesticides, which have proven to be effective [36]. However, reports indicate that *S. frugiperda* has developed resistance to various insecticides, including bromides, neonicotinoids, and pyrethroids [37–39]. Insect resistance to insecticides is mainly attributed to enhanced metabolism and alterations in target site structures [40–42].

To date, some miRNAs have been identified in *S. frugiperda*, with functional studies primarily focusing on antiviral immune defense and adaptive evolution [43,44]. However, systematic identification and functional analysis of miRNAs related to detoxification metabolism in *S. frugiperda* remain limited [45]. Using high-throughput sequencing of miRNA libraries, this study investigated miRNAs in *S. frugiperda* at different developmental stages, analyzed differentially expressed miRNAs in larvae, and identified target genes and functional predictions of these miRNAs across four stages. This research aims to elucidate the regulatory mechanisms of *S. frugiperda* growth and development, analyze detoxification-related target genes, and provide new strategies for the effective management of *S. frugiperda* outbreaks and damage.

2. Materials and Methods

2.1. Insect Collection and Rearing

The larvae of *S. frugiperda* were collected from a cornfield at the Xindu Base of the Sichuan Academy of Agricultural Sciences, China. They were continuously raised in the insect laboratory of Plant Conservation Institute of Sichuan Academy of Agricultural Sciences. The larvae were reared continuously in a light incubator under controlled conditions (16/8 hr light/dark (L/D) cycle, 65% humidity, 25 °C) and fed with an artificial diet [46], which consisted of the following components: wheat germ powder 280 g, soy protein powder 90 g, yeast powder 35 g, agar 25 g, vitamin B complex 0.2 g, cholesterol 12 g, sorbate 2 g, ascorbic acid 12 g, methyl p-hydroxybenzoate 5 g, formaldehyde 4 mL, penicillin 0.2 g, and distilled water 1500 mL. Sampling periods at different developmental stages were as follows: 500 eggs were collected within 24 h post-laying, 20–30 third instar larvae were collected within 12 h post-molting, 3–5 pupae were collected within 36 h post-pupation, and 2–3 adults were collected within 12 h post-emergence. Each sample contained 0.3–0.5 g and had three biological replicates. Samples were placed in 2.0 mL RNase-free centrifuge tubes with TRIzol reagent (Invitrogen Life Technologies, Carlsbad, CA, USA) for temporary storage. After collection, samples were stored in an ultra-low temperature freezer at −80 °C.

2.2. Small RNA Library Preparation and Sequencing

Total RNA was extracted using TRIzol reagent (Invitrogen Life Technologies, Carlsbad, CA, USA) following the manufacturer's instructions. RNA concentrations were measured using a NanoDrop spectrophotometer (Thermo Scientific, Sunnyvale, CA, USA), and RNA quality and integrity were assessed with an Agilent 2100 Bioanalyzer (Agilent Technologies, Waldbronn, Germany). Total RNA was used to construct small RNA (sRNA) libraries with the NEBNext Multiplex Small RNA Library Prep Set for Illumina (New England Biolabs, Ipswich, MA, USA).

One microgram of total RNA from *S. frugiperda* samples was ligated to 3′ and 5′ adapters using Ligation Enzyme Mix. The samples were then reverse-transcribed using Superscript II reverse transcriptase. The sRNA libraries were subjected to quality control, and the average insert size was approximately 140 to 150 bp. cDNA was prepared using random primers and a cDNA Synthesis Kit (Invitrogen, CA, USA). The sequencing library was quantified using the Agilent High Sensitivity DNA assay on an Agilent Bioanalyzer (Agilent Technologies, Germany) and sequenced on a NovaSeq 6000 (Illumina) in PE150 mode at Personal Biotechnology Co., Ltd. (Shanghai, China).

2.3. Unigene Assembly, Analysis, and Annotation

Raw sequencing reads underwent quality control using FastQC program (Babraham Bioinformatics, Cambridge, UK) [47], and then the raw data were filtered using the Personalbio company's self-developed script. Clean reads ranging from 18 to 36 nt were filtered, and deduplication was performed to obtain unique reads for subsequent analysis. The reference genome index was built using Bowtie2 v2.5.1 [48], and de-duplicated clean reads were mapped to the reference genome using miRDeep2 v2.0 (https://sourceforge.net/projects/mireap/, accessed on 1 October 2023) [49]. Unique reads were aligned to known miRNAs in the miRBase R.22. database (http://www.mirbase.org/, accessed on 1 October 2023) and annotated with other non-coding RNAs. Sequences without annotations were analyzed using mireap v0.2 for novel miRNA prediction. Annotation results were organized according to the priority: known miRNA > piRNA > rRNA > tRNA > snRNA > snoRNA > novel miRNA, ensuring each small RNA had a unique annotation.

2.4. Differential Expression Analysis of miRNAs

The reads count of miRNAs was determined based on sequences aligned to mature miRNAs. The highest abundance among miRNAs of the same name was used for subsequent analysis. Differentially expressed miRNAs were identified using DESeq v1.39.0 [50], with transcripts showing a log2 fold change > 1 and p-value < 0.05 considered significant. Bidirectional cluster analysis of all miRNAs and samples was performed using the R package Pheatmap v1.0.12 [51], with the Euclidean method for distance calculation and Complete Linkage for hierarchical clustering.

2.5. Quantitative Reverse Transcription PCR

Eight differentially expressed miRNAs were selected for qRT-PCR using initial RNA samples. Stem-loop RT primers and gene-specific primers were designed (Table 1). Reverse transcription was performed with the PrimeScript RT reagent kit with gDNA Eraser (Takara, Tokyo, Japan), the reference gene Actin reverse transcription was performed using the general primers of the kit, and the specific primers of miRNAs were reversed. We used 1 μg of total RNA under the following conditions: 16 °C for 30 min, 42 °C for 30 min, and 85 °C for 5 min. The reference gene qRT-PCR reaction contained 2 μL of diluted cDNA (15 ng), 10 μL of 2× AceQ qPCR SYBR Green Master mix (Vazyme, Nanjing, China), and 10 μM of each primer in a 20 μL total volume [52]. The miRNAs qRT-PCR reaction contained 2 μL of diluted cDNA (15 ng), 10 μL of 2× AceQ Universal U^+ Probe Master mix (Vazyme, China), 10 μM probe, and 10 μM of each primer in a 20 μL total volume [53]. The conditions used were: 95 °C for 5 min, followed by 40 cycles of 95 °C for 10 s, 60 °C for 30 s, and a melting curve from 68 °C to 95 °C. The Applied Biosystems QuantStudio 6 Flex system (Thermo Scientific, Sunnyvale, CA, USA) was used with Sf-actin as an endogenous control. The $2^{-\Delta\Delta CT}$ method was used to calculate the relative expression levels of miRNAs [54]. The qRT-PCR included three technical and biological replicates.

Table 1. Primers used for the qRT-PCR.

Primer	Sequence (from 5′ End to 3′ End)
miR-6094-5P-RT	GTCGTATCCAGTGCAGGGTCCGAGGTAT TCGCACTGGATACGACAGGTAC
miR-6094-5P-F	TCAGCGGTGGCCTGGG
miR-6094-3P-RT	GTCGTATCCAGTGCAGGGTCCGAGGTATTCGCACTGGATACGACAGGATC
miR-6094-3P-F	CGCGTATTCGAGACCTCTGCT
miR-10505-3p-RT	GTCGTATCCAGTGCAGGGTCCGAGGTATTCGCACTGGATACGACGAGCCA
miR-10505-3p-F	CGCGCGTAGGGTTAGAAACT
miR-14-5p-RT	GTCGTATCCAGTGCAGGGTCCGAGGTATTCGCACTGGATACGACTCGAGT
miR-14-5p-F	GCGGGGGGAGGAATTG
miR-2765-5p-RT	GTCGTATCCAGTGCAGGGTCCGAGGTATTCGCACTGGATACGACGCCAAC
miR-2765-5p-F	CGCGCGTGGTAACTCCACCACC
miR-277-3p-RT	GTCGTATCCAGTGCAGGGTCCGAGGTATTCGCACTGGATACGACTGTCGT
miR-277-3p-F	CGCGCGTAAATGCACTATCTGGT
miR-307-3p-RT	GTCGTATCCAGTGCAGGGTCCGAGGTATTCGCACTGGATACGACGCTCAC
miR-307-3p-F	CGCGCGTCACAACCTCCTTGA
miR-34-5p-RT	GTCGTATCCAGTGCAGGGTCCGAGGTATTCGCACTGGATACGACACAACC
miR-34-5p-F	CGCGCGTGGCAGTGTGGTTAGCT
miR-R	AGTGCAGGGTCCGAGGTATT
Sf-actin(nei)-F	CGAGAAGATGACCCAGAT
Sf-actin(nei)-R	GATAGCACAGCCTGGATA
probe	FAM-CGCACTGGATACGAC-MGB

2.6. Gene Ontology (GO) and Kyoto Encyclopedia of Genes and Genomes (KEGG) Enrichment Analysis

Target genes of differentially expressed miRNAs were predicted using miRanda v3.3a (https://bioweb.pasteur.fr/packages/pack@miRanda@3.3a/, accessed on 13 December 2023), considering the 3′ untranslated region (UTR) sequences of *S. frugiperda* mRNAs. GO (http://geneontology.org/, accessed on 13 December 2023) and KEGG (http://www.kegg.jp/, accessed on 13 December 2023) enrichment analyses were performed on the predicted target genes. GO enrichment was conducted using topGO v2.50.0 [55], with significant enrichment determined by p-value < 0.05 using the hypergeometric distribution method. KEGG pathway enrichment analysis was performed using clusterProfiler v4.6.0 [56], focusing on pathways with p-values < 0.05.

2.7. Statistical and Data Analysis

Statistical analyses were conducted using Statistical Product and Service Solutions (SPSS) v20.0 (SPSS Inc., Chicago, IL, USA). Comparisons were performed using Student's t-test, with a p-value < 0.05 considered statistically significant.

3. Results

3.1. Small RNA Sequencing Data in *S. frugiperda*

High-throughput sequencing generated twelve sRNA libraries for *S. frugiperda* at four different developmental stages: egg, larva, pupa, and adult. The transcriptome data for these stages contain 20,128,612, 16,315,920, 23,520,239, and 17,593,809 raw reads, respectively (Table 2). The total raw sequences for the four developmental stages exceeded 10 Mb, with high-quality reads constituting 91–97%, indicating good sequencing quality. After filtering clean reads (≥18 nt), the number of valid sequences obtained for egg, larva, pupa and adult stages were 17,231,905, 3,731,980, 17,141,575, and 12,107,130 reads, respectively.

Table 2. Summary of sequencing reads.

Sample	Flag	Raw Reads	Average Raw Reads	Clean Reads	Average Clean Reads
Egg-1	E01	23,489,347	20,128,612	20,154,610	17,231,905
Egg-2	E02	24,772,159	20,128,612	20,586,799	17,231,905
Egg-3	E03	12,124,331	20,128,612	10,954,307	17,231,905
Larva-1	L01	10,837,072	16,315,920	1,502,965	3,731,980
Larva-2	L02	17,172,714	16,315,920	2,494,144	3,731,980
Larva-3	L03	20,937,973	16,315,920	7,198,831	3,731,980
Pupa-1	P01	20,479,951	23,520,239	14,980,746	17,141,575
Pupa-2	P02	25,798,866	23,520,239	19,742,017	17,141,575
Pupa-3	P03	24,281,901	23,520,239	16,701,963	17,141,575
Adult-1	A01	23,403,358	17,593,809	13,367,017	12,107,130
Adult-2	A02	12,435,180	17,593,809	10,254,039	12,107,130
Adult-3	A03	16,942,888	17,593,809	12,700,333	12,107,130

The sRNA lengths for the four developmental stages of *S. frugiperda* ranged from 18 to 36 nt (Figure 1). In larvae, pupae, and adults, the 22 nt sequences were the most abundant, accounting for 43.20%, 48.18%, and 38.99% of the total, respectively. In eggs, the 27 nt sequences were the most abundant, making up 30.72% of the total.

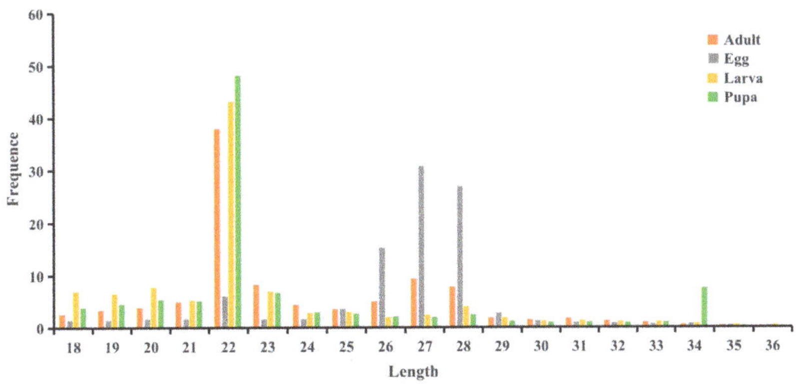

Figure 1. Length distribution of sRNAs in different stages of *S. frugiperda*. X-axis indicates sequence length, while Y-axis shows the frequence in percentage.

3.2. Small RNA Classification and Annotation

To identify known non-coding RNAs, the precursor and mature sequences of miRNAs for the species were downloaded from miRBase. The deduplicated sequences were aligned to these miRNAs to annotate sRNAs at different developmental stages of *S. frugiperda* (Figure 2A, Table S1). Among the annotated sRNAs, ribosomal RNA (rRNA) was the most abundant, constituting 32.89% of the total. Larvae and pupae had higher rRNA quantities than adults and eggs. tRNA, snoRNA, and snRNA were significant components of sRNAs, with maximum proportions of 0.40%, 0.12%, and 0.14%, respectively. A significant proportion of sRNAs (31.94–81.26%) remained unannotated across all developmental stages.

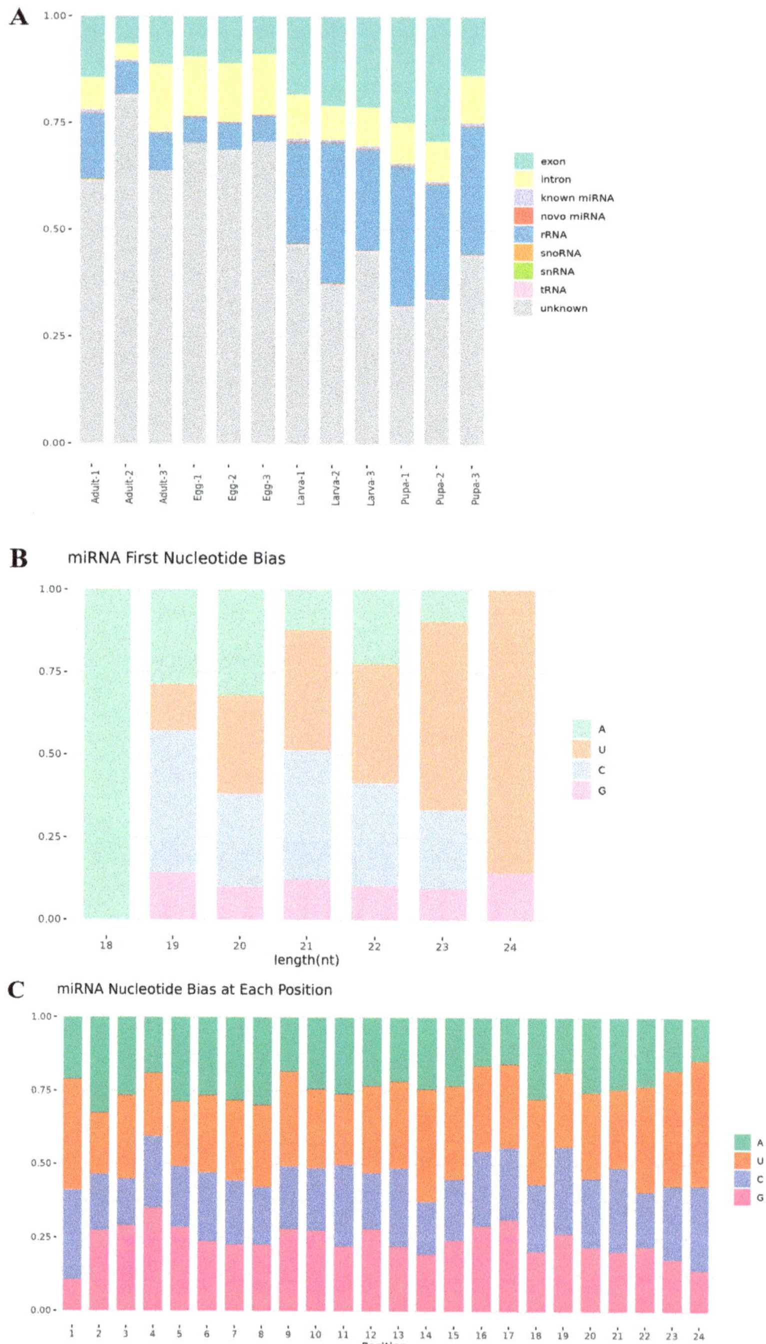

Figure 2. Composition of sRNAs in *S. frugiperda*. (**A**) Annotation of sRNAs. X-axis shows the sample number, whereas Y-axis shows the proportion of the de-duplicated sequences annotated to each sRNA

to the total de-duplicated sequences. (**B**) First base bias of the miRNAs in *S. frugiperda*. X-axis shows the miRNAs length, and Y-axis shows the frequency in percentage. (**C**) Base bias of the miRNAs at each position. X-axis shows the nucleotide position, and Y-axis shows the frequency in percentage.

The nucleotide preference at the first base of miRNAs in different developmental stages of *S. frugiperda* was analyzed (Figure 2B), with most sRNAs showing a preference for U (uracil) at the first base. In all samples, the first base of 18 bp sRNAs preferred A (adenine). Different nucleotide positions exhibited different preferences (Figure 2C). The preference for U at the 5′ end is a conserved characteristic of miRNAs.

3.3. Identification of Known and Novel miRNAs in *S. frugiperda*

Precursor and mature sequences of miRNAs were downloaded from miRBase, and deduplicated sequences were aligned to these miRNAs for annotation. A total of 135, 149, 168, and 164 known miRNAs were identified in eggs, larvae, pupae, and adults, respectively. Among these, 115 known miRNAs were present across all four developmental stages (Figure 3A). Additionally, novel miRNAs were predicted from sequences not annotated with any information using mireap analysis, resulting in 147, 83, 122, and 138 novel miRNAs identified in eggs, larvae, pupae, and adults, respectively. Among these, 32 novel miRNAs were common across all four stages (Figure 3B). The ten most highly expressed miRNAs were detected (Table S2), with sfr-miR-2766-3p, sfr-miR-279a-3p, and sfr-miR-10-5p being the most abundantly expressed.

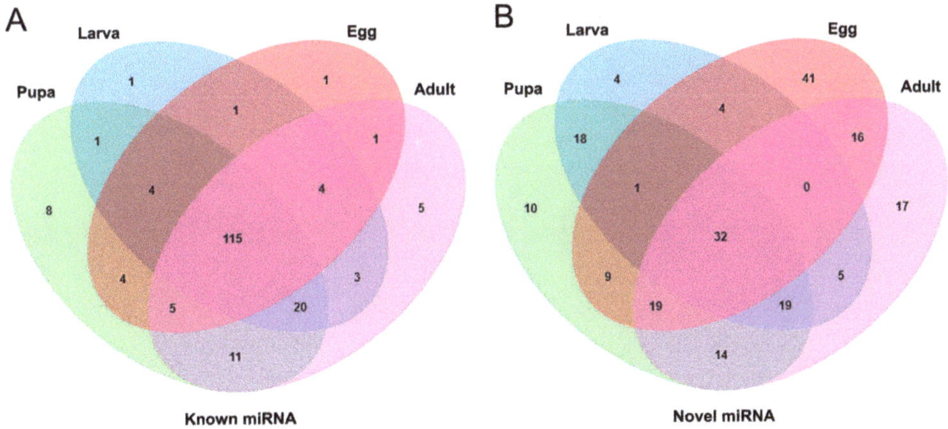

Figure 3. Venn diagram of differential expression in known (**A**) and novel (**B**) miRNAs of *S. frugiperda*. Numbers indicate shared miRNAs.

3.4. Identification of Differentially Expressed miRNAs in Larvae

miRNAs expressed in both larvae and other developmental stages (eggs, pupae, and adults) were selected based on differential expression (log2 fold change > 1) and significance (p-value < 0.05). The results showed 67 differentially expressed miRNAs across the four developmental stages. The number of differentially expressed miRNAs between larvae and eggs, pupae, and adults were 54, 15, and 18, respectively (Table 3).

Table 3. The number of differentially expressed miRNAs in different stages of *S. frugiperda*.

Control	Case	Upregulated Genes	Downregulated Genes	Total DEGs
Egg	Larva	32	22	54
Adult	Larva	9	9	18
Pupa	Larva	6	9	15

Eight miRNAs showed differential expression in *S. frugiperda* larvae compared to the other three stages (Figure 4). Among them, sfr-miR-14-5p, sfr-miR-6094-5p, and sfr-miR-6094-3p were significantly upregulated in larvae compared to the other stages. The most significantly upregulated miRNA in larvae was sfr-miR-6094-3p, with expression levels increasing by 223.41, 482.67, and 79.76 times compared to eggs, pupae, and adults, respectively. sfr-miR-6094-5p and sfr-miR-14-5p were also highly upregulated, with sfr-miR-6094-5p showing increases of 209.54, 253.92, and 87.57 times, and sfr-miR-14-5p showing increases of 23.89, 3.69, and 5.21 times, compared to eggs, pupae, and adults, respectively.

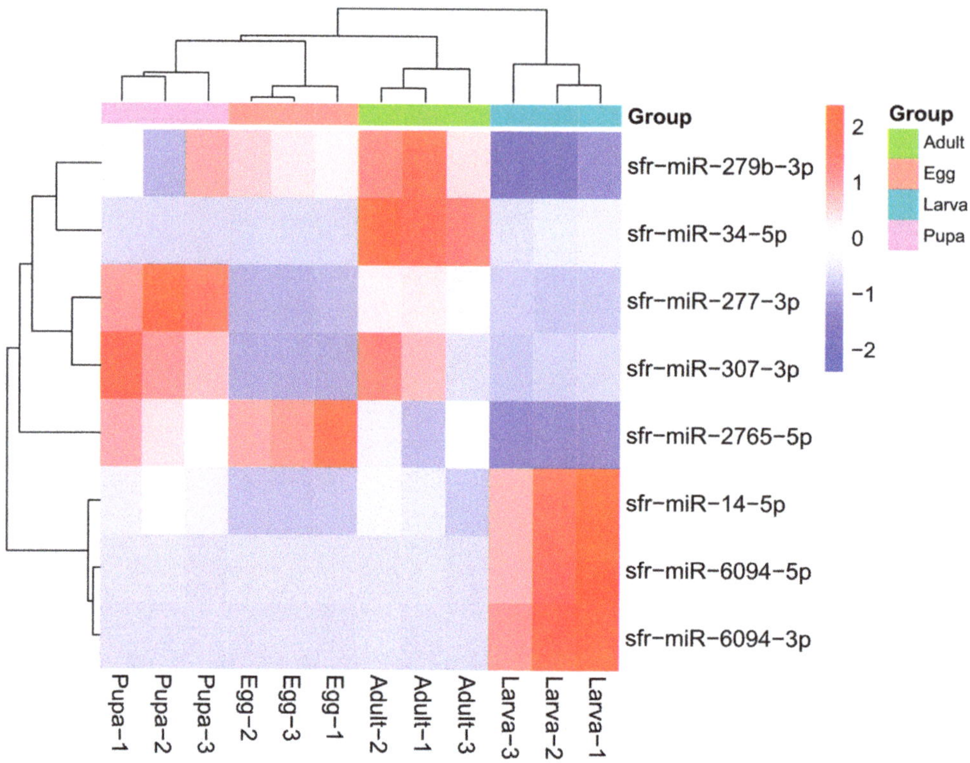

Figure 4. Heatmap of differentially expressed miRNAs in different stages of *S. frugiperda*. Red indicates upregulation; blue means downregulation.

Conversely, the most significantly downregulated miRNA in larvae was sfr-miR-2765-5p, with expression decreasing by 74.20, 61.11, and 37.06 times compared to eggs, pupae, and adults, respectively. sfr-miR-279b-3p was also significantly downregulated, with decreases of 3.89, 3.77, and 6.97 times compared to eggs, pupae, and adults, respectively. Additionally, sfr-miR-277-3p and sfr-miR-307-3p were downregulated in larvae compared to pupae and adults, and sfr-miR-34-5p was downregulated in larvae compared to adults.

3.5. qRT-PCR Validation of Differentially Expressed miRNAs in Larvae

To further confirm the miRNA sequencing results, eight miRNAs (miR-6094-5p, miR-6094-3p, miR-14-5p, miR-10505-3p, miR-2765-5p, miR-277-3p, miR-307-3p, and miR-34-5p) were randomly selected for qRT-PCR based on their differential expression in larvae compared to the other three developmental stages. The qRT-PCR results showed consistent expression patterns with RNA-seq. Compared with the egg stage, the expression of miR-2765-5p was downregulated, and the expression of other miRNAs was upregulated (Figure 5A). Compared with the pupa stage, the expressions of miR-6094-5p, miR-6094-

3p, miR-10505-3p, miR-14-5p, and miR-34-5p were upregulated, while the expressions of miR-2765-5p, miR-277-3p, and miR-307-3p were downregulated (Figure 5B). Compared with the adult stage, the expressions of miR-6094-5p, miR-6094-3p, and miR-14-5p were upregulated, while the expressions of miR-2765-5p, miR-277-3p, miR-307-3p, and miR-34-5p were downregulated (Figure 5C). This suggests that the orientation of regulation and expression pattern of these miRNAs validated by RT-qPCR was consistent with the results from the miRNA sequencing, thus confirming the reliability and repeatability of the miRNA sequencing method.

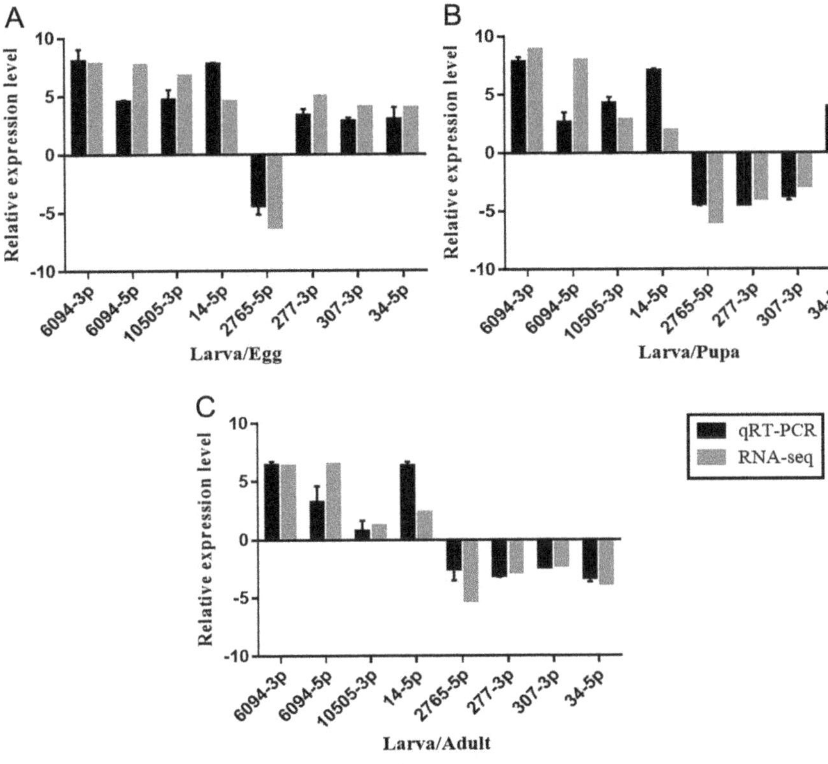

Figure 5. Relative expression level of selected differentially expressed miRNAs of *S. frugiperda* between qRT-PCR and RNA-seq in (**A**) larvae vs. eggs, (**B**) larvae vs. pupae, and (**C**) larvae vs. adults.

3.6. Prediction of Targeted Genes

Using miRanda, the 3′ UTR sequences of the species' mRNAs were targeted for the prediction of differentially expressed miRNA target genes. A total of 2019 target genes and 66,004 target sites were predicted for the 67 differentially expressed miRNAs in *S. frugiperda* larvae.

To further investigate the functions of miRNAs in the *S. frugiperda* transcriptome, the top twenty significantly enriched KEGG pathways were analyzed. In larvae vs. adults (Figure 6A), the most significantly enriched pathways were chemokine signaling pathway, gonadotropin-releasing hormone (GnRH) secretion, cyclic adenosine monophosphate (cAMP) signaling pathway, GnRH signaling pathway, and axon guidance. In larvae vs. eggs (Figure 6B), the top pathways were endocytosis, apelin signaling pathway, chemokine signaling pathway, circadian entrainment, and phospholipase D signaling pathway. In larvae vs. pupae (Figure 6C), the most significantly enriched pathways were oxytocin signaling pathway, vascular smooth muscle contraction, serotonergic synapse, axon guidance, and GnRH secretion.

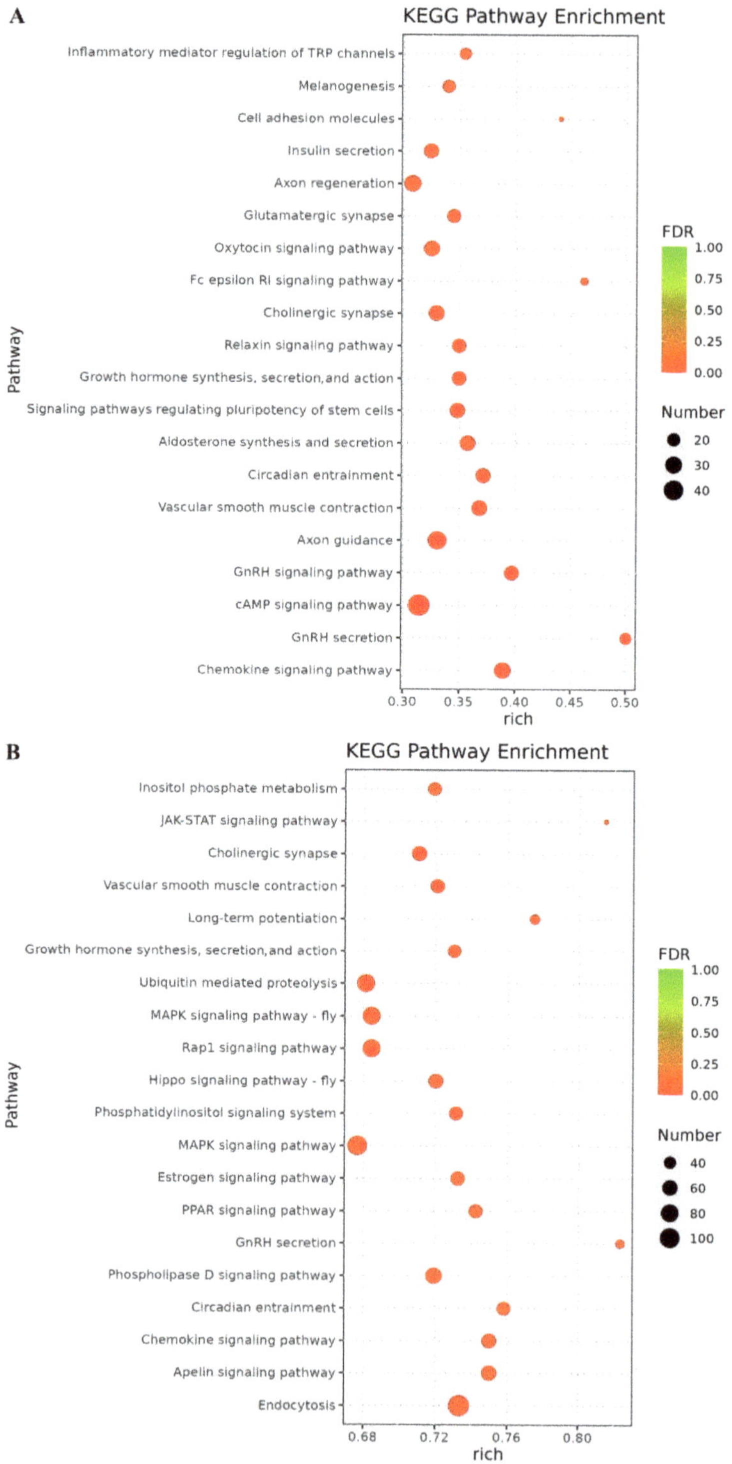

Figure 6. *Cont.*

C

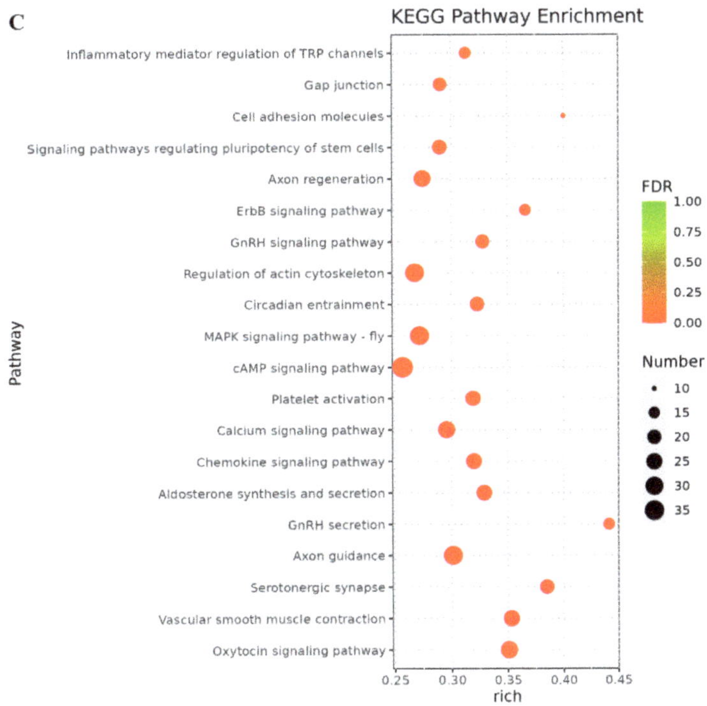

Figure 6. The top 20 significantly enriched KEGG pathways of *S. frugiperda* in (**A**) larvae vs. adults, (**B**) larvae vs. eggs, and (**C**) larvae vs. pupae. X-axis shows the enrichment factor, and Y-axis shows the pathway names. Definitions of the pathways are available at https://www.kegg.jp/kegg/pathway.html. The enrichment degree was measured by Rich factor value and the number of miRNA target genes enriched on this pathway. Rich factor refers to the ratio of the number of differential miRNA target genes enriched in this pathway to the number of differential miRNA target genes annotated. The greater the Rich factor, the greater the degree of enrichment.

GO enrichment analysis of the target genes of differentially expressed miRNAs showed that the functions of differentially expressed genes in larvae compared to the other three stages were mainly distributed across cellular components, molecular functions, and biological processes. In larvae, adults, and pupae, the G protein-coupled receptor signaling pathway was a significantly enriched subgroup (Figures 7 and 8). In larvae and eggs, localization was a significantly enriched subgroup (Figure 9).

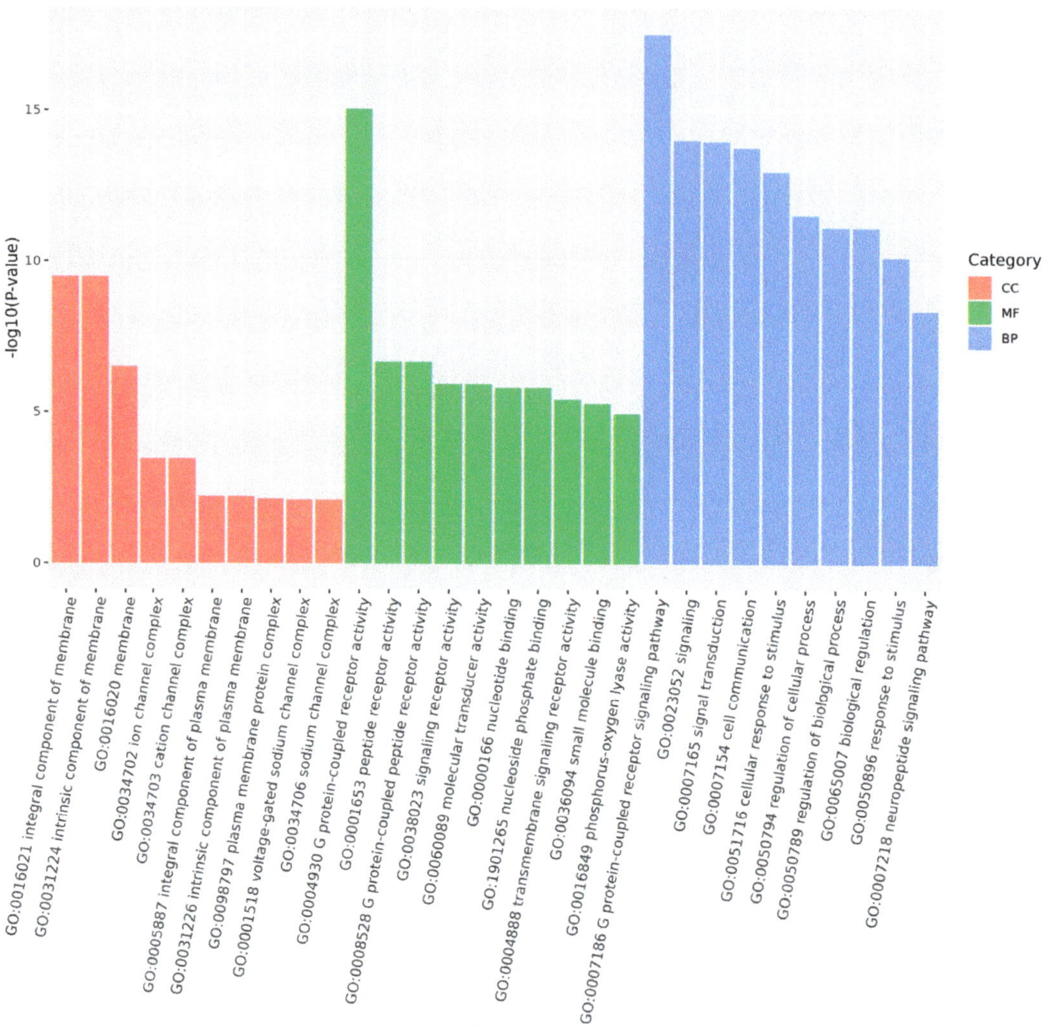

Figure 7. The top 30 GO terms of the target genes of differentially expressed miRNAs of *S. frugiperda* in larvae vs. adults. The enrichment degree was measured using the false discovery rate (FDR) value and the number of miRNA target genes enriched on this pathway. FDR generally ranges from 0 to 1, and the closer it is to zero, the more significant the enrichment. The top 30 pathways with the smallest FDR value, that is, the most significant enrichment, were selected for display.

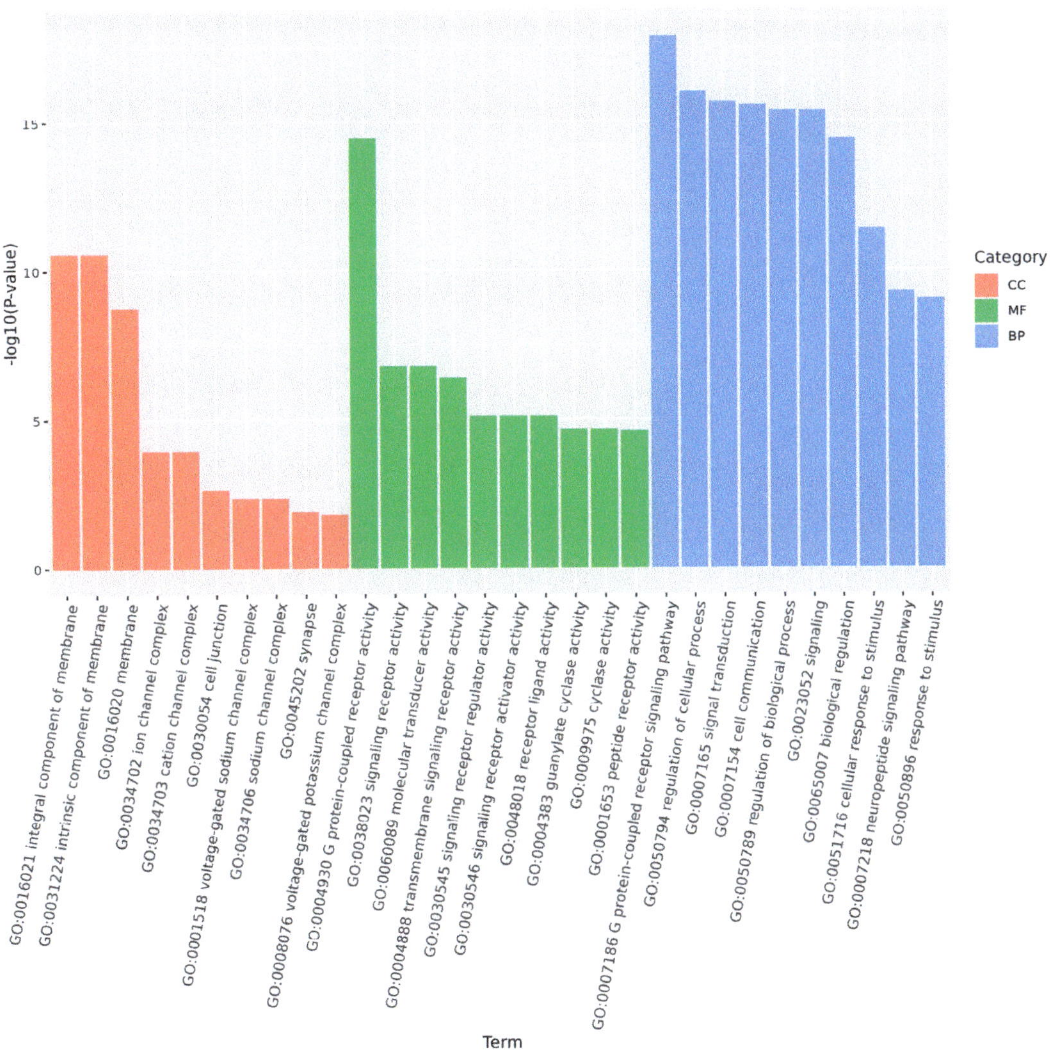

Figure 8. The top 30 GO terms of the target genes of differentially expressed miRNAs of *S. frugiperda* in larvae vs. pupae. The enrichment degree was measured by FDR value and the number of miRNA target genes enriched on this pathway. FDR generally ranges from 0 to 1, and the closer it is to zero, the more significant the enrichment. The top 30 pathways with the smallest FDR value, that is, the most significant enrichment, were selected for display.

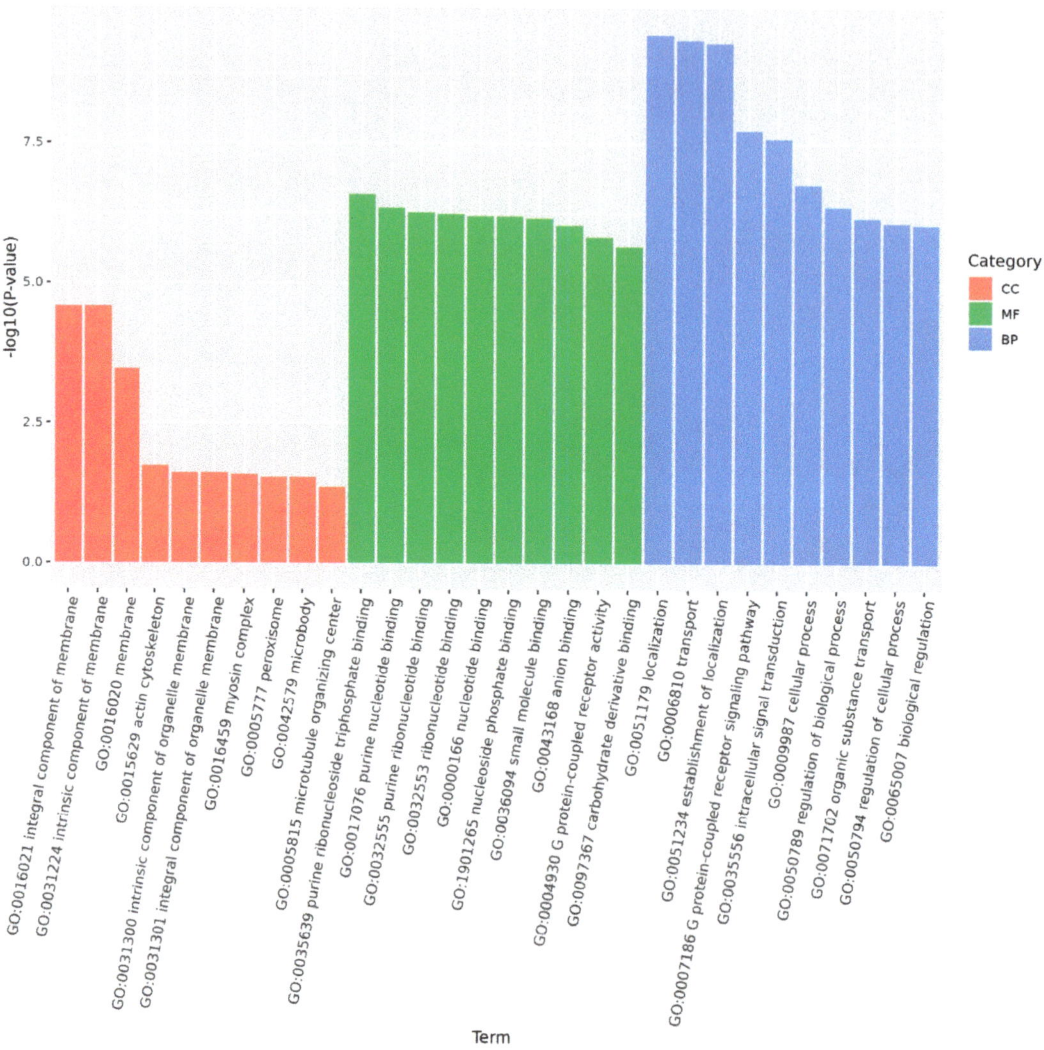

Figure 9. The top 30 GO terms of the target genes of differentially expressed miRNAs of *S. frugiperda* in larvae vs. eggs. X-axis shows GO terms, and Y-axis shows −log10 (*p*-value). The enrichment degree was measured by FDR value and the number of miRNA target genes enriched on this pathway. FDR generally ranges from 0 to 1, and the closer it is to zero, the more significant the enrichment. The top 30 pathways with the smallest FDR value, that is, the most significant enrichment, were selected for display.

To better understand the interactions between miRNAs and their potential target genes, eight of the most significantly differentially expressed miRNAs and detoxification metabolism-related target mRNAs were selected to construct a miRNA–mRNA network (Figure 10). The eight miRNAs included the four most upregulated miRNAs in larvae (sfr-miR-14-5p, sfr-miR-6094-5p, sfr-miR-6094-3p, and sfr-miR-34-5p) and the four most downregulated miRNAs (sfr-miR-2765-5p, sfr-miR-279b-3p, sfr-miR-277-3p, and sfr-miR-307-3p). A total of 82 target genes were identified, targeting 10 gene families involved in detoxification metabolism. These included twenty-nine genes in the P450 gene family, twenty-four in the ABC transporters gene family, thirteen in the GSTs gene family, four in the CarEs gene family, nine in the Acetylcholine gene family, twenty-three in the Sodium channel gene family, seven in the Ryanodine gene family, eight in the Cadherin gene family, seven in the Alkaline phosphatase gene family, and eleven in the Aminopeptidase gene family. These results suggest that detoxification metabolism-related pathways play a critical role in insecticide resistance in *S. frugiperda*.

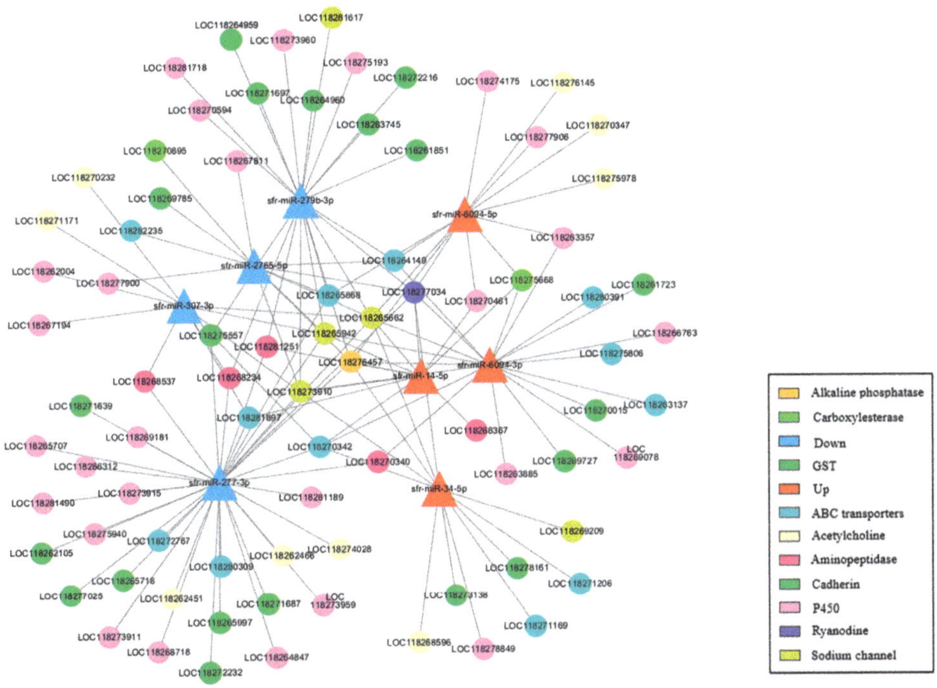

Figure 10. A miRNA–mRNA network of 8 differentially expressed miRNAs and their target genes related to detoxification metabolism in *S. frugiperda*. Red triangles represent upregulated miRNAs while blue triangles represent downregulated miRNAs. Other circles represent different target genes related to detoxification metabolism.

4. Discussion

In recent years, increasing numbers of miRNAs have been discovered in various eukaryotic organisms [57]. In insects, miRNAs regulate gene expression at the post-transcriptional level by either repressing translation or degrading mRNA, with their functions mainly focused on various physiological processes in insects [58–61]. The larval stage is the primary period during which *S. frugiperda* infests crops like corn [62]. It is also a crucial period for rapid growth and preparation for pupation. Studying miRNAs preferentially expressed in the larvae of the fall armyworm can aid in identifying miRNAs involved in regulating growth and development, as well as analyzing detoxification-related target genes, providing new insights and methods for effectively controlling outbreaks and damage caused by this pest.

Chemical control of *S. frugiperda* has a long history, so a variety of resistance mechanisms have emerged [63]. In the United States and Brazil, chemical control of *S. frugiperda* is mainly used before planting Bt transgenic maize [39]. The metabolic capacity of carbaryl -resistant strains is five times higher than that of sensitive strains, and the resistance is mainly caused by oxidative metabolism such as P450 hydroxylation and epoxidation [64]. The activities of various detoxification metabolic enzymes such as multifunctional oxidase (MFO), GSTs, and esterase (ESTs) of laboratory-resistant strains are significantly higher than those of sensitive strains [37]. The sensitivity of acetylcholinesterase (AChE), a molecular target of carbamate and organophosphorus insecticides, to menafarb is also significantly reduced [65]. All these indicate that the resistance of *S. frugiperda* in the field population is caused by several mechanisms, so the study of insecticide resistance is particularly important in chemical control.

This study also provides comprehensive insights into the sRNA landscape of *S. frugiperda* across its developmental stages, with a particular focus on identifying and predicting differentially expressed miRNAs associated with detoxification pathways in larvae. The results revealed critical patterns and potential regulatory roles of miRNAs, which could be crucial for understanding and managing this pest's resistance to insecticides.

Our analysis identified a substantial number of known and novel miRNAs across the developmental stages of *S. frugiperda*, including 135, 149, 168, and 164 known miRNAs in eggs, larvae, pupae, and adults, respectively. Additionally, we discovered 147, 83, 122, and 138 novel miRNAs in these stages. These findings are consistent with previous studies that have identified numerous miRNAs in various insect species [31,43,66–69], underscoring the conserved nature of miRNA-mediated regulation in insect development and physiology.

Notably, miR-2766-3p, miR-279a-3p, and miR-10-5p were among the most abundantly expressed miRNAs in *S. frugiperda*, with significant expression in the larval stage. These miRNAs have been implicated in crucial biological processes such as development, stress response, and resistance mechanisms in other insect species. For instance, miR-2766 regulates tyrosine hydroxylase in *H. armigera*, influencing larval–pupal metamorphosis [70]. Similarly, miR-279a-3p modulates CYP325BB1 expression, affecting insecticide resistance in mosquitoes [71]. In addition, miR-10 is also the most abundantly expressed miRNA in larvae of *H. armigera* when parasitized by the wasp *Diadegma semiclausum* [72].

The differential expression analysis highlighted eight miRNAs with significant changes in expression levels in *S. frugiperda* larvae compared to other developmental stages. These include upregulated miRNAs such as sfr-miR-6094-3p, sfr-miR-6094-5p, and sfr-miR-34-5p, and downregulated miRNAs such as sfr-miR-2765-5p and sfr-miR-279b-3p. The high expression of sfr-miR-34-5p in larvae and its decrease in eggs and pupae suggest its pivotal role in larval development. This pattern aligns with previous findings in other insects, where miRNAs were also highly expressed in larvae compared to other stages [73–78].

The KEGG and GO enrichment analyses of target genes predicted for differentially expressed miRNAs revealed their involvement in critical biological pathways, including detoxification metabolism and G protein-coupled receptor (GPCR) signaling. GPCR signaling has been linked to insecticide resistance in various insects, including *S. frugiperda* [79]. Our study identified 82 potential detoxification-related target genes regulated by differentially expressed miRNAs, including genes from the P450, GST, and ABC transporter families. These gene families are well-known for their roles in insecticide detoxification and resistance [80].

Previous studies have demonstrated that miRNAs can modulate insecticide sensitivity by targeting detoxification genes. For instance, miR-2b-3p and miR-14-5p in *P. xylostella* larvae were shown to influence the expression of CYP9F2 and CYP307a1, affecting the larval detoxification capacity [81]. Similarly, regulating novel_miR-1517 expression in *Bemisia tabaci* affected its sensitivity to imidacloprid by modulating CYP6CM1 expression [82]. The chlorpyrifos- and cypermethrin-resistant nature of *S. frugiperda* strains showed that the A201S, G227A, and F290V point mutations of AChE are resistant to chlorpyrifos [39]. The T929I, L932F, and L1014F point mutations in sodium ion channels lead to resistance to cyhalothrin. The increased expressions of P450, GSTs, and ESTs are involved in resistance to chlorpyrifos and cypermethrin [83]. Our findings suggest that miRNAs in *S. frugiperda* may similarly regulate detoxification genes, contributing to its resistance to various insecticides.

The findings of this study have significant implications for understanding the molecular mechanisms underlying insecticide resistance in *S. frugiperda*. The identified miRNAs and their target genes involved in detoxification pathways offer potential targets for developing novel pest management strategies. By manipulating specific miRNAs or their targets, it might be possible to enhance the efficacy of insecticides or develop new control methods that circumvent existing resistance mechanisms. Future research should focus on functional validation of these miRNAs and their targets using techniques such as miRNA mimics and inhibitors, gene knockout or knockdown experiments, and biochemical assays to confirm their roles in detoxification and resistance. Additionally, exploring the environmental and physiological factors that influence miRNA expression and activity will further elucidate their regulatory networks in *S. frugiperda*.

5. Conclusions

In summary, eight miRNAs showed differential expression in *S. frugiperda* larvae compared to the other three stages. A total of 82 target genes were identified, targeting 10 gene families involved in detoxification metabolism. These results provide a theoretical basis for further exploring the mechanism of miRNAs regulating detoxification metabolism genes.

Supplementary Materials: The following supporting information can be downloaded at: https://www.mdpi.com/article/10.3390/genes15081021/s1, Table S1: Annotation of sRNAs in *S. frugiperda*. Table S2: Ten of the most highly expressed miRNAs in *S. frugiperda*. Numbers indicate absolute read counts.

Author Contributions: Conceptualization, H.-L.L. and S.C.; methodology, C.-Y.Y. and C.-H.L.; software, C.-Y.Y. and C.-H.L.; validation, C.-C.Z. and Z.-Z.W.; formal analysis, Y.-P.W. and X.-Y.C.; investigation, J.-W.G. and W.-J.Y.; data curation, Y.-P.W.; writing—original draft preparation, Y.-P.W. and X.-Y.C.; writing—review and editing, D.-Q.P.; project administration, H.-L.L. and D.-Q.P.; funding acquisition, H.-L.L. and Y.-P.W. All authors have read and agreed to the published version of the manuscript.

Funding: This research was funded by "Identification and functional prediction of preferred expression microRNAs in *S. frugiperda* larvae" (grant number 2023NSFSC1258), "Research on the mechanism of the identification of *Telenomus remus* Nixon and the parasitism of *S. frugiperda*" (grant number 2023NSFSC0152), "Study on the disaster law and key control technology of main pests in Sichuan characteristic fruits" (2021XKJS084), and "Sichuan Fruit Innovation Team of National Modern Agricultural Industry Technology System" (grant number SCCXTD-2024-4).

Institutional Review Board Statement: Not applicable.

Informed Consent Statement: Not applicable.

Data Availability Statement: The original data presented in the study are openly available in FigShare at https://doi.org/10.6084/m9.figshare.26181866.v1.

Acknowledgments: We thank the anonymous reviewers for valuable comments.

Conflicts of Interest: The authors declare no conflicts of interest.

References

1. Ambros, V. The functions of animal microRNAs. *Nature* **2004**, *431*, 350–355. [CrossRef] [PubMed]
2. Asgari, S. Venom proteins from polydnavirus-producing endoparasitoids: Their role in host-parasite interactions. *Arch. Insect Biochem.* **2006**, *61*, 146–156. [CrossRef]
3. Ibanez-Ventoso, C.; Vora, M.; Driscoll, M. Sequence relationships among *C. elegans, D. melanogaster* and human microRNAs highlight the extensive conservation of microRNAs in biology. *PLoS ONE* **2008**, *3*, e2818. [CrossRef] [PubMed]
4. Gomez-Orte, E.; Belles, X. MicroRNA-dependent metamorphosis in hemimetabolan insects. *Proc. Natl. Acad. Sci. USA* **2009**, *106*, 21678–21682. [CrossRef]
5. Asgari, S. Role of MicroRNAs in insect host-microorganism interactions. *Front. Physiol.* **2011**, *2*, 48. [CrossRef]
6. Freitak, D.; Knorr, E.; Vogel, H.; Vilcinskas, A. Gender- and stressor-specific microRNA expression in *Tribolium castaneum*. *Biol. Lett.* **2012**, *8*, 860–863. [CrossRef]
7. Fullaondo, A.; Lee, S.Y. Identification of putative miRNA involved in *Drosophila melanogaster* immune response. *Dev. Comp. Immunol.* **2012**, *36*, 267–273. [CrossRef] [PubMed]
8. Yu, X.; Zhou, Q.; Cai, Y.; Luo, Q.; Lin, H.; Hu, S.; Yu, J. A discovery of novel microRNAs in the silkworm (*Bombyx mori*) genome. *Genomics* **2009**, *94*, 438–444. [CrossRef]

9. Legeai, F.; Rizk, G.; Walsh, T.; Edwards, O.; Gordon, K.; Lavenier, D.; Leterme, N.; Mereau, A.; Nicolas, J.; Tagu, D.; et al. Bioinformatic prediction, deep sequencing of microRNAs and expression analysis during phenotypic plasticity in the pea aphid, *Acyrthosiphon pisum*. *BMC Genom.* **2010**, *11*, 281. [CrossRef]
10. Kloosterman, W.P.; Plasterk, R.H. The diverse functions of microRNAs in animal development and disease. *Dev. Cell* **2006**, *11*, 441–450. [CrossRef]
11. Krol, J.; Loedige, I.; Filipowicz, W. The widespread regulation of microRNA biogenesis, function and decay. *Nat. Rev. Genet.* **2010**, *11*, 597–610. [CrossRef]
12. Zhang, Y.; Zhou, X.; Ge, X.; Jiang, J.; Li, M.; Jia, S.; Yang, X.; Kan, Y.; Miao, X.; Zhao, G.; et al. Insect-Specific microRNA involved in the development of the silkworm *Bombyx mori*. *PLoS ONE* **2009**, *4*, e4677. [CrossRef]
13. Etebari, K.; Asgari, S. Conserved microRNA miR-8 blocks activation of the Toll pathway by upregulating Serpin 27 transcripts. *RNA Biol.* **2013**, *10*, 1356–1364. [CrossRef]
14. Lucas, K.J.; Zhao, B.; Liu, S.; Raikhel, A.S. Regulation of physiological processes by microRNAs in insects. *Curr. Opin. Insect Sci.* **2015**, *11*, 1–7. [CrossRef]
15. Ling, L.; Kokoza, V.A.; Zhang, C.; Aksoy, E.; Raikhel, A.S. MicroRNA-277 targets insulin-like peptides 7 and 8 to control lipid metabolism and reproduction in *Aedes aegypti* mosquitoes. *Proc. Natl. Acad. Sci. USA* **2017**, *114*, E8017–E8024. [CrossRef]
16. Zhang, Z.; Liu, X.; Shiotsuki, T.; Wang, Z.; Xu, X.; Huang, Y.; Li, M.; Li, K.; Tan, A. Depletion of juvenile hormone esterase extends larval growth in *Bombyx mori*. *Insect Biochem. Mol. Biol.* **2017**, *81*, 72–79. [CrossRef]
17. Song, J.; Li, W.; Zhao, H.; Gao, L.; Fan, Y.; Zhou, S. The microRNAs let-7 and miR-278 regulate insect metamorphosis and oogenesis by targeting the juvenile hormone early-response gene Kruppel-homolog 1. *Development* **2018**, *145*, 170670. [CrossRef]
18. Shang, F.; Niu, J.; Ding, B.Y.; Zhang, W.; Wei, D.D.; Wei, D.; Jiang, H.B.; Wang, J.J. The miR-9b microRNA mediates dimorphism and development of wing in aphids. *Proc. Natl. Acad. Sci. USA* **2020**, *117*, 8404–8409. [CrossRef]
19. Lin, L. Functional Analysis of microRNA let-7 and miR-2 Regulating Development in *Bombyx mori*. Ph.D. Thesis, Northwestern Polytechnical University, Xi'an, China, 2017.
20. Wang, Y.L.; Yang, M.L.; Jiang, F.; Zhang, J.Z.; Kang, L. MicroRNA-dependent development revealed by RNA interference-mediated gene silencing of LmDicer1 in the migratory locust. *Insect Sci.* **2013**, *20*, 53–60. [CrossRef]
21. Lozano, J.; Montanez, R.; Belles, X. MiR-2 family regulates insect metamorphosis by controlling the juvenile hormone signaling pathway. *Proc. Natl. Acad. Sci. USA* **2015**, *112*, 3740–3745. [CrossRef]
22. Chen, X.; Fu, J. The microRNA miR-14 regulates egg-laying by targeting EcR in honeybees (*Apis mellifera*). *Insects* **2021**, *12*, 351. [CrossRef]
23. He, K.; Xiao, H.; Sun, Y.; Ding, S.; Situ, G.; Li, F. Transgenic microRNA-14 rice shows high resistance to rice stem borer. *Plant Biotechnol. J.* **2019**, *17*, 461–471. [CrossRef]
24. Puthiyakunnon, S.; Yao, Y.; Li, Y.; Gu, J.; Peng, H.; Chen, X. Functional characterization of three MicroRNAs of the Asian tiger mosquito, *Aedes albopictus*. *Parasite Vector* **2013**, *6*, 230. [CrossRef]
25. Peng, W.; Zheng, W.W.; Tariq, K.; Yu, S.N.; Zhang, H.Y. MicroRNA Let-7 targets the ecdysone signaling pathway E75 gene to control larval-pupal development in *Bactrocera dorsalis*. *Insect Sci.* **2019**, *26*, 229–239. [CrossRef]
26. Yang, M.; Wei, Y.; Jiang, F.; Wang, Y.; Guo, X.; He, J.; Kang, L. MicroRNA-133 inhibits behavioral aggregation by controlling dopamine synthesis in locusts. *PLoS Genet.* **2014**, *10*, e1004206. [CrossRef]
27. Zha, W.; Chang, M.; You, A. Regulation of microRNA on insect growth and development. *Hubei Agric. Sci.* **2019**, *58*, 198–202. [CrossRef]
28. Deng, C.; He, Q. Advanced research of microRNA in insects. *J. Heilongjiang Bayi Agric. Univ.* **2021**, *33*, 15–21.
29. Li, X.; Guo, L.; Zhou, X.; Gao, X.; Liang, P. miRNAs regulated overexpression of ryanodine receptor is involved in chlorantraniliprole resistance in *Plutella xylostella* (L.). *Sci. Rep.* **2015**, *5*, 14095. [CrossRef]
30. Zhu, B.; Sun, X.; Nie, X.; Liang, P.; Gao, X. MicroRNA-998-3p contributes to Cry1Ac-resistance by targeting ABCC2 in lepidopteran insects. *Insect. Biochem. Mol. Biol.* **2020**, *117*, 103283. [CrossRef]
31. Mao, K.; Jin, R.; Ren, Z.; Zhang, J.; Li, Z.; He, S.; Ma, K.; Wan, H.; Li, J. miRNAs targeting CYP6ER1 and CarE1 are involved in nitenpyram resistance in *Nilaparvata lugens*. *Insect Sci.* **2022**, *29*, 177–187. [CrossRef]
32. Ma, K.; Li, F.; Tang, Q.; Liang, P.; Liu, Y.; Zhang, B.; Gao, X. CYP4CJ1-mediated gossypol and tannic acid tolerance in *Aphis gossypii* Glover. *Chemosphere* **2019**, *219*, 961–970. [CrossRef]
33. Van den Berg, J.; Brewer, M.J.; Reisig, D.D. A special collection: *Spodoptera frugiperda* (fall armyworm): Ecology and management of its world-scale invasion outside of the Americas. *J. Econ. Entomol.* **2022**, *115*, 1725–1728. [CrossRef]
34. Goergen, G.; Kumar, P.L.; Sankung, S.B.; Togola, A.; Tamo, M. First report of outbreaks of the fall armyworm *Spodoptera frugiperda* (JE Smith) (Lepidoptera, Noctuidae), a new alien invasive pest in west and central Africa. *PLoS ONE* **2016**, *11*, e165632. [CrossRef]
35. Cock, M.; Beseh, P.K.; Buddie, A.G.; Cafa, G.; Crozier, J. Molecular methods to detect *Spodoptera frugiperda* in Ghana, and implications for monitoring the spread of invasive species in developing countries. *Sci. Rep.* **2017**, *7*, 4103. [CrossRef]
36. Belay, D.K.; Huckaba, R.M.; Foster, J.E. Susceptibility of the fall armyworm, *Spodoptera frugiperda* (lepidoptera: Noctuidae), at Santa Isabel, Puerto Rico, to different insecticides. *Fla. Entomol.* **2012**, *95*, 476–478. [CrossRef]
37. Yu, S.J.; McCord, E.J. Lack of cross-resistance to indoxacarb in insecticide-resistant *Spodoptera frugiperda* (Lepidoptera: Noctuidae) and *Plutella xylostella* (Lepidoptera: Yponomeutidae). *Pest Manag. Sci.* **2007**, *63*, 63–67. [CrossRef] [PubMed]

38. Sharon, G.; Segal, D.; Ringo, J.M.; Hefetz, A.; Zilber-Rosenberg, I.; Rosenberg, E. Commensal bacteria play a role in mating preference of *Drosophila melanogaster*. *Proc. Natl. Acad. Sci. USA* **2010**, *107*, 20051–20056. [CrossRef]
39. Carvalho, R.A.; Omoto, C.; Field, L.M.; Williamson, M.S.; Bass, C. Investigating the molecular mechanisms of organophosphate and pyrethroid resistance in the fall armyworm *Spodoptera frugiperda*. *PLoS ONE* **2013**, *8*, e62268. [CrossRef]
40. Nascimento, A.R.B.D.; Farias, J.R.; Bernardi, D.; Horikoshi, R.J.; Omoto, C. Genetic basis of *Spodoptera frugiperda* (Lepidoptera: Noctuidae) resistance to the chitin synthesis inhibitor lufenuron. *Pest Manag. Sci.* **2016**, *72*, 810–815. [CrossRef]
41. Bolzan, A.; Padovez, F.E.; Nascimento, A.R.; Kaiser, I.S.; Lira, E.C.; Amaral, F.S.; Kanno, R.H.; Malaquias, J.B.; Omoto, C. Selection and characterization of the inheritance of resistance of *Spodoptera frugiperda* (Lepidoptera: Noctuidae) to chlorantraniliprole and cross-resistance to other diamide insecticides. *Pest Manag. Sci.* **2019**, *75*, 2682–2689. [CrossRef]
42. Lira, E.C.; Bolzan, A.; Nascimento, A.R.; Amaral, F.S.; Kanno, R.H.; Kaiser, I.S.; Omoto, C. Resistance of *Spodoptera frugiperda* (Lepidoptera: Noctuidae) to spinetoram: Inheritance and cross-resistance to spinosad. *Pest Manag. Sci.* **2020**, *76*, 2674–2680. [CrossRef]
43. Mone, Y.; Nhim, S.; Gimenez, S.; Legeai, F.; Seninet, I.; Parrinello, H.; Negre, N.; D'Alencon, E. Characterization and expression profiling of microRNAs in response to plant feeding in two host-plant strains of the lepidopteran pest *Spodoptera frugiperda*. *BMC Genom.* **2018**, *19*, 804. [CrossRef]
44. Karamipour, N.; Fathipour, Y.; Talebi, A.A.; Asgari, S.; Mehrabadi, M. The microRNA pathway is involved in *Spodoptera frugiperda* (Sf9) cells antiviral immune defense against *Autographa californica* multiple nucleopolyhedrovirus infection. *Insect Biochem. Mol. Biol.* **2019**, *112*, 103202. [CrossRef]
45. Mahalle, R.M.; Sun, W.; Posos-Parra, O.A.; Jung, S.; Mota-Sanchez, D.; Pittendrigh, B.R.; Seong, K.M. Identification of differentially expressed miRNAs associated with diamide detoxification pathways in *Spodoptera frugiperda*. *Sci. Rep.* **2024**, *14*, 4308. [CrossRef]
46. He, L.; Wang, T.; Chen, Y.; Ge, S.; Wyckhuys, K.; Wu, K. Larval diet affects development and reproduction of East Asian strain of the fall armyworm, *Spodoptera frugiperda*. *J. Integr. Agric.* **2021**, *20*, 736–744. [CrossRef]
47. Andrews, S.A. Quality Control Tool for High Throughput Sequence Data. Available online: http://www.bioinformatics.bbsrc.ac.uk/projects/fastqc/ (accessed on 7 December 2016).
48. Langmead, B.; Salzberg, S.L. Fast gapped-read alignment with bowtie 2. *Nat. Methods* **2012**, *9*, 357–359. [CrossRef] [PubMed]
49. Friedlander, M.R.; Chen, W.; Adamidi, C.; Maaskola, J.; Einspanier, R.; Knespel, S.; Rajewsky, N. Discovering microRNAs from deep sequencing data using miRDeep. *Nat. Biotechnol.* **2008**, *26*, 407–415. [CrossRef]
50. Anders, S.; Huber, W. Differential expression analysis for sequence count data. *Genome Biol.* **2010**, *11*, R106. [CrossRef]
51. Hu, K. Become competent in generating RNA-seq heat maps in one day for novices without prior R experience. *Methods Mol. Biol.* **2021**, *2239*, 269–303. [CrossRef]
52. Peng, X.; Nguyen, A.; Ghosh, D. Quantification of M13 and T7 bacteriophages by Taqman and SYBR green qPCR. *J. Virol. Methods* **2018**, *252*, 100–107. [CrossRef]
53. Sriram, K.; Wiley, S.Z.; Moyung, K.; Gorr, M.W.; Salmerón, C.; Marucut, J.; French, R.P.; Lowy, A.M.; Insel, P.A. Detection and quantification of GPCR mRNA: An assessment and implications of data from high-content methods. *ACS Omega* **2019**, *4*, 17048–17059. [CrossRef]
54. Livak, K.J.; Schmittgen, T.D. Analysis of relative gene expression data using real-time quantitative PCR and the 2(-Delta Delta C(T)) Method. *Methods* **2001**, *25*, 402–408. [CrossRef]
55. Yang, K.; Sablok, G.; Qiao, G.; Nie, Q.; Wen, X. IsomiR2Function: An integrated workflow for identifying microRNA variants in plants. *Front. Plant Sci.* **2017**, *8*, 322. [CrossRef]
56. Yu, G.; Wang, L.G.; Han, Y.; He, Q.Y. Clusterprofiler: An R package for comparing biological themes among gene clusters. *Omics* **2012**, *16*, 284–287. [CrossRef]
57. Asgari, S. MicroRNA functions in insects. *Insect Biochem. Mol. Biol.* **2013**, *43*, 388–397. [CrossRef]
58. Belles, X. MicroRNAs and the evolution of insect metamorphosis. *Annu. Rev. Entomol.* **2017**, *62*, 111–125. [CrossRef] [PubMed]
59. Lucas, K.; Raikhel, A.S. Insect microRNAs: Biogenesis, expression profiling and biological functions. *Insect Biochem. Mol. Biol.* **2013**, *43*, 24–38. [CrossRef]
60. Azzam, G.; Smibert, P.; Lai, E.C.; Liu, J.L. *Drosophila* Argonaute 1 and its miRNA biogenesis partners are required for oocyte formation and germline cell division. *Dev. Biol.* **2012**, *365*, 384–394. [CrossRef]
61. Wightman, B.; Ha, I.; Ruvkun, G. Posttranscriptional regulation of the heterochronic gene lin-14 by lin-4 mediates temporal pattern formation in *C. elegans*. *Cell* **1993**, *75*, 855–862. [CrossRef]
62. Gouin, A.; Bretaudeau, A.; Nam, K.; Gimenez, S.; Aury, J.M.; Duvic, B.; Hilliou, F.; Durand, N.; Montagne, N.; Darboux, I.; et al. Two genomes of highly polyphagous lepidopteran pests (*Spodoptera frugiperda*, Noctuidae) with different host-plant ranges. *Sci. Rep.* **2017**, *7*, 11816. [CrossRef]
63. Rios-Diez, J.D.; Saldamando-Benjumea, C. Susceptibility of *Spodoptera frugiperda* (lepidoptera: Noctuidae) strains from central Colombia to two insecticides, methomyl and lambda-cyhalothrin: A study of the genetic basis of resistance. *J. Econ. Entomol.* **2011**, *104*, 1698–1705. [CrossRef] [PubMed]
64. Mccord, E.; Yus, J. The mechanisms of carbaryl resistance in the fall armyworm, *Spodoptera frugiperda* (J.E. Smith). *Pestic. Biochem. Physiol.* **1987**, *27*, 114–122. [CrossRef]

65. Gutiérrez-Moreno, R.; Mota-Sanchez, D.; A Blanco, C.; E Whalon, M.; Terán-Santofimio, H.; Rodriguez-Maciel, J.C.; DiFonzo, C. Field-evolved resistance of the fall armyworm (Lepidoptera: Noctuidae) to synthetic insecticides in Puerto Rico and Mexico. *J. Econ. Entomol.* **2019**, *112*, 792–802. [CrossRef] [PubMed]
66. Yang, Y.; Zhang, Y.; Wang, A.; Duan, A.; Xue, C.; Wang, K.; Zhao, M.; Zhang, J. Four microRNAs, miR-13b-3p, miR-278-5p, miR-10483-5p, and miR-10485-5p, mediate insecticide tolerance in *Spodoptera frugiperda*. *Front. Genet.* **2021**, *12*, 820778. [CrossRef] [PubMed]
67. Zhu, B.; Li, X.; Liu, Y.; Gao, X.; Liang, P. Global identification of microRNAs associated with chlorantraniliprole resistance in diamondback moth *Plutella xylostella* (L.). *Sci. Rep.* **2017**, *7*, 40713. [CrossRef]
68. Li, J.M.; Zhou, Y.R.; Sun, Z.T.; Wang, X.; Xie, L.; Chen, J.P. Identification and profiling of conserved and novel microRNAs in *Laodelphax striatellus* in response to rice black-streaked dwarf virus (RBSDV) infection. *Genom. Data* **2015**, *3*, 63–69. [CrossRef] [PubMed]
69. Zhang, B.Z.; Hu, G.L.; Lu, L.Y.; Hu, S.F.; Li, Y.S.; Su, X.; Dong, W.Y.; Zhen, C.A.; Liu, R.Q.; Kong, F.B.; et al. Identification of differentially expressed microRNAs under imidacloprid exposure in *Sitobion miscanthi*. *Pestic. Biochem. Phys.* **2021**, *177*, 104885. [CrossRef]
70. Shen, Z.J.; Zhu, F.; Liu, Y.J.; Li, Z.; Moural, T.W.; Liu, X.M.; Liu, X. MicroRNAs miR-14 and miR-2766 regulate tyrosine hydroxylase to control larval-pupal metamorphosis in *Helicoverpa armigera*. *Pest Manag. Sci.* **2022**, *78*, 3540–3550. [CrossRef]
71. Li, X.; Hu, S.; Zhang, H.; Yin, H.; Wang, H.; Zhou, D.; Sun, Y.; Ma, L.; Shen, B.; Zhu, C. MiR-279-3p regulates deltamethrin resistance through CYP325BB1 in *Culex pipiens pallens*. *Parasite Vector* **2021**, *14*, 528. [CrossRef]
72. Etebari, K.; Hussain, M.; Asgari, S. Identification of microRNAs from *Plutella xylostella* larvae associated with parasitization by *Diadegma semiclausum*. *Insect Biochem. Mol. Biol.* **2013**, *43*, 309–318. [CrossRef]
73. Bai, J.; Lin, G.; Wang, Y.; Zhao, X.; Peng, L.; Xie, M.; Yang, J.; He, W.; You, M. Identification and functional characterization of larval-based microRNAs in the diamondback moth, *Plutella xylostella*. *J. Biosaf.* **2022**, *31*, 64–74. [CrossRef]
74. Yang, S. Function Analysis of Protein Coding Gene and miRNA in Pupation and Pupae Development of in *Chilo suppressalis*. Ph.D. Thesis, Nanjing Agricultural University, Nanjing, China, 2015.
75. Liang, Z.; Yang, Y.; Sun, X.; Du, J.; Wang, Q.; Zhang, G.; Zhang, J.; Yin, X.; Singh, D.; Su, P.; et al. Integrated analysis of microRNA and mRNA expression profiles in the fat bodies of MbMNPV-Infected *Helicoverpa armigera*. *Viruses* **2022**, *15*, 19. [CrossRef] [PubMed]
76. Wang, C.; Guo, X.; Li, Y.; Zhang, J.; Fu, Y. miR-34-5p, encoded by *Spodoptera frugiperda*, participates in anti-baculovirus by regulating innate immunity in the insect host. *Int. J. Biol. Macro. Mol.* **2022**, *222*, 2190–2199. [CrossRef]
77. Liu, Z.; Xu, J.; Ling, L.; Luo, X.; Yang, D.; Yang, X.; Zhang, X.; Huang, Y. miR-34 regulates larval growth and wing morphogenesis by directly modulating ecdysone signaling and cuticle protein in *Bombyx mori*. *RNA Biol.* **2020**, *17*, 1342–1351. [CrossRef]
78. Wu, Y.; Guo, Y.; Fan, X.; Zhao, H.; Zhang, Y.; Guo, S.; Jing, X.; Liu, Z.; Feng, P.; Liu, X.; et al. ame-miR-34 modulates the larval body weight and immune response of Apis mellifera Workers to *Ascosphara apis* Invasion. *Int. J. Mol. Sci.* **2023**, *24*, 1214. [CrossRef] [PubMed]
79. Li, T.; Liu, N. Role of the G-protein-coupled receptor signaling pathway in insecticide resistance. *Int. J. Mol. Sci.* **2019**, *20*, 4300. [CrossRef] [PubMed]
80. Hilliou, F.; Chertemps, T.; Maïbèche, M.; Le Goff, G. Resistance in the genus *Spodoptera*: Key insect detoxification genes. *Insects* **2021**, *12*, 544. [CrossRef] [PubMed]
81. Etebari, K.; Afrad, M.H.; Tang, B.; Silva, R.; Furlong, M.J.; Asgari, S. Involvement of microRNA miR-2b-3p in regulation of metabolic resistance to insecticides in *Plutella xylostella*. *Insect Mol. Biol.* **2018**, *27*, 478–491. [CrossRef]
82. Gong, P.P.; Wei, X.G.; Liu, S.N.; Yang, J.; Fu, B.L.; Liang, J.J.; Huang, M.J.; Du, T.H.; Yin, C.; Ji, Y.; et al. Novel_miR-1517 mediates CYP6CM1 to regulate imidacloprid resistance in *Bemisia tabaci* (Hemiptera: Gennadius). *Pestic. Biochem. Phys.* **2023**, *194*, 105469. [CrossRef]
83. Giraudo, M.; Hilliou, F.; Fricaux, T.; Audant, P.; Feyereisen, R.; Le Goff, G. Cytochrome p450s from the fall armyworm (*Spodoptera frugiperda*): Responses to plant allelochemicals and pesticides. *Insect Mol. Biol.* **2015**, *24*, 115–128. [CrossRef]

MDPI AG
Grosspeteranlage 5
4052 Basel
Switzerland
Tel.: +41 61 683 77 34
www.mdpi.com

Genes Editorial Office
E-mail: genes@mdpi.com
www.mdpi.com/journal/genes

www.ingramcontent.com/pod-product-compliance
Lightning Source LLC
Chambersburg PA
CBHW061448210926
51850CB00008B/1648

* 9 7 8 3 7 2 5 8 1 9 3 1 7 *